Group Theoretical Techniques in Quantum Chemistry

THEORETICAL CHEMISTRY

A Series of Monographs

Volume 1 T. E. PEACOCK: Electronic Properties of Aromatic and Heterocyclic Molecules

Volume 2 R. McWEENY and B. T. SUTCLIFFE: Methods of Molecular Quantum Mechanics

Volume 3 J. LINDERBERG and Y. ÖHRN: Propagators in Quantum Chemistry

Volume 4 M. S. CHILD: Molecular Collision Theory

Volume 5 C. D. H. CHISHOLM: Group Theoretical Techniques in Quantum Chemistry

Forthcoming Volumes

A. C. HURLEY: The Electronic Theory of Small Molecules

Group Theoretical Techniques in Quantum Chemistry

C. D. H. CHISHOLM,
Department of Chemistry,
The University of Sheffield,
Sheffield, England

1976

Academic Press
London . New York . San Francisco

A Subsidiary of Harcourt Brace Jovanovich, Publishers

ACADEMIC PRESS INC. (LONDON) LTD.
24/28 Oval Road,
London NW1

United States Edition published by
ACADEMIC PRESS INC.
111 Fifth Avenue
New York, New York 10003

Library of Congress Catalog Card Number: 75-27236
ISBN: 0-12-172950-8

Filmset by The Universities Press, Belfast

Printed in Great Britain by
J. W. Arrowsmith Ltd., Bristol

PREFACE

This book has been written primarily for postgraduate students studying quantum chemistry. However parts of it should also be of interest to general research workers in theoretical chemistry who are not familiar with the more advanced aspects of group theoretical techniques. With this in mind the basic concepts and results of quantum mechanics are assumed known by the reader. As far as mathematics is concerned only a knowledge of elementary calculus and matrix theory is assumed. Any other mathematics which is required is either specifically included or else referenced. The most important area of mathematics needed is linear algebra and since many readers will not be familiar with this a whole chapter is devoted to the elements of linear algebra. This comes in Chapter 2 after a simple introductory chapter on symmetry.

Chapters 3, 4 and 5 are fairly straightforward presentations of the theory and some applications of finite molecular symmetry groups. Since there are already many books dealing with this no great detail is presented. Chapter 6 deals with permutational symmetry and the symmetric group. In this chapter and indeed throughout the book no general theorems are proven. Instead the aim is to present the results with examples in such a way that the reader can understand and use the theorems without detailed proofs which can be found elsewhere in the literature. Chapter 7 uses the symmetric group briefly to discuss spin-free quantum chemistry. There are not many books for chemists which treat the important continuous groups in detail. It is the hope that this situation will be remedied here. Thus Chapters 8 and 9 deal with the theory and representations of continuous groups firstly using ideas based on permutational symmetry and secondly using the concept of infinitesimal transformations. Chapter 10 then gives some applications of continuous groups in the study of the quantum mechanics of simple systems.

Chapter 11 is a straightforward account of the inclusion of spin within a group theoretical framework. Chapter 12 introduces the important and useful coupling and recoupling coefficients associated with the concept of direct product. Chapter 13 is concerned with the useful branching results. In Chapter 14 group theoretical concepts are used to classify the states of many-electron systems. Chapter 15 deals with the theory of fractional parentage while the final chapter is devoted to tensor operators and some of their important applications. This last part is not considered in great detail as there are already many excellent books available on the subject of tensor operators.

Two appendices are included. Appendix 1 deals with the elegant but somewhat abstract general theory of Lie groups and their algebras. Many readers may be content to ignore this and simply accept the results which are given in Chapter 9. Other readers however, like the author, may well find that a study of this material is well worth the effort. Appendix 2 is a collection of some basic tables which, if included in the main text, could destroy the continuity of reading.

I would like to thank particularly Professor Roy McWeeny for reading the whole manuscript, correcting errors, and making very helpful suggestions for general improvement of presentation. I would also like to thank Dr. F. R. Innes for helpful discussions on recoupling and fractional parentage. My thanks also to Mrs. S. P. Rogers for painstakingly and skilfully typing an extremely difficult manuscript. Finally my thanks to Academic Press for their cooperation at all times.

C. D. H. CHISHOLM

Sheffield
September 1975

CONTENTS

Preface v

Chapter 1. Molecular Symmetry Groups

1.1 Molecular symmetry elements and transformations . . 1
1.2 Finite groups and molecular symmetry groups . . . 2
1.3 Classification of molecular symmetry groups . . . 7

Chapter 2. Elements of Linear Algebra

2.1 Introduction 10
2.2 Linear spaces 10
2.3 Linear subspaces 12
2.4 Linear operators 13
2.5 Invariant subspaces 15
2.6 The Dual space 16
2.7 Cogredience and contragredience 17
2.8 Vector spaces 18
2.9 Contravariant and covariant components of a vector . . 21
2.10 Complex vectors 23
2.11 Tensor spaces 24
2.12 The metric tensor 26
2.13 Hilbert spaces 27
2.14 Algebras 28
2.15 Euclidean 3-space 29

Chapter 3. Group Representations

3.1 Symmetry groups and linear spaces 34
3.2 Character 37
3.3 Irreducible representations 38
3.4 Representation theory of finite groups 39
3.5 Group representations and quantum mechanics . . . 42

Chapter 4. Irreducible Representations of Finite Molecular Symmetry Groups

4.1 Spherical harmonics 46
4.2 Notation for irreducible representations 49
4.3 Irreducible representations of the groups $\mathscr{C}_n$ . . . 50

4.4 Irreducible representations of the groups $\mathscr{D}_n$ . . . 52
4.5 Irreducible representations of the octahedral group $\mathscr{O}$. 54

Chapter 5. Applications of Irreducible Representations
5.1 Projection operators and molecular orbitals 58
5.2 Direct products 65
5.3 Selection rules 67
5.4 Branching rules 72

Chapter 6. The Symmetric Group
6.1 Permutations 75
6.2 Partitions and conjugate classes 77
6.3 Young Tableaux 78
6.4 The standard irreducible representations of $\mathscr{S}_n$ 84
6.5 Standard Young operators 88
6.6 Dual representations 88
6.7 Direct products 89
6.8 Outer product representations 92
6.9 The Antisymmetrizer 94

Chapter 7. Spin-free Quantum Chemistry
7.1 Spin values in a many-electron system 98
7.2 Permutational symmetry of spin functions. 99
7.3 The spin-free Hamiltonian 101
7.4 Examples of the use of the spin-free formalism . . . 103

Chapter 8. Continuous Groups. The Method of Irreducible Tensors
8.1 The full linear group $GL(M)$ 106
8.2 The special unitary group $SU(M)$ 116
8.3 The rotation group $R(M)$ 119
8.4 Scalar invariants 124
8.5 The Symplectic group $Sp(M)$ 126
8.6 Direct products 129

Chapter 9. Continuous Groups. The Method of Lie Algebras
9.1 A one-parameter Lie group of transformations. . . . 131
9.2 The molecular symmetry groups $\mathscr{C}_{\infty v}$ and $\mathscr{D}_{\infty h}$ . . . 137
9.3 Invariant integration 139
9.4 The rotation group $R(3)$ 142
9.5 The group $SU(2)$ 148

9.6 The rotation group in four dimensions $R(4)$ . . . 153
9.7 The unitary unimodular group in three dimensions $SU(3)$. 157

Chapter 10. The Quantum Mechanics of Simple Systems
10.1 Harmonic oscillators 161
10.2 The rigid rotator 164
10.3 The hydrogen atom 164

Chapter 11. Spinors and Double Groups
11.1 The concept of a spinor 167
11.2 The double groups $\mathscr{C}_n^*$ 168
11.3 The double groups $\mathscr{D}_n^*$ 169
11.4 The octahedral double group $\mathscr{O}^*$ 170

Chapter 12. Direct Products and Coupling Coefficients
12.1 Introduction 173
12.2 Coupling of two angular momenta 173
12.3 Coupling of representations 176
12.4 Recoupling of angular momenta 181
12.5 Recoupling of representations 185

Chapter 13. Subgroups and Branching Rules
13.1 Introduction 188
13.2 Branching rules for the reduction $R^+(3) \to \mathscr{G}$ . . . 188
13.3 Branching rules for the reduction $SU(M) \to R(M)$ . . 190
13.4 Branching rules for the reduction $SU(M) \to R(3)$ . . 192
13.5 Branching rules for the reduction $R(M) \to R(3)$ . . 195

Chapter 14. Classification of Many-Electron States
14.1 Coupling schemes for atomic systems 198
14.2 Terms in LS coupling 198
14.3 Seniority 201
14.4 Terms from configurations involving inequivalent electrons 203
14.5 Multiplet structure 204
14.6 jj coupling 205
14.7 Molecular terms 208
14.8 Multiplet structure and $\gamma\gamma$ coupling 210

Chapter 15. Fractional Parentage
15.1 Introduction 212
15.2 Substantive coefficients of fractional parentage . . . 213
15.3 CFP in the spin-free formalism 218

15.4 Factorization of CFP . . . 220
15.5 Explicit expressions for substantive CFP . . . 220
15.6 Fractional parentage and molecular configurations . . 223
15.7 Adjective coefficients of fractional parentage . . . 224

Chapter 16. Tensor Operator Analysis
16.1 Introduction . . . 229
16.2 Irreducible tensor operators . . . 229
16.3 The Wigner-Eckart theorem . . . 230
16.4 Products of tensor operators . . . 232
16.5 Simplification of matrix elements . . . 233
16.6 The Coulomb interaction . . . 234
16.7 First order energies for N-electron atomic states . . 237

Appendix 1. General Results in the Theory of Lie Groups
A1.1 Lie groups and Lie algebras . . . 242
A1.2 The Lie algebra as a vector space . . . 244
A1.3 The classification of simple or semi-simple Lie algebras . 245
A1.4 Some detailed examples for particular Lie groups . . 252
A1.5 Canonical parameters . . . 253
A1.6 Conjugate classes in linear groups . . . 254
A1.7 Representations of Lie algebras and Lie groups . . 256
A1.8 The irreducible representations of $R^*(4)$ . . . 259
A1.9 The irreducible representations of $SU(3)$ . . . 260
A1.10 The Casimir Operator . . . 261

Appendix 2. Some Basic Tables Relating to Classification of States

Subject Index . . . 269

CHAPTER 1

Molecular Symmetry Groups

1.1 Molecular symmetry elements and transformations

We begin by considering the molecule H_2O which is shown in Fig. 1 along with a set of coordinate axes. It is clear that this molecule possesses symmetry. Thus if we rotate the molecule by 180° about the Z-axis in a clockwise direction as viewed from 0 we cannot distinguish between the initial and final configurations of the atoms. We say that the Z-axis is a symmetry element and that the rotation by 180° is a symmetry transformation. Other symmetry transformations of H_2O are evidently mirror reflections in the planes YZ and XZ. There also exists the trivial symmetry transformation in which each atom is left unchanged. This is called the identity transformation and is denoted by E.

The above concepts can be generalized to any molecular system. If the atoms comprising the molecule are brought into an identical configuration by a clockwise rotation about a given axis ξ by an angle $(2\pi/n)$, where n is a positive integer, then we call the axis ξ an n-fold rotation axis and denote the transformation by $C_n(\xi)$. When $n = 1$ we have the identity transformation E. If we carry out the transformation C_n twice in succession it is clear that we obtain a further symmetry transformation of the molecule. This transformation is denoted by $C_n \,.\, C_n = C_n^2$. Such a procedure can be repeated to generate the symmetry transformations $C_n^3, C_n^4, \ldots, C_n^n \equiv E$. We note that the smallest angle of rotation is used to characterize the axis. If there is more than one axis of rotational symmetry then the axes with the largest value of n are called principal axes. When there is only one principal axis it is conventional to choose this as the Z-axis.

Suppose that the atoms in a molecule can be brought into an identical configuration by mirror reflection in a plane Π. We denote this symmetry transformation by $\sigma(\Pi)$. It is clear that

$$\sigma^2(\Pi) \equiv E$$

It is usual to denote mirror reflection in a plane perpendicular to the principal axis by σ_h and to denote mirror reflection in a plane passing through the principal axis by σ_v. If there exist 2-fold axes perpendicular to the principal axis such that the plane Π contains the principal axis and bisects the angle between two of the 2-fold axes then reflection in this plane is denoted by σ_d.

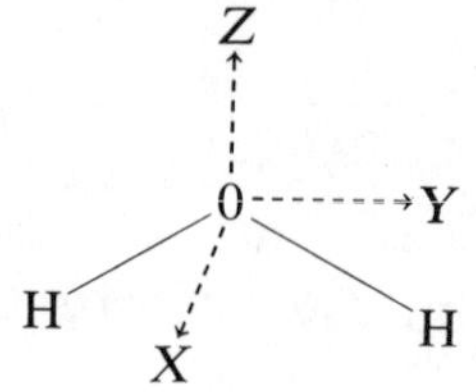

FIG. 1. The water molecule.

Sometimes there exists a symmetry transformation which consists of the combined operation of rotation C_n followed by reflection σ_h. This rotation-reflection transformation is denoted by $S_n = \sigma_h C_n$. Of particular importance is the symmetry transformation S_2. This clearly corresponds to inversion in a centre of symmetry and we write $I \equiv S_2$.

As we shall presently see, the set of all symmetry transformations of a non-linear molecule forms a finite mathematical group. It is therefore useful to consider some of the results contained in the theory of finite groups.

1.2 Finite groups and molecular symmetry groups

Let $\mathcal{G} = \{A, B, \ldots\}$ be a set of g abstract elements in which a single law of composition called multiplication is defined. Thus any two elements A, B of $\mathcal{G}$ possess a unique product $C = AB$. The set $\mathcal{G}$ is called a group of order g if the following conditions are satisfied:

I. Closure: to every ordered pair of elements A, B of $\mathcal{G}$ there belongs a unique element C of $\mathcal{G}$

$$C = AB$$

II. Associative law: if A, B, C are any three elements of $\mathcal{G}$ then

$$(AB)C = A(BC) = ABC$$

III. Unit element: $\mathcal{G}$ contains an element E called the unit element or identity such that for every element A of $\mathcal{G}$

$$AE = EA = A$$

IV. Inverse element: corresponding to every element A of $\mathcal{G}$ there exists in $\mathcal{G}$ an element B such that

$$AB = BA = E$$

It is usual to denote B by A^{-1}.

A group which has the additional property that for every two if its elements

$$AB = BA$$

is called an Abelian group.

Let $\mathcal{G} = \{A, B, C, \ldots\}$ be the set of all symmetry transformations of a molecule. We define a law of combination

$$C = AB \tag{1.2.1}$$

to mean the following. Apply the transformation B to the molecule and then apply the transformation A to the result. Note that although we read

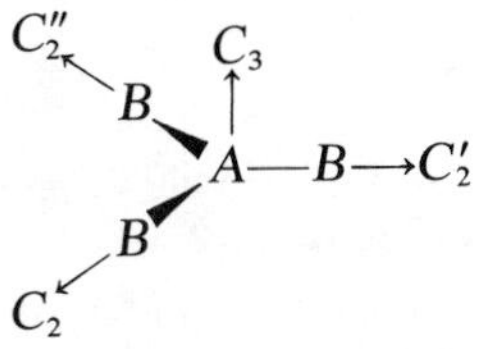

FIG. 2. Rotational symmetry of AB_3.

AB from left to right we apply it in the reverse order. It is clear that for all A and B in $\mathcal{G}$ the element C is a further symmetry transformation of the molecule. Hence C is in $\mathcal{G}$. The associative law is evidently satisfied. In $\mathcal{G}$ we can always find the identity transformation E which is the unit element. Finally for each symmetry transformation A in $\mathcal{G}$ there exists another symmetry transformation A^{-1} in $\mathcal{G}$ which reverses the effect of A. Thus the inverse of each symmetry transformation exists. The set of all symmetry transformations of a molecule now constitutes a group of transformations called the symmetry group of the molecule.

A finite group is completely specified when all possible products AB are known or can be determined. These products are usually displayed in a multiplication table called the group table. The product AB is that element which is found at the intersection of the row marked A and the column marked B.

When we wish to illustrate general results for molecular symmetry groups by means of an example we shall frequently use the group $\mathcal{G} = \{E, C_3, C_3^2, C_2, C_2', C_2''\}$ where C_3, C_2, C_2' and C_2'' are shown in Fig. 2. This group is denoted by $\mathcal{D}_3$ and it describes the purely rotational symmetry transformations of a planar AB_3 molecule. The complete symmetry group of such a molecule must of course also include the mirror reflections. By using relations of the type (1.2.1) we find the group table for $\mathscr{D}_3$ to be

$\mathscr{D}_3$	E	C_3	C_3^2	C_2	C_2'	C_2''
E	E	C_3	C_3^2	C_2	C_2'	C_2''
C_3	C_3	C_3^2	E	C_2''	C_2	C_2'
C_3^2	C_3^2	E	C_3	C_2'	C_2''	C_2
C_2	C_2	C_2'	C_2''	E	C_3	C_3^2
C_2'	C_2'	C_2''	C_2	C_3^2	E	C_3
C_2''	C_2''	C_2	C_2'	C_3	C_3^2	E

It is often unnecessary to write out the complete group table. We note that all the elements of the group $\mathscr{D}_3$ can be expressed in terms of C_3 and C_2. Thus

$$C_3^2 = C_3 \,.\, C_3 \qquad E = C_3 \,.\, C_3 \,.\, C_3 \tag{1.2.2}$$
$$C_2' = C_3 \,.\, C_3 \,.\, C_2 \qquad C_2'' = C_3 \,.\, C_2$$

In this case the elements C_3 and C_2 are called *generators* and the equations (1.2.2) are called *defining relations*.

In general suppose that each of the elements of a finite group can be expressed in terms of products of elements in a subset of the group. We say that this subset constitutes a set of generators for the group. The group can now be specified by means of this set of independent generators, (for which the associative law is assumed), together with a system of defining relations which are sufficient to enable the complete group table to be constructed.

Consider next an abstract group $\mathscr{G}$ of order g and a subset $\mathscr{H}$ of $\mathscr{G}$ containing h elements. If $\mathscr{H}$ is itself a group under the same law of composition of the elements as in $\mathscr{G}$ then we say that $\mathscr{H}$ is a *subgroup* of $\mathscr{G}$ of order h. A *proper* subgroup is any subgroup other than the whole group $\mathscr{G}$ itself or the group consisting of the unit element alone. An important theorem states that the order of a subgroup of a finite group is a divisor of the order of the group. Subgroups of $\mathscr{D}_3$ are readily spotted from the group table. Thus $\mathscr{C}_3 = \{E, C_3, C_3^2\}$ is a subgroup of $\mathscr{D}_3$ of order 3 while $\mathscr{C}_2 = \{E, C_2\}$ is a subgroup of $\mathscr{D}_3$ of order 2.

An element B of a group $\mathscr{G}$ is said to be *conjugate* to the element A of $\mathscr{G}$ if there exists an element S of $\mathscr{G}$ such that

$$B = SAS^{-1}$$

All those elements of a group which are conjugate to one another constitute what is called a *class* of conjugate elements. For the group $\mathscr{D}_3$ we find quite straightforwardly three classes of conjugate elements. They are

$$\mathscr{C}^{(1)} = \{E\} \qquad \mathscr{C}^{(2)} = \{C_3, C_3^2\} \qquad \mathscr{C}^{(3)} = \{C_2, C_2', C_2''\}$$

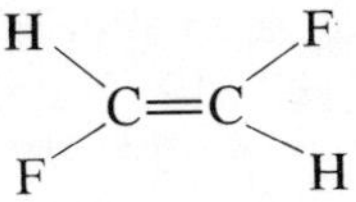

FIG. 3. Transdifluoroethene.

If $\mathcal{H}$ is a subgroup of $\mathcal{G}$ we can construct the set of elements AHA^{-1} where A is a fixed element of $\mathcal{G}$ and H runs through all the elements of $\mathcal{H}$. This set which can be denoted $A\mathcal{H}A^{-1}$ is also a subgroup of $\mathcal{G}$. If for all elements A in $\mathcal{G}$ we have

$$A\mathcal{H}A^{-1} = \mathcal{H}$$

then $\mathcal{H}$ is called an *invariant subgroup* in $\mathcal{G}$. Thus $\mathcal{C}_3$ is an invariant subgroup in $\mathcal{D}_3$ but $\mathcal{C}_2$ is not. A group which has no proper invariant subgroups is called *simple* and a group which has no proper abelian invariant subgroups is called *semi-simple*. The group $\mathcal{D}_3$ is neither simple nor semi-simple.

We consider next the molecule shown in Fig. 3. The symmetry transformations of this molecule are evidently E, C_2, σ_h and I. This symmetry group is denoted by $\mathcal{C}_{2h}$. Consider the subgroups $\mathcal{C}_2 = \{E, C_2\}$ and $\mathcal{C}_i = \{E, I\}$. These subgroups have only the identity E in common and all the elements of $\mathcal{C}_2$ commute with all the elements of $\mathcal{C}_i$. Furthermore any element of $\mathcal{C}_{2h}$ can be expressed as the product of an element of $\mathcal{C}_2$ and an element of $\mathcal{C}_i$ since $\sigma_h = C_2 I$. In general we say that $\mathcal{G}$ is the *direct product* of subgroups $\mathcal{H}_1$ and $\mathcal{H}_2$ if all the elements of $\mathcal{H}_1$ commute with all the elements of $\mathcal{H}_2$, $\mathcal{H}_1$ and $\mathcal{H}_2$ have only the identity in common, and every element of $\mathcal{G}$ can be expressed as the product of an element of $\mathcal{H}_1$ and an element of $\mathcal{H}_2$. We write $\mathcal{G} = \mathcal{H}_1 \otimes \mathcal{H}_2$. Thus $\mathcal{C}_{2h} = \mathcal{C}_2 \otimes \mathcal{C}_i$.

We now consider the set of all permutations of three objects denoted by the numerals 1, 2 and 3. By a permutation (of degree 3) we mean the operation of re-arranging three distinct objects 1, 2 and 3. The identity $1 \to 1, 2 \to 2, 3 \to 3$ is denoted by E, the transposition $1 \to 2, 2 \to 1, 3 \to 3$ is denoted by (12), and the cycle $1 \to 2, 2 \to 3, 3 \to 1$ is denoted by (123). The set of six permutations $\{E, (12), (23), (13), (123), (132)\}$ forms a group called the *symmetric group* on three objects and denoted $\mathcal{S}_3$. The group table is readily found to be

$\mathcal{S}_3$	E	(12)	(13)	(23)	(123)	(132)
E	E	(12)	(13)	(23)	(123)	(132)
(12)	(12)	E	(132)	(123)	(23)	(13)
(13)	(13)	(123)	E	(132)	(12)	(23)
(23)	(23)	(132)	(123)	E	(13)	(12)
(123)	(123)	(13)	(23)	(12)	(132)	E
(132)	(132)	(23)	(12)	(13)	E	(123)

If we rename the elements of $\mathcal{S}_3$ as follows $E \leftrightarrow E$; $C_2 \leftrightarrow (12)$; $C_2'' \leftrightarrow (23)$; $C_2' \leftrightarrow (13)$; $C_3^2 \leftrightarrow (123)$; $C_3 \leftrightarrow (132)$ then we obtain the group table for $\mathcal{D}_3$. In general let $\mathcal{G} = \{A, B, \ldots\}$ and $\mathcal{G}' = \{A', B', \ldots\}$ be two groups of order g. These groups are said to be *isomorphic* ($\mathcal{G} \sim \mathcal{G}'$) if a one-to-one correspondence

$$A \leftrightarrow A', \qquad B \leftrightarrow B', \qquad \ldots$$

can be set up between their elements such that if $AB = C$ then $A'B' = C'$. Isomorphic groups clearly have the same structure although they may differ in the nature of their elements. Any information about $\mathcal{G}'$ which does not depend on the particular nature of its elements applies equally to $\mathcal{G}$. Thus $\mathcal{S}_3$ and $\mathcal{D}_3$ are isomorphic groups; $\mathcal{S}_3 \sim \mathcal{D}_3$.

If we have a correspondence between the elements of two groups $\mathcal{G}$ and $\mathcal{G}'$ which is not one to one (so that two different elements in $\mathcal{G}$ may give rise to the same element in $\mathcal{G}'$) and if further $AB = C$ implies $A'B' = C'$, though not the reverse, then we say that the correspondence is a *homomorphism*. If we have a one to one correspondence of a group $\mathcal{G}$ onto itself which associates with every element A of $\mathcal{G}$ a unique element A_t in $\mathcal{G}$ such that $(AB)_t = A_t B_t$ then the correspondence is called an *automorphism*. In particular if X is a fixed element of $\mathcal{G}$ the correspondence $A \leftrightarrow X^{-1}AX$ for all A in $\mathcal{G}$ is called an *inner automorphism*.

Finally we consider a very important result called Lagrange's Theorem. Let $\mathcal{H}$ be a subgroup of order h of a finite group $\mathcal{G}$ of order g. The theorem states that h is a factor of g, i.e.

$$g = nh$$

We call the integer n the *index* of $\mathcal{H}$ in $\mathcal{G}$. Furthermore there exists a set of n elements $Q_1, Q_2, \ldots, Q_n$ in $\mathcal{G}$ such that $\mathcal{G}$ can be written in the form

$$\mathcal{G} = \{Q_1\mathcal{H}, Q_2\mathcal{H}, \ldots, Q_n\mathcal{H}\} \tag{1.2.3}$$

We call the set $\{Q_i\mathcal{H}\}$ the ith *coset* of $\mathcal{G}$ relative to $\mathcal{H}$. For some element Q_i given by (1.2.3) any element R of $\mathcal{G}$ can be expressed as

$$R = Q_iS \tag{1.2.4}$$

where S is an element of $\mathcal{H}$.

As an illustration we take the group $\mathcal{G} = \mathcal{D}_3$. The first part of the theorem shows that only subgroups of orders 2 and 3 can be found. We have already seen that $\mathcal{C}_2$ and $\mathcal{C}_3$ are such subgroups. For the subgroup $\mathcal{H} = \mathcal{C}_3$ the two cosets may be taken as $E\mathcal{C}_3$ and $C_2\mathcal{C}_3$. Thus

$$\mathcal{D}_3 = \{E\mathcal{C}_3, C_2\mathcal{C}_3\}$$

and $Q_1 = E$, $Q_2 = C_2$. The elements of $\mathcal{D}_3$ can be written as

$$E = Q_1E \qquad C_3 = Q_1C_3 \qquad C_3^2 = Q_1C_3^2$$
$$C_2 = Q_2E \qquad C_2' = Q_2C_3 \qquad C_2'' = Q_2C_3^2$$

For the subgroup $\mathcal{H} = \mathcal{C}_2$ the three cosets may be taken as $E\mathcal{C}_2$, $C_3\mathcal{C}_2$ and $C_3^2\mathcal{C}_2$. Thus

$$\mathcal{D}_3 = \{E\mathcal{C}_2, C_3\mathcal{C}_2, C_3^2\mathcal{C}_2\}$$

amd in this case

$$Q_1 = E \qquad Q_2 = C_3 \qquad Q_3 = C_3^2$$

The elements of $\mathcal{D}_3$ can now be written as

$$E = Q_1E \qquad C_2 = Q_1C_2$$
$$C_3 = Q_2E \qquad C_2'' = Q_2C_2$$
$$C_3^2 = Q_3E \qquad C_2' = Q_3C_2$$

1.3 Classification of molecular symmetry groups

The various molecular symmetry groups can be classified according to the types of symmetry transformation which are included in them. The first main class of molecular symmetry groups possess a unique principal axis of symmetry. The types of group belonging to this class are denoted $\mathcal{C}_n$, $\mathcal{C}_{nv}$, $\mathcal{C}_{nh}$, $\mathcal{S}_n$, $\mathcal{D}_n$, $\mathcal{D}_{nd}$, $\mathcal{D}_{nh}$. They are specified in detail in Table 1. The group $\mathcal{S}_2$ is usually denoted $\mathcal{C}_i$ and the group $\mathcal{C}_{1h}$ is usually denoted $\mathcal{C}_s$.

The second main class of molecular symmetry groups have no unique principal axis but instead have several n-fold axes with $n > 2$. The most important of these groups, at least for our purposes, are the cubic groups $\mathcal{T}$, $\mathcal{O}$, $\mathcal{T}_d$ and $\mathcal{O}_h$. Their symmetry transformations are illustrated in Fig. 4. It is not hard to see that the centres of the faces of a regular octahedron may

TABLE 1. Molecular symmetry groups possessing a unique principal axis (the Z-axis)

Group	Characterization by symmetry transformations
$\mathcal{C}_n$	C_n
$\mathcal{C}_{nv}$	C_n and σ_v
$\mathcal{C}_{nh}$	C_n and σ_h
$\mathcal{S}_n$	S_n
$\mathcal{D}_n$	C_n and n 2-fold axes C_2', C_2'', . . . perpendicular to C_n.
$\mathcal{D}_{nd}$	C_n, C_2', C_2'', . . . and σ_d
$\mathcal{D}_{nh}$	C_n, C_2', C_2'', . . . and σ_h

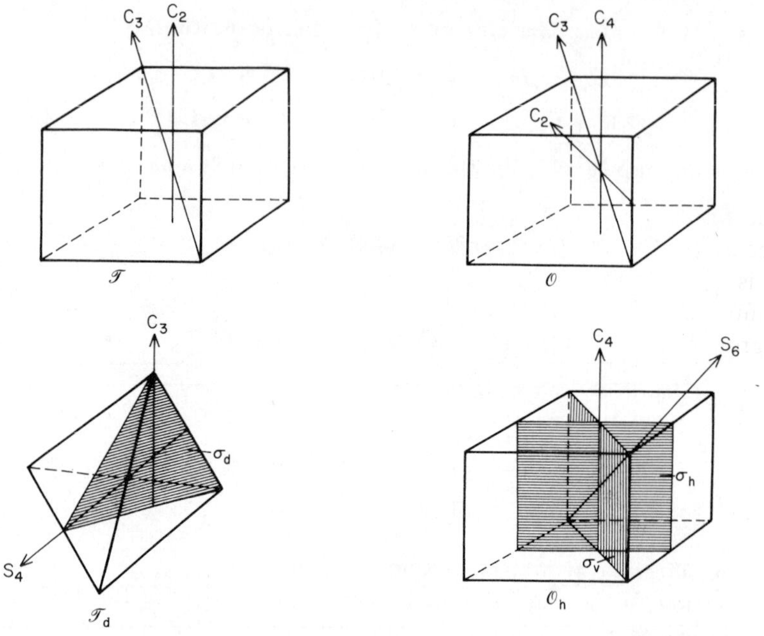

FIG. 4. The cubic groups.

be taken as the vertices of a cube. Conversely to every cube we can inscribe an octahedron whose vertices lie at the centres of the faces of the cube. It follows that if a cube is transformed into itself so is the octahedron and vice-versa. Thus the symmetry groups of a cube and regular octahedron are identical. For most purposes it is more convenient to consider a cube rather than an octahedron.

By examining the symmetry transformations in subgroups of the various molecular symmetry groups we readily find the direct product relationships shown in Table 2. It should also be clear that amongst the molecular symmetry groups there are many isomorphisms. Some of these are shown in Table 3.

TABLE 2. Symmetry groups as direct products

$\mathscr{C}_{nh} = \mathscr{C}_n \otimes \mathscr{C}_i$	n even
$\mathscr{C}_{nh} = \mathscr{C}_n \otimes \mathscr{C}_s$	n odd
$\mathscr{D}_{nh} = \mathscr{D}_n \otimes \mathscr{C}_i$	n even
$\mathscr{D}_{nh} = \mathscr{C}_n \otimes \mathscr{C}_s$	n odd
$\mathscr{D}_{nd} = \mathscr{D}_n \otimes \mathscr{C}_i$	n odd
$\mathscr{O}_h = \mathscr{O} \otimes \mathscr{C}_i$	

TABLE 3. Isomorphic symmetry groups

$\mathscr{D}_{nd} \sim \mathscr{D}_{2n}$	n even
$\mathscr{S}_{2n} \sim \mathscr{C}_{2n}$	
$\mathscr{C}_{nv} \sim \mathscr{D}_{n}$	
$\mathscr{T}_{d} \sim \mathscr{O}$	

The third class of molecular symmetry groups is appropriate to linear molecules. Since a rotation by any angle whatsoever about the molecular axis is a symmetry transformation it follows that there is at least a continuously infinite number of symmetry transformations. Such continuous groups will be dealt with later.

CHAPTER 2

Elements of Linear Algebra

2.1 Introduction

In general the equations we meet in quantum chemistry are linear differential equations. By a linear differential equation we mean that if $\psi_1, \psi_2, \ldots, \psi_n$ are n solutions of the equation then any linear combination

$$\psi = c_1\psi_1 + c_2\psi_2 + \ldots + c_n\psi_n$$

is also a solution. Thus we are immediately led to a study of linear algebra.

2.2 Linear spaces

Consider the totality of directed line segments which can be drawn from a point in a plane. Such directed line segments are called *plane vectors.* Any two vectors α and β in the plane can be added to give a vector $\gamma = \alpha + \beta$ which is also in the plane. Any vector α in the plane can be multiplied by a scalar c to give a vector $c\alpha$ which is also in the plane. We say that the plane constitutes a *linear space.*

Consider next the totality of solutions of a homogeneous linear second order differential equation. Clearly the sum of any two such solutions is again a solution and if we multiply any solution by a scalar c the resulting function is a solution. These properties are identical to those of plane vectors and we say that the totality of solutions constitutes a linear space. In particular if E is a doubly degenerate energy level of some quantum mechanical system there correspond to E two eigenfunctions ψ_1 and ψ_2. Any linear combination of ψ_1 and ψ_2 is also an eigenfunction belonging to E and we say that the totality of such linear combinations constitutes a linear space.

The above ideas can be generalized as follows. Consider a totality $\mathscr{L}$ of entities $\{\alpha, \beta, \gamma, \ldots\}$ called *vectors.* We suppose that these vectors can be added together and multiplied by an arbitrary scalar c. If $(\alpha + \beta)$ and $c\alpha$ are vectors in $\mathscr{L}$ for all α and β in $\mathscr{L}$ then we say that $\mathscr{L}$ is a *linear space.*

A set of vectors $e_1, e_2, \ldots, e_n$ in $\mathscr{L}$ is said to be *linearly independent* if and only if, for all scalars c_i the equation

$$c_1e_1 + c_2e_2 + \ldots + c_ne_n = 0$$

implies

$$c_1 = c_2 = \ldots = c_n = 0$$

Vectors which are not linearly independent are called *linearly dependent.* If in $\mathscr{L}$ we can find at most n vectors $e_1, e_2, \ldots, e_n$ which are linearly independent then any vector α in $\mathscr{L}$ can be expressed as a linear combination

$$\alpha = \alpha^1 e_1 + \alpha^2 e_2 + \ldots + \alpha^n e_n \equiv (\alpha^1, \alpha^2, \ldots, \alpha^n) \tag{2.2.1}$$

and we say that $\mathscr{L}$ is an n-dimensional linear space. It should be noted that in α^i the index i is used as a superscript and not as an exponent.

If the α^i and the scalars c are complex numbers then we say that $\mathscr{L}$ is a complex linear space. If the α^i and the scalars c are real numbers then $\mathscr{L}$ is a real linear space. In what follows a linear space is assumed complex unless otherwise stated. The vectors $e_1, e_2, \ldots, e_n$ are said to constitute a *basis* in $\mathscr{L}$. Alternatively we say that $\mathscr{L}$ is *spanned* by the basis vectors $e_1, e_2, \ldots, e_n$. The scalars $\alpha^1, \alpha^2, \ldots, \alpha^n$ are called the *components* of α in the basis $e_1, e_2, \ldots, e_n$. Clearly the basis vectors $e_1, e_2, \ldots, e_n$ are not unique since any n linearly independent vectors can be chosen to span the space.

If we adopt the convention of writing a set of basis vectors as a row matrix and the components of a vector as a column matrix then (2.2.1) can be written in matrix notation as

$$\alpha = \mathbf{e}\boldsymbol{\alpha} \tag{2.2.2}$$

where

$$\mathbf{e} = [e_1 e_2 \ldots e_n]$$

and

$$\boldsymbol{\alpha} = \begin{bmatrix} \alpha^1 \\ \alpha^2 \\ \cdot \\ \cdot \\ \cdot \\ \alpha^n \end{bmatrix}$$

Suppose now that $e'_1, e'_2, \ldots, e'_n$ is a second basis in $\mathscr{L}$. We have

$$\alpha = \sum_{i=1}^{n} (\alpha^i) e_i = \sum_{j=1}^{n} (\alpha^j)' e'_j \tag{2.2.3}$$

where the $(\alpha^j)'$ are the components of α in the basis $e'_1, \ldots, e'_n$. Clearly the basis vectors e'_i can be expressed linearly in terms of the basis vectors e_k, i.e.

$$e'_j = \sum_{i=1}^{n} e_i a_{ij} \qquad j = 1, 2, \ldots, n \tag{2.2.4}$$

In fact (2.2.4) is essentially the definition of the second basis. The scalars a_{ij} can be regarded as the elements of a non-singular $n \times n$ matrix $\mathbf{A}$. Thus in matrix notation we can write (2.2.4) as

$$\mathbf{e}' = \mathbf{eA} \tag{2.2.5}$$

It readily follows from (2.2.3) and (2.2.4) that

$$\alpha^j = \sum_{i=1}^{n} a_{ji}(\alpha^i)' \qquad j = 1, 2, \ldots, n \tag{2.2.6}$$

The above set of equations can be written in matrix form as

$$\boldsymbol{\alpha} = \mathbf{A}\boldsymbol{\alpha}' \tag{2.2.7}$$

It is convenient, as we shall see later, to write

$$\mathbf{R} = \mathbf{A}^{-1} \tag{2.2.8}$$

In terms of **R** we now have

$$\boldsymbol{\alpha}' = \mathbf{R}\boldsymbol{\alpha} \tag{2.2.9}$$

$$\mathbf{e}' = \mathbf{e}\mathbf{R}^{-1} \tag{2.2.10}$$

2.3 Linear subspaces

If there exists within $\mathscr{L}$ a subset $\mathscr{M}$ of vectors such that $\mathscr{M}$ is itself a linear space (under the same law of addition and scalar multiplication as in $\mathscr{L}$) then we call $\mathscr{M}$ a *linear subspace* of $\mathscr{L}$. In ordinary three-dimensional space V a plane A and a line L are linear subspaces of dimensions two and one respectively. Furthermore any vector V can be expressed uniquely as the sum of a vector in A and a vector in L.

In general we say that a linear space $\mathscr{L}$ is the *direct sum* of two subspaces $\mathscr{L}_1$ and $\mathscr{L}_2$ if every vector α in $\mathscr{L}$ can be written uniquely in the form

$$\alpha = \beta + \gamma$$

with β in $\mathscr{L}_1$ and γ in $\mathscr{L}_2$. This direct sum is written as

$$\mathscr{L} = \mathscr{L}_1 \oplus \mathscr{L}_2 \tag{2.3.1}$$

It is not hard to show that if $e_1, e_2, \ldots, e_{n_1}$ is a basis for the n_1-dimensional subspace $\mathscr{L}_1$ and $e_{n_1+1}, e_{n_1+2}, \ldots, e_{n_1+n_2}$ is a basis for the n_2-dimensional subspace $\mathscr{L}_2$ then $e_1, e_2, \ldots, e_{n_1}, \; e_{n_1+1}, \ldots, e_{n_1+n_2}$ is a basis for the (n_1+n_2)-dimensional space $\mathscr{L} = \mathscr{L}_1 \oplus \mathscr{L}_2$. If we change to new bases

$$e'_k = \sum_{i=1}^{n_1} e_i a_{ik} \qquad k = 1, 2, \ldots, n_1$$

and

$$e'_l = \sum_{j=n_1+1}^{n_1+n_2} e_j b_{jl} \qquad l = n_1+1, \ldots, n_1+n_2$$

in $\mathscr{L}_1$ and $\mathscr{L}_2$ respectively then the new basis in $\mathscr{L}$ is given by

$$e'_k = \sum_{i=1}^{n_1+n_2} e_i c_{ik} \qquad k = 1, 2, \ldots, n_1 + n_2$$

where the matrix $\mathbf{C} = [c_{ij}]$ has the block diagonal form

$$\mathbf{C} = \left[\begin{array}{c|c} \mathbf{A} & \mathbf{O} \\ \hline \mathbf{O} & \mathbf{B} \end{array}\right] \tag{2.3.2}$$

This matrix is called the *direct sum* of **A** and **B** and we write (2.3.2) as

$$\mathbf{C} = \mathbf{A} \oplus \mathbf{B}$$

We can also construct the *direct product* $\mathscr{L}_1 \otimes \mathscr{L}_2$ of two linear spaces $\mathscr{L}_1$ and $\mathscr{L}_2$. Suppose that $\mathscr{L}_1$ is an m-dimensional space spanned by $e_1, e_2, \ldots, e_m$ and $\mathscr{L}_2$ is an n-dimensional space spanned by $f_1, f_2, \ldots, f_n$. The space $\mathscr{L}_1 \otimes \mathscr{L}_2$ is defined to the mn-dimensional space spanned by the mn products $g_{ij} = e_i f_j$ $(i = 1, 2, \ldots, m;\ j = 1, 2, \ldots, n)$. If we change to new bases

$$e'_i = \sum_{k=1}^{m} e_k a_{ki} \qquad i = 1, 2, \ldots, m$$

and

$$f'_j = \sum_{l=1}^{n} f_l b_{lj} \qquad j = 1, 2, \ldots, n$$

in $\mathscr{L}_1$ and $\mathscr{L}_2$ respectively then the new basis in $\mathscr{L}_1 \otimes \mathscr{L}_2$ is given by

$$g'_{ij} = e'_i f'_j = \sum_{k,l} e_k f_l a_{ki} b_{lj} = \sum_{k,l} g_{kl} a_{ki} b_{lj} \tag{2.3.3}$$

The transformation matrix between g_{kl} and g'_{ij} has elements $a_{ki} b_{lj}$. This matrix is called the *direct product* of **A** and **B**. It is denoted by $\mathbf{A} \otimes \mathbf{B}$ and is given by

$$\mathbf{A} \otimes \mathbf{B} = \begin{bmatrix} a_{11}\mathbf{B} & a_{12}\mathbf{B} & \cdots & a_{1m}\mathbf{B} \\ a_{21}\mathbf{B} & & \cdots & a_{2m}\mathbf{B} \\ \cdots & \cdots & \cdots & \\ a_{m1}\mathbf{B} & & \cdots & a_{mm}\mathbf{B} \end{bmatrix} \tag{2.3.4}$$

2.4 Linear operators

Let $\mathscr{L}$ be an n-dimensional linear space. We define a *mapping* A of the linear space $\mathscr{L}$ onto itself as a rule by which we associate with each vector α in $\mathscr{L}$ another β in $\mathscr{L}$. The vector β is called the *image* of α under A. The mapping is said to be *one to one* if no two vectors have the same image and if every image vector arises from the mapping of one and only one vector. The mapping

$$A:\alpha\to\beta$$

can be written as

$$\beta=\hat{A}\alpha \tag{2.4.1}$$

where $\hat{A}$ is an *operator* which acts upon vectors α in $\mathscr{L}$ to give vectors β in $\mathscr{L}$. If for all α and β in $\mathscr{L}$ we have

$$\begin{aligned}\hat{A}(\alpha+\beta)&=\hat{A}\alpha+\hat{A}\beta\\ \hat{A}(c\alpha)&=c\hat{A}\alpha\end{aligned} \tag{2.4.2}$$

where c is a scalar then we say that $\hat{A}$ is a *linear operator.*

Now let $e_1, e_2, \ldots, e_n$ be a basis in $\mathscr{L}$. We have

$$\alpha=\sum_i \alpha^i e_i \qquad \beta=\sum_i \beta^i e_i$$

and

$$\beta=\hat{A}\alpha=\sum_i \alpha^i(\hat{A}e_i)$$

The vectors $\hat{A}e_i=e_i'$ can be regarded as forming a new basis in $\mathscr{L}$. We have

$$\sum_i \beta^i e_i=\sum_j \alpha^j(\hat{A}e_j)=\sum_j \alpha^j e_j'$$

According to (2.2.4) we can write

$$\sum_i \beta^i e_i=\sum_j \alpha^j \sum_k e_k a_{kj}$$

Now the basis vectors e_i are linearly independent and it follows that

$$\beta^i=\sum_j a_{ij}\alpha^j$$

This can be written in matrix notation as

$$\boldsymbol{\beta}=\mathbf{A}\boldsymbol{\alpha} \tag{2.4.3}$$

Comparison of (2.4.1) and (2.4.3) allows us to call $\mathbf{A}$ an *n-dimensional matrix representation* of the linear operator $\hat{A}$ in the basis e_i. If we use a different basis e_i' which is related to the basis e_i through a linear transformation with matrix $\mathbf{B}$ then we have (using (2.2.7))

$$\mathbf{e}'=\mathbf{eB} \qquad \boldsymbol{\alpha}=\mathbf{B}\boldsymbol{\alpha}' \qquad \boldsymbol{\beta}=\mathbf{B}\boldsymbol{\beta}'$$

and it follows that

$$\boldsymbol{\beta}'=(\mathbf{B}^{-1}\mathbf{AB})\boldsymbol{\alpha}' \tag{2.4.4}$$

Thus matrix representations of a linear operator $\hat{A}$ in different bases are transforms of one another. The matrix $\mathbf{A}$ and its transform $(\mathbf{B}^{-1}\mathbf{AB})$ are called *equivalent representations* of $\hat{A}$ since they only differ through a choice of basis.

2.5 Invariant subspaces

Let $\mathscr{A}=\{A_1, A_2, \ldots\}$ be a set of linear mappings of an n-dimensional linear space $\mathscr{L}$ onto itself. In particular this set may constitute a group of mappings. Suppose there exists a linear subspace $\mathscr{L}_1$ of $\mathscr{L}$ such that under a set of linear mappings $\mathscr{A}=\{A_1, A_2, \ldots\}$ of $\mathscr{L}$ onto $\mathscr{L}$ every vector on $\mathscr{L}_1$ is mapped onto another vector in $\mathscr{L}_1$. When this is the case $\mathscr{L}_1$ is called an *invariant subspace* of $\mathscr{L}$ under the set $\mathscr{A}$. The set of mappings $\mathscr{A}$ is also said to be *reducible*. If no such invariant subspace of $\mathscr{L}$ exists then the set of mappings $\mathscr{A}$ is said to be *irreducible*. If there exists a second invariant subspace $\mathscr{L}_2$ of $\mathscr{L}$ such that

$$\mathscr{L}=\mathscr{L}_1 \oplus \mathscr{L}_2$$

then the set of mappings $\mathscr{A}$ is said to be *decomposable* or *completely irreducible*. It follows from (2.3.2) that when $\mathscr{A}$ is decomposable all the matrix representations $\mathbf{A}_1, \mathbf{A}_2, \ldots$ can be expressed in block diagonal form by a suitable ordering of the basis vectors in $\mathscr{L}$.

In connection with the irreducibility of sets (or groups) of mappings there are two very important results which we now state without proof. They are usually referred to collectively as Schur's Lemma (see Hamermesh, 1962).

I. Let $\mathscr{A}=\{A_1, A_2, \ldots\}$ and $\mathscr{B}=\{B_1, B_2, \ldots\}$ be two irreducible sets of linear mappings in two linear spaces $\mathscr{L}_1$ and $\mathscr{L}_2$ respectively. Let the dimensions of $\mathscr{L}_1$ and $\mathscr{L}_2$ be n and m respectively. Consider the matrix representation $\mathbf{A}_i$ and $\mathbf{B}_i$ of the sets $\mathscr{A}$ and $\mathscr{B}$ respectively. If there exists a matrix $\mathbf{C}$ such that

$$\mathbf{CA}_i=\mathbf{B}_i\mathbf{C} \qquad i=1, 2, \ldots$$

then either $\mathbf{C}=\mathbf{O}$ or else $\mathscr{A}$ and $\mathscr{B}$ are equivalent in the sense of (2.4.4). In that case $n=m$.

II. If a matrix $\mathbf{C}$ commutes with all the matrices $\mathbf{A}_1, \mathbf{A}_2, \ldots$ of an irreducible set $\mathscr{A}$ then $\mathbf{C}$ is a numerical multiple $\lambda\mathbf{I}$ of the unit matrix

$$\mathbf{C}=\lambda\mathbf{I} \tag{2.5.1}$$

2.6 The dual space

Just as functions of scalars can be defined so too can functions of vectors. In particular we define ϕ as a *linear function* on an n-dimensional linear space $\mathscr{L}$ by the properties

$$\phi(\alpha+\beta)=\phi(\alpha)+\phi(\beta) \tag{2.6.1}$$

$$\phi(k\alpha)=k\phi(\alpha) \tag{2.6.2}$$

where α and β are vectors in $\mathscr{L}$ and k is a scalar. It can be shown (Jacobson, 1953) that the totality of all linear functions $\{\phi, \psi, \ldots\}$ forms an n-dimensional linear space. This new linear space is called the *dual space* of $\mathscr{L}$ and we denote it by $\tilde{\mathscr{L}}$. It can further be shown (Jacobson, 1953) that

$$(\tilde{\tilde{\mathscr{L}}})=\mathscr{L}$$

A specially important linear function is given by

$$\phi(\alpha)=c \tag{2.6.3}$$

where c is a scalar. This linear function is essentially a mapping of the n-dimensional space $\mathscr{L}$ onto the 1-dimensional space of scalars. If $(\alpha^1, \alpha^2, \ldots, \alpha^n)$ are the components of α in the basis $\{e_i\}$ then

$$\alpha=\sum_i e_i\alpha^i \tag{2.6.4}$$

and

$$\phi(\alpha)=\sum_i \phi(e_i)\alpha^i=\sum_i c_i\alpha^i=c \tag{2.6.5}$$

where

$$c_i=\phi(e_i) \qquad i=1, 2, \ldots, n \tag{2.6.6}$$

It is now apparent that ϕ is completely determined by the scalars $c_1, c_2, \ldots, c_n$. We can in fact regard $(c_1, c_2, \ldots, c_n)$ as the components of ϕ relative to the *dual basis* $\{\tilde{e}_i\}$ in $\tilde{\mathscr{L}}$ defined by (see Jacobson, 1953)

$$\tilde{e}_i(e_j)=\delta_{ij} \tag{2.6.7}$$

This can be clarified as follows. If

$$\phi=\sum_{j=1}^{n} \tilde{e}_j c_j \tag{2.6.8}$$

then

$$\phi(e_i)=\sum_j \tilde{e}_j(e_i)c_j=c_i$$

which is in accord with (2.6.6).

From (2.6.5) we see that it is possible to express (2.6.3) in matrix form as

$$\tilde{\boldsymbol{\phi}}\boldsymbol{\alpha} = c \tag{2.6.9}$$

where $\tilde{\boldsymbol{\phi}}$ is the matrix

$$\tilde{\boldsymbol{\phi}} = [c_1 c_2 \ldots c_n]$$

and c is a scalar. The transposition of ϕ is necessary since we have agreed to write the components of a vector as a column matrix and only a row times column matrix product can give a scalar.

2.7 Cogredience and contragredience

Let $e_1, e_2, \ldots, e_n$ be a basis for an n-dimensional linear space $\mathscr{L}$. Let $e_1', e_2', \ldots, e_n'$ be a second basis for $\mathscr{L}$ given by

$$\mathbf{e}' = \mathbf{e}\mathbf{A} \tag{2.7.1}$$

If a set of quantities $x_1, x_2, \ldots, x_n$ which, written as a row matrix $\mathbf{x}$, transform under the basis change (2.7.1) according to

$$\mathbf{x}' = \mathbf{x}\mathbf{A} \tag{2.7.2}$$

we say $\mathbf{x}$ transforms *cogrediently* to $\mathbf{e}$. If a set of quantities $\alpha_1, \alpha_2, \ldots, \alpha_n$ which, written as a column matrix $\boldsymbol{\alpha}$, transform under the basis change (2.7.1) according to

$$\boldsymbol{\alpha}' = \mathbf{A}^{-1}\boldsymbol{\alpha} = \mathbf{R}\boldsymbol{\alpha} \tag{2.7.3}$$

we say $\boldsymbol{\alpha}$ transforms *contragrediently* to $\mathbf{e}$. The transformations (2.7.2) and (2.7.3) are said to be *contragredient*. We note that for contragredient transformations we have the invarient

$$\mathbf{x}'\boldsymbol{\alpha}' = \mathbf{x}\mathbf{A}\mathbf{A}^{-1}\boldsymbol{\alpha} = \mathbf{x}\boldsymbol{\alpha} \tag{2.7.4}$$

The standard contragredient law (2.7.3) is

$$\alpha_i' = \sum_j R_{ij}\alpha_j \tag{2.7.5}$$

Written in the same form (2.7.1) becomes

$$e_i' = \sum_j R_{ij}^{\#} e_j \tag{2.7.6}$$

where $\mathbf{R}^{\#}$ is the contragredient matrix of $\mathbf{R}$ defined by

$$\mathbf{R}^{\#} = \tilde{\mathbf{R}}^{-1} \tag{2.7.7}$$

According to (2.2.9) the components α^i of a vector α in $\mathscr{L}$ transform contragrediently. We say that the α^i form a *contravariant set* or more briefly that vectors in $\mathscr{L}$ are *contravariant vectors.*

Consider next vectors ϕ in the dual space $\tilde{\mathscr{L}}$. Since $\phi(\alpha)=c$ is a scalar it follows that a transformation A on $\mathscr{L}$ induces a transformation B on $\tilde{\mathscr{L}}$

$$\boldsymbol{\phi}'=\mathbf{B}\boldsymbol{\phi} \tag{2.7.8}$$

such that

$$\tilde{\boldsymbol{\phi}}'\boldsymbol{\alpha}'=\tilde{\boldsymbol{\phi}}\mathbf{B}\mathbf{A}^{-1}\boldsymbol{\alpha}=c$$

and $\mathbf{B}=\tilde{\mathbf{A}}$. We have

$$\tilde{\boldsymbol{\phi}}'=\tilde{\boldsymbol{\phi}}\mathbf{A} \tag{2.7.9}$$

and it follows that the components c_i of ϕ transform cogrediently. We say that the c_i form a *covariant set* or that vectors in $\tilde{\mathscr{L}}$ are *covariant vectors.*

We now have

$$\boldsymbol{\alpha}'=\mathbf{R}\boldsymbol{\alpha} \tag{2.7.10}$$

for contravariant vectors and

$$\boldsymbol{\phi}'=\mathbf{R}^{\#}\boldsymbol{\phi} \tag{2.7.11}$$

for covariant vectors. Alternatively if the covariant vectors undergo the transformation

$$\boldsymbol{\phi}'=\mathbf{S}\boldsymbol{\phi} \tag{2.7.12}$$

then the contravariant vectors (in the dual space) undergo the contragredient transformation

$$\boldsymbol{\alpha}'=\mathbf{S}^{\#}\boldsymbol{\alpha} \tag{2.7.13}$$

We note that for *unitary* transformations

$$\mathbf{R}^{-1}=\tilde{\mathbf{R}}^{*}=\mathbf{R}^{\dagger} \tag{2.7.14}$$

and then

$$\mathbf{R}^{\#}=\mathbf{R}^{*} \tag{2.7.15}$$

For *orthogonal* transformations

$$\mathbf{R}^{-1}=\tilde{\mathbf{R}} \tag{2.7.16}$$

and then

$$\mathbf{R}^{\#}=\mathbf{R} \tag{2.7.17}$$

2.8 Vector spaces

A linear space $\mathscr{L}$ becomes a *vector space* when a *scalar product* (inner product) is defined in it. Sometimes what we call a linear space is called an affine vector space. In such cases what we call a vector space is called a metric space or unitary space or inner product space. In 3-dimensional

Euclidean space the definitions of concepts such as length (magnitude) of vectors and angles between vectors are obvious. In order to take these concepts over to a general n-dimensional linear space $\mathscr{L}$ it is necessary to define a *metric* and a scalar product in $\mathscr{L}$. This is achieved as follows.

We associate with every pair of vectors α and β in $\mathscr{L}$ a complex number (scalar) which is denoted by $\langle\alpha \mid \beta\rangle$ and is called the scalar product of α and β. The scalar product $\langle\alpha \mid \beta\rangle$ can be defined in any way we choose so long as the definition satisfies the following axioms.

$$\langle\alpha \mid \beta\rangle=\langle\beta \mid \alpha\rangle^* \tag{2.8.1}$$

$$\langle\lambda_1\alpha+\lambda_2\beta \mid \gamma\rangle=\lambda_1^*\langle\alpha \mid \gamma\rangle+\lambda_2^*\langle\beta \mid \gamma\rangle \tag{2.8.2}$$

$$\langle\alpha \mid \alpha\rangle\geqslant 0 \qquad \text{and} \qquad \langle\alpha \mid \alpha\rangle=0 \qquad \text{if and only if } \alpha=0 \tag{2.8.3}$$

In (2.8.2) λ_1 and λ_2 are scalars and α, β, γ are any three vectors in $\mathscr{L}$. The quantity $\langle\alpha \mid \alpha\rangle$ is called the metric in $\mathscr{L}$. It is clear that $\langle\alpha \mid \alpha\rangle$ is a real number. The quantity $+\sqrt{\langle\alpha \mid \alpha\rangle}$ is called the *magnitude* of the vector α. In a real vector space the *angle* ω between two non-zero vectors is defined by

$$\cos\omega=\frac{\langle\alpha \mid \beta\rangle}{\sqrt{\langle\alpha \mid \alpha\rangle\langle\beta \mid \beta\rangle}} \tag{2.8.4}$$

One of the most convenient ways of defining a scalar product in an n-dimensional linear space is to consider the components of the base vectors, defined by

$$\left.\begin{aligned}\varepsilon_1&=(1,0,\ldots,0)\\ \varepsilon_2&=(0,1,0,\ldots,0)\\ &\cdots\cdots\cdots\cdots\\ \varepsilon_n&=(0,0,\ldots,0,1)\end{aligned}\right\} \tag{2.8.5}$$

A scalar product in $\mathscr{L}$ is now defined by setting

$$\begin{aligned}\langle\varepsilon_i \mid \varepsilon_j\rangle=\delta_{ij}&=1 \qquad \text{if} \qquad i=j\\ &=0 \qquad \text{if} \qquad i\neq j\end{aligned} \tag{2.8.6}$$

If

$$\alpha=\sum_{i=1}^{n} a^i\varepsilon_i \qquad \text{and} \qquad \beta=\sum_{j=1}^{n} b^j\varepsilon_j$$

are any two vectors in $\mathscr{L}$ then we have

$$\langle\alpha \mid \beta\rangle=\sum_{i=1}^{n}(a^i)^*b^i \tag{2.8.7}$$

and

$$\langle\alpha \mid \alpha\rangle=\sum_{i}|a^i|^2\geqslant 0 \tag{2.8.8}$$

Now let $e_1, e_2, \ldots, e_n$ be an arbitrary basis in $\mathscr{L}$. The scalar product can be defined in terms of the *metric matrix* **G** whose elements g_{ij} are given by

$$g_{ij} = \langle e_i \mid e_j \rangle \tag{2.8.9}$$

If $\alpha = \sum_i \alpha^i e_i$ and $\beta = \sum_j \beta^j e_j$ then we have

$$\langle \alpha \mid \beta \rangle = \sum_{i,j} (\alpha^i)^* g_{ij} \beta^j \tag{2.8.10}$$

This can be written in matrix notation as

$$\langle \boldsymbol{\alpha} \mid \boldsymbol{\beta} \rangle = \tilde{\boldsymbol{\alpha}}^* \mathbf{G} \boldsymbol{\beta} = \boldsymbol{\alpha}^\dagger \mathbf{G} \boldsymbol{\beta} \tag{2.8.11}$$

If the basis vectors e_i are related to the vectors ε_i through the linear transformation

$$e_i = \sum_k \varepsilon_k a_{ki} \tag{2.8.12}$$

then we find

$$g_{ij} = \sum_k a^\dagger_{ik} a_{kj} \qquad \mathbf{G} = \mathbf{A}^\dagger \mathbf{A} \tag{2.8.13}$$

If the vectors e_i constitute an *orthonormal basis* so that

$$\langle e_i \mid e_j \rangle = g_{ij} = \delta_{ij}$$

then $\mathbf{A}^\dagger \mathbf{A} = \mathbf{I}$ and the matrix **A** is unitary. It follows that transition from one orthonormal basis to another orthonormal basis is accomplished by means of a unitary transformation. When $\mathbf{G} = \mathbf{I}$ we speak of a *unitary* vector space.

Consider next a linear mapping of a vector space $\mathscr{L}$ onto itself given by the linear operator $\hat{A}$. If α and β are vectors in $\mathscr{L}$ then the image vectors are

$$\alpha' = \hat{A}\alpha \qquad \beta' = \hat{A}\beta$$

Suppose further that $\mathbf{A} = [a_{ij}]$ is a unitary matrix representation of $\hat{A}$. We have

$$\boldsymbol{\alpha}' = \mathbf{A}\boldsymbol{\alpha} \qquad \boldsymbol{\beta}' = \mathbf{A}\boldsymbol{\beta}$$

Within the basis $\{\varepsilon_i\}$ we can write

$$\langle \alpha \mid \beta \rangle = \boldsymbol{\alpha}^\dagger \boldsymbol{\beta}$$

$$\langle \alpha' \mid \beta' \rangle = (\boldsymbol{\alpha}')^\dagger \boldsymbol{\beta}' = \langle \hat{A}\alpha \mid \hat{A}\beta \rangle$$

Since **A** is a unitary matrix it follows that

$$\langle \hat{A}\alpha \mid \hat{A}\beta \rangle = \langle \alpha \mid \beta \rangle \tag{2.8.14}$$

An operator $\hat{A}$ which satisfies (2.8.14) is called a *unitary operator.*

Consider now a basis change in $\mathscr{L}$ given by

$$\mathbf{e}' = \mathbf{eA}$$

We know that the components α^i of a vector α in $\mathscr{L}$ undergo the contragredient transformation

$$\boldsymbol{\alpha}' = \mathbf{A}^{-1}\boldsymbol{\alpha}$$

If α and β are vectors on $\mathscr{L}$ the scalar product in the $\mathbf{e}$ basis is by (2.8.11)

$$\langle \alpha \mid \beta \rangle = \boldsymbol{\alpha}^\dagger \mathbf{G} \boldsymbol{\beta}$$

In the $\mathbf{e}'$ basis the scalar product is

$$\langle \alpha \mid \beta \rangle = \boldsymbol{\alpha}'^\dagger \mathbf{G}' \boldsymbol{\beta}'$$

where $\mathbf{G}'$ has matrix elements

$$g'_{ij} = \langle e'_i \mid e'_j \rangle$$

Now

$$g'_{ij} = \left\langle \sum_k e_k A_{ki} \,\middle|\, \sum_l e_l A_{lj} \right\rangle$$

$$= \sum_{k,l} A^*_{ki} g_{kl} A_{lj} = \sum_{k,l} A^\dagger_{ik} g_{kl} A_{lj}$$

and

$$\mathbf{G}' = \mathbf{A}^\dagger \mathbf{G} \mathbf{A}$$

It is then at once evident that

$$\boldsymbol{\alpha}'^\dagger \mathbf{G}' \boldsymbol{\beta}' = \boldsymbol{\alpha}^\dagger \mathbf{A}^{-1\dagger} \mathbf{A}^\dagger \mathbf{G} \mathbf{A} \mathbf{A}^{-1} \boldsymbol{\beta} = \boldsymbol{\alpha}^\dagger \mathbf{G} \boldsymbol{\beta}$$

and the alternative scalar product expressions are identically equal as they must be since $\langle \alpha \mid \beta \rangle$ is a scalar associated with two vectors α, β and does not depend on basis.

2.9 Contravariant and covariant components of a vector

We have seen that for a linear space there exists two types of vector: the contravariant vector in the space itself and the covariant vector in the dual space (or vice-versa). In this section we show that when a metric is defined so that the linear space becomes a vector space then a connection exists between contravariant vectors and covariant vectors. This implies that for a vector space there is only one type of vector, which can be described either by a contravariant set of components or by a covariant set of components.

Let $\mathscr{L}$ be an n-dimensional vector space. Let $\{e_i\}$ be a basis in $\mathscr{L}$ and α a vector in $\mathscr{L}$ given by

$$\alpha = \sum_{i=1}^{n} \alpha^i e_i \tag{2.9.1}$$

We have already seen that the components α^i-transform contragrediently to the base vectors. We call α^i the *contravariant components* of α.

Consider now a set of n scalars $\alpha_1, \alpha_2, \ldots, \alpha_n$ defined by

$$\alpha_j = \langle \alpha \mid e_j \rangle \qquad j = 1, 2, \ldots, n \tag{2.9.2}$$

If $\{e_i'\}$ is a second basis in $\mathscr{L}$ such that

$$e_j' = \sum_k e_k a_{ki} \qquad j = 1, 2, \ldots, n \tag{2.9.3}$$

then in this basis we have

$$\alpha_j' = \langle \alpha \mid e_j' \rangle = \sum_k \alpha_k a_{ki} \tag{2.9.4}$$

The set of scalars $\{\alpha_j\}$ thus transform cogrediently to the base vectors. It is also clear that this set of scalars completely determines α just as the set $\{\alpha^i\}$ does. We call α_j the *covariant components* of α. If we substitute (2.9.1) into (2.9.2) we find

$$\alpha_j = \sum_i (\alpha^i)^* g_{ij} \tag{2.9.5}$$

where as before

$$g_{ij} = \langle e_i \mid e_j \rangle$$

If we denote by g^{ij} the elements of the matrix $\mathbf{G}^{-1}$ then from (2.9.5) we find

$$\alpha^j = \sum_i \alpha_i^* (g^{ij})^* \qquad j = 1, 2, \ldots, n \tag{2.9.6}$$

The scalar product (2.8.10) can now be written in the alternative forms

$$\langle \alpha \mid \beta \rangle = \sum_{ij} (\alpha^i)^* g_{ij} \beta^j = \sum_j \alpha_j \beta^j = \sum_{i,j} \beta_i^* (g^{ij})^* \alpha_j \tag{2.9.7}$$

The metric matrix $\mathbf{G}$ clearly has the property

$$\tilde{\mathbf{G}} = \mathbf{G}^* \tag{2.9.8}$$

In an orthonormal basis $g_{ij} = \delta_{ij}$ and we have

$$\alpha_j = (\alpha^i)^* \qquad \alpha^i = \alpha_j^* \tag{2.9.9}$$

For such a basis the metric is given by

$$\langle \alpha \mid \alpha \rangle = \sum_j \alpha_j \alpha^j = \sum_j |\alpha^j|^2 = \sum_j |\alpha_j|^2 \tag{2.9.10}$$

In many applications the space $\mathscr{L}$ is a real vector space and

$$\mathbf{G}^* = \mathbf{G} = \tilde{\mathbf{G}} \tag{2.9.11}$$

It follows immediately from (2.9.5), (2.9.6) and (2.9.7) that when $\mathscr{L}$ is real we have the relations

$$\alpha_i = \sum_j g_{ij}\alpha^j \qquad i = 1, 2, \ldots, n \tag{2.9.12}$$

$$\alpha^i = \sum_j g^{ij}\alpha_j \qquad i = 1, 2, \ldots, n \tag{2.9.13}$$

$$\langle \alpha \mid \beta \rangle = \sum_i \alpha_i\beta^i = \sum_{i,j} g_{ij}\alpha^i\beta^j = \sum_{i,j} g^{ij}\alpha_i\beta_j \tag{2.9.14}$$

In an orthonormal basis (2.9.12) reduces to

$$\alpha_i = \alpha^i \tag{2.9.15}$$

and even the distinction between covariant and contravariant components vanishes. This is of course the situation which prevails in elementary vector analysis.

2.10 Complex vectors

We begin by reminding ourselves that the components of a vector describe the relationship of the vector to a given basis. When the basis is changed the components change but the vector is invariant. In a vector space there is a second invariant, namely the scalar product.

We have already discussed in Section 2.6 the dual of a linear space. Here we consider the dual of a unitary vector space $\mathscr{L}$. The scalar product of two vectors α and β in $\mathscr{L}$ given by

$$\langle \beta \mid \alpha \rangle = \sum_{i=1}^{n} (\beta^i)^*\alpha^i = c \tag{2.10.1}$$

is evidently a mapping of $\mathscr{L}$ onto the 1-dimensional space of scalars c. If we compare (2.10.1) with (2.6.5) we see that the vector whose components are $(\beta^i)^*$ can be regarded as a vector in the dual space. Thus in a unitary vector space the distinction between contravariant and covariant vectors has shrunk to the distinction between a vector and its complex conjugate. This complex conjugate vector has components, relative to the dual basis, which are the complex conjugates of the components of the vector. Thus if

$$\alpha = \sum_j \alpha^i e_i$$

then the dual or complex conjugate vector is defined by

$$\alpha = \alpha^* = \sum_j (\alpha^j)^* \tilde{e}_j$$

Since $(\alpha^*)^* = \alpha$ we have

$$\alpha = \sum_j \alpha^i \tilde{e}_i^*$$

and it follows that we can write

$$\tilde{e}_j = e_j^*$$

It must be realized that the vector α and its complex conjugate α^* belong to different spaces. One is in the dual space of the other. Since no definition exists for the addition of a vector in one space to a vector in another space it is clear that a complex vector is essentially different from a complex number. Complex vectors cannot in general be separated into real and imaginary parts since the procedure for obtaining the real part of an ordinary complex quantity is to take half the sum of the quantity and its conjugate. For this reason Dirac (1958) prefers to call α^* the conjugate imaginary of α. He denotes α by the ket $|\alpha\rangle$ and α^* by the bra $\langle\alpha|$.

2.11 Tensor spaces

Let $\mathscr{L}$ be an n-dimensional linear space and let $\mathscr{G}$ by a set (or group) of linear mappings of $\mathscr{L}$ onto itself. When the covariant vectors (in $\mathscr{L}$) with components $(\alpha_1, \alpha_2, \ldots, \alpha_n)$ transform under the linear transformation R in $\mathscr{G}$ according to

$$\alpha_i' = \sum_{j=1}^{n} R_{ij}\alpha_j \qquad i = 1, 2, \ldots, n \tag{2.11.1}$$

then by (2.7.13) the contravariant vectors (in the dual space $\tilde{\mathscr{L}}$) with components $(\alpha^1, \alpha^2, \ldots, \alpha^n)$ are required to transform according to

$$(\alpha^i)' = \sum_{j=1}^{n} R_{ij}^{\#}\alpha^j \qquad i = 1, 2, \ldots, n \tag{2.11.2}$$

where $\mathbf{R}^{\#}$ is the contragredient matrix defined in (2.7.7). If $\{\beta_i\}$ and $\{\beta^i\}$ are other sets of covariant and contravariant vector components in $\mathscr{L}$ and $\tilde{\mathscr{L}}$ respectively we have the transformations

$$\beta_k' = \sum_l R_{kl}\beta_l \tag{2.11.3}$$

$$(\beta^k)' = \sum_l R_{kl}^{\#}\beta^l \tag{2.11.4}$$

We now consider certain direct products (outer products or tensor products). These products are the three types of n^2 quantities namely $\alpha_j\beta_l$, $\alpha^j\beta^l$ and $\alpha_j\beta^l$.

In passing it is worth mentioning that when $\mathscr{L}$ is not simply a linear space but a vector space then it is sometimes necessary to introduce further types of products such as $\alpha_j^*\beta_l$ where the set $\{\alpha_i^*\}$ undergoes the complex conjugate transformation law (see McWeeny, 1963)

$$(\alpha_i^*)' = \sum_{j=1}^{n} R_{ij}^* \alpha_j^* \qquad (2.11.5)$$

In later chapters we shall be concerned with linear spaces together with their duals and it is unnecessary to consider explicitly transformations like (2.11.5). This is because we use unitary bases and the four transformation laws reduce to two.

From (2.11.1) and (2.11.3) we have

$$\alpha_i'\beta_k' = \sum_{j,l} R_{ij}R_{kl}\alpha_j\beta_l \qquad (2.11.6)$$

From (2.11.2) and (2.11.4) we have

$$(\alpha^i)'(\beta^k)' = \sum_{j,l} R_{ij}^{\#} R_{kl}^{\#} \alpha^j\beta^l \qquad (2.11.7)$$

From (2.11.1) and (2.11.4) we have

$$\alpha_i'(\beta^k)' = \sum_{j,l} R_{ij}R_{kl}^{\#}\alpha_j\beta^l \qquad (2.11.8)$$

Consider next any set of n^2 quantities T_{jl} whose transformation law is (2.11.6) i.e.

$$T_{ik}' = \sum_{j,l} R_{ij}R_{kl}T_{jl} \qquad (2.11.9)$$

We say that the n^2 quantities T_{jl} constitute the components of a *covariant tensor of rank two* with respect to the linear transformations in $\mathscr{G}$. The totality of all such second rank tensors constitutes a *tensor space* of rank two. Likewise any set of n^2 quantities T^{jl} whose transformation law is (2.11.7) i.e.

$$(T^{ik})' = \sum_{j,l} R_{ij}^{\#} R_{kl}^{\#} T^{jl} \qquad (2.11.10)$$

are said to constitute the components of a *contravariant tensor of rank two* with respect to $\mathscr{G}$. A set of n^2 quantities T_j^l whose transformation law is

(2.11.8) i.e.

$$(T_i^k)' = \sum_{j,l} R_{ij} R^{\#}_{kl} T_j^l \tag{2.11.11}$$

are said to constitute the components of a *mixed tensor of rank two* with respect to $\mathscr{G}$.

The above concepts can readily be generalized. Thus a set of n^r quantities $T^{l_1 \ldots l_s}_{j_1 \ldots j_{r-s}}$ whose transformation law is given by

$$(T^{k_1 \ldots k_s}_{i_1 \ldots i_{r-s}})' = \sum_{\substack{j_1, \ldots, j_{r-s} \\ l_1, \ldots, l_s}} R_{i_1 j_1} \ldots R_{i_{r-s} j_{r-s}} R^{\#}_{k_1 l_1} \ldots R^{\#}_{k_s l_s} T^{l_1 \ldots l_s}_{j_1 \ldots j_{r-s}} \tag{2.11.12}$$

are said to constitute the components of a *mixed tensor of rank r* with respect to $\mathscr{G}$. We say that the tensor is of type $(s, r-s)$ meaning that its contravariant rank is s and its covariant rank is $r-s$.

2.12 The metric tensor

In this section we consider the transformation properties of the metric matrix elements in more detail.

Let $\mathscr{L}$ be an n dimensional vector space with metric matrix elements given by

$$g_{ij} = \langle e_i \mid e_j \rangle \tag{2.12.1}$$

When the contravariant components $\{\alpha^i\}$ of a vector α in $\mathscr{L}$ undergo the transformation (2.11.2) the covariant components $\{\alpha_i\}$ undergo the transformation (2.11.1) and these transformations are induced by the basis change

$$e_i' = \sum_k e_k R^{\#-1}_{ki} = \sum_k e_k \tilde{R}_{ki} \tag{2.12.2}$$

Now in the basis $\{e_i'\}$ the metric matrix elements are given by

$$g_{ij}' = \langle e_i' \mid e_j' \rangle \tag{2.12.3}$$

Substitution of (2.12.2) in (2.12.3) leads to the transformation law

$$g_{ij}' = \left\langle \sum_k e_k \tilde{R}_{ki} \;\middle|\; \sum_l e_l \tilde{R}_{lj} \right\rangle$$

$$= \sum_{k,l} \tilde{R}^*_{ki} g_{kl} \tilde{R}_{lj} = \sum_{k,l} R^*_{ik} R_{jl} g_{kl} \tag{2.12.4}$$

Comparison of (2.12.4) and (2.11.9) shows that the metric matrix elements do not form the components of a rank two covariant tensor unless the vector space is real. In a complex vector space it is possible to attach an extended tensorial meaning to the metric by introducing explicitly further transformation of the type (2.11.5) (see McWeeny, 1963).

In a real vector space the g_{ij} form the covariant components of a rank two tensor called the *metric tensor.* Likewise the inverse metric matrix elements g^{ij} form the contravariant components of the metric tensor.

In conventional tensor analysis a process called *contraction* is defined which reduces the rank of a given tensor by two. In this process two tensor indices are set equal and summation over the index is carried out. Thus $g^{il}g_{jk}$ form the components of a fourth rank mixed tensor. If we set $l = j$ and sum over j we have

$$\sum_j g^{ij}g_{jk} = \delta^i_k$$

and the Kronecker delta is a mixed second rank tensor.

2.13 Hilbert spaces

We consider the totality $\mathscr{L}$ of all continuous single valued complex functions $f(x)$ of a real variable x which is defined in the range $a \leqslant x \leqslant b$. We further suppose that the integral

$$\int_a^b |f(x)|^2 \, dx$$

exists. If $f(x)$ and $g(x)$ are any two functions in $\mathscr{L}$ then it can be shown that a scalar product defined by

$$\langle f \mid g \rangle = \int_a^b f^*(x) g(x) \, dx \tag{2.13.1}$$

exists. It follows that the totality of functions $\mathscr{L}$ constitutes a vector space. Such a vector space is called a *Hilbert space.* We now assume the existence of a complete set of orthonormal functions $\xi_1(x)$, $\xi_2(x), \ldots, \xi_n(x), \ldots$. Since the functions are orthonormal we have

$$\langle \xi_i \mid \xi_j \rangle = \int_a^b \xi_i^*(x) \xi_j(x) \, dx = \delta_{ij}$$

Since the set of functions is assumed to be complete we can expand any vector $f(x)$ in $\mathscr{L}$ as

$$f(x) = \sum_{i=1}^{\infty} f_i \xi_i(x)$$

where f_i are scalars. It is clear that we can regard the set $\{\xi_i(x)\}$ as basis vectors for $\mathscr{L}$. Hilbert space is thus an example of an infinite dimensional vector space.

We can evidently generalize the above concepts to construct Hilbert spaces with functions of any number of variables. The wave functions which occur in quantum mechanics are vectors in a Hilbert space.

Consider now a finite set of functions $\phi_1(x), \phi_2(x), \ldots, \phi_n(x)$ where $\phi_i(x)$ is a vector in a Hilbert space. We suppose that

$$\langle\phi_i \mid \phi_j\rangle = \int_a^b \phi_i^*(x)\phi_j(x)\, dx = \delta_{ij}$$

The vectors $\phi_1, \phi_2, \ldots, \phi_n$ are linearly independent and span an n-dimensional subspace of Hilbert space. A general vector in this subspace is

$$\phi(x) = \sum_{i=1}^{n} a_i\phi_i(x)$$

If $\psi(x) = \sum_{i=1} b_i\phi_i(x)$ is another vector in the subspace then

$$\langle\phi \mid \psi\rangle = \int_a^b \phi^*(x)\psi(x)\, dx = \sum_{i=1}^{n} a_i^* b_i$$

and the properties of finite dimensional vector spaces are recovered.

An extremely important case of an n-dimensional subspace which consists of vectors from a Hilbert space is given by the wave functions which belong to an n-fold degenerate energy eigenvalue.

2.14 Algebras

An *algebra* is a linear space in which there is also defined a binary operation of multiplication which combines any two vectors of the space to form a third vector of the space.

Let us consider a finite group $\mathscr{G}$ of order g. We can label the elements by $R(a) = R_a$ where a is a parameter which can assume g values $a = 1, 2, 3, \ldots, g$. These values of a can be represented as points and we call the set of points the *group manifold*. This group manifold can formally be regarded as a g-dimensional linear space which is spanned by the basis vectors $R_1, R_2, \ldots, R_g$. A general vector in the space is given by

$$\alpha = \sum_{i=1}^{g} \alpha_i R_i$$

where the α_i are scalars. If $\beta = \sum_{j=1}^{g} \beta_j R_j$ is a second vector in the space then

$$\alpha\beta = \sum_{i=1}^{g} \sum_{j=1}^{g} \alpha_i\beta_j(R_iR_j)$$

Since R_i and R_j are group elements the product R_iR_j is a group element R_k and it follows that $\alpha\beta$ is a vector in the g-dimensional space. This g-dimensional linear space is thus an algebra which is called the group algebra.

A *subalgebra* $\mathscr{B}$ of an algebra $\mathscr{A}$ is a subspace of $\mathscr{A}$ such that any two vectors of the subspace when combined by the binary operation of multiplication form another vector of the subspace.

If a subalgebra $\mathscr{B}$ has the property that for β in $\mathscr{B}$, $\alpha\beta$ is also in $\mathscr{B}$ for any element α of the whole algebra $\mathscr{A}$, then B is called an *ideal* in $\mathscr{A}$. An algebra which has no proper ideals is called a *simple* algebra.

An algebra $\mathscr{A}$ can be expressed as the direct sum of two (or more) subalgebras $\mathscr{B}_1$ and $\mathscr{B}_2$ if the linear space $\mathscr{A}$ is the direct sum of $\mathscr{B}_1$ and $\mathscr{B}_2$ and if further $\alpha\beta = 0$ for all vectors α in $\mathscr{B}_1$ and all vectors β in $\mathscr{B}_2$. We write $\mathscr{A} = \mathscr{B}_1 \oplus \mathscr{B}_2$. If $\mathscr{B}_1$ and $\mathscr{B}_2$ are simple subalgebras then $\mathscr{A}$ is called a *semi-simple* algebra.

2.15 Euclidean 3-space

The space within which all our observations take place is ordinary Euclidean space. This space, which we shall denote by E_3, is evidently a real 3-dimensional vector space within which an orthonormal basis $e = (e_1, e_2, e_3)$ can always be chosen. Within such a basis the distinction between covariant and contravariant components of a vector vanishes. It is customary to denote a vector in E_3 by r and to write

$$r = xe_1 + ye_2 + ze_3 \tag{2.15.1}$$

$$x = \langle r \mid e_1 \rangle \qquad y = \langle r \mid e_2 \rangle \qquad z = \langle r \mid e_3 \rangle \tag{2.15.2}$$

We call (x, y, z) the Cartesian components of r. Alternatively we say that (x, y, z) are the Cartesian coordinates of a point $P(x, y, z)$ in E_3.

If R is a mapping of E_3 onto itself then the image vector r' arising from r is given by (2.4.1) as

$$r' = \hat{R}r = x'e_1 + y'e_2 + z'e_3 \tag{2.15.3}$$

Now

$$r' = \hat{R}(xe_1 + ye_2 + ze_3) = x(\hat{R}e_1) + y(\hat{R}e_2) + z(\hat{R}e_3)$$

The vectors

$$e_i' = \hat{R}e_i \qquad i = 1, 2, 3 \tag{2.15.4}$$

can be regarded as forming a new basis in E_3. If $\hat{R}$ is unitary (e.g. a rotation) then $e' = (e_1', e_2', e_3')$ constitutes a new set of orthonormal basis vectors (see (2.8.13)). We now have

$$r' = xe_1' + ye_2' + ze_3' = x'e_1 + y'e_2 + z'e_3 \tag{2.15.5}$$

According to (2.2.4) we can express the new basis vectors in terms of the old basis vectors by

$$e_j' = e_1 A_{1j} + e_2 A_{2j} + e_3 A_{3j} \qquad j = 1, 2, 3 \tag{2.15.6}$$

where A_{ij} are the elements of a matrix **A**. It readily follows from (2.15.6) and (2.15.4) that

$$\hat{\mathbf{R}}\mathbf{e} = \mathbf{e}' = \mathbf{eA} \tag{2.15.7}$$

We turn our attention next to scalar valued functions $\psi(r)$ of the vectors r in E_3. Such functions are called functions of position. Given an orthonormal basis e and a point $P(x, y, z)$ the function $\psi(r) = \psi(x, y, z)$ determines a scalar ψ_P called the value of the function ψ at the point P. As an example we have the function

$$\psi(r) = \psi(x, y, z) = xy$$

In order to make it clear that we are working in the basis e we write

$$\psi_P = \psi_e(r)$$

where r is the position vector of P.

Consider now a new orthonormal basis e'. The point P (assumed fixed in space) is assigned new coordinates (x', y', z') given by

$$x' = \langle r \mid e_1' \rangle \qquad y' = \langle r \mid e_2' \rangle \qquad z' = \langle r \mid e_3' \rangle$$

The position vector r of P is

$$r = x'e_1' + y'e_2' + z'e_3'$$

and ψ_P is the value of a function $\psi_{e'}(r)$. Now the basis e' can be related to the original basis e. Thus for example the new basis may have been obtained by rotating the original basis by 45° about the z axis in an anticlockwise direction as viewed from the origin. In that case

$$e_1' = \frac{1}{\sqrt{2}}(e_1 - e_2)$$

$$e_2' = \frac{1}{\sqrt{2}}(e_1 + e_2)$$

$$e_3' = e_3$$

The position vector r is given by

$$r = \frac{1}{\sqrt{2}}(x - y)e_1' + \frac{1}{\sqrt{2}}(x + y)e_2' + ze_3' = x'e_1' + y'e_2' + z'e_3'$$

and we see that the function $\psi(r)=xy$ in the new basis is

$$\psi_{e'}(r)=x'y'=\tfrac{1}{2}(x^2-y^2)$$

This function is a different function from xy. More generally when the basis is changed we have

$$\psi_e(r)\neq\psi_{e'}(r)$$

Since $\psi_{e'}$ is a new function we denote it by ψ'_e and say that a transformation

$$\mathbf{e}'=\hat{\mathbf{R}}\mathbf{e}=\mathbf{eA} \tag{2.15.8}$$

in E_3 induces the transformation

$$\psi'_e(r)\equiv\hat{R}\psi_e(r)=\psi_{e'}(r) \tag{2.15.9}$$

in the space of scalar valued functions ψ_e. We call ψ'_e the *transformed function.*

We could alternatively regard (x', y', z') as the coordinates of a new point P' relative to the original basis e. In that case

$$\begin{aligned}\psi_P&=\psi_e(r)=\psi_e(x, y, z)\\ \psi'_P&=\psi_{e'}(r)=\psi_e(x', y', z')\neq\psi_P\end{aligned} \tag{2.15.10}$$

From this point of view we have kept a fixed basis in E_3 and transformed the vectors according to

$$r'=\hat{R}r=x'e_1+y'e_2+z'e_3 \tag{2.15.11}$$

Thus

$$\psi'_P=\psi_e(r')=\psi_e(\hat{R}r) \tag{2.15.12}$$

We can also simultaneously transform the vectors in E_3 and the basis by means of an operator $\hat{R}$. Since

$$r'=xe'_1+ye'_2+ze'_3$$

we see that

$$\psi'_{P'}=\psi_{e'}(r')=\psi_e(r)=\psi_P \tag{2.15.13}$$

and the value of the transformed function at the transformed point is the same as the value of the original function at the original point. We have

$$\psi_e(r)=\psi_{e'}(r')=\psi_{e'}(\hat{R}r)=\psi'_e(\hat{R}r)=\hat{R}\psi_e(\hat{R}r) \tag{2.15.14}$$

When r is replaced by $\hat{R}^{-1}r$ we obtain the relation

$$\hat{R}\psi_e(r)=\psi_e(\hat{R}^{-1}r) \tag{2.15.15}$$

We now suppose the transformations to be rotations and we shall speak of rotated bases and rotated functions. Consider the three elementary functions

$$\begin{aligned} X &= X_e(r) = x = \langle e_1 \mid r\rangle \\ Y &= Y_e(r) = y = \langle e_2 \mid r\rangle \\ Z &= Z_e(r) = z = \langle e_3 \mid r\rangle \end{aligned} \tag{2.15.16}$$

The rotated functions are given by (2.15.9) as

$$\begin{aligned} \hat{R}X_e(r) &= X_{e'}(r) = \langle e'_1 \mid r\rangle \\ \hat{R}Y_e(r) &= Y_{e'}(r) = \langle e'_2 \mid r\rangle \\ \hat{R}Z_e(r) &= Z_{e'}(r) = \langle e'_3 \mid r\rangle \end{aligned} \tag{2.15.17}$$

From (2.15.16) we readily see that

$$\hat{R}[X \quad Y \quad Z] = [X \quad Y \quad Z]\mathbf{A} \tag{2.15.18}$$

and the functions transform cogrediently. It follows that when x, y, z are regarded as three functions of position they do not in general transform like the components of a vector (see McWeeny, 1963).

The rotated functions can also be defined by (2.15.15). Thus

$$\hat{R}X_e(r) = X_e(\hat{R}^{-1}r) = \langle e_1 \mid \hat{R}^{-1}r\rangle \tag{2.15.19}$$

Since $\mathbf{A}^{-1} = \tilde{\mathbf{A}}$ for rotations and since the basis e is orthonormal it is straightforward to show that (2.15.19) is the same as (2.15.17).

We are often required to find the effect of an operator $\hat{R}$ on a product of two functions of position. If ψ and ϕ are any two vectors in function space it can be seen from (2.15.15) that

$$\hat{R}[\psi_e(r)\phi_e(r)] = [\hat{R}\psi_e(r)][\hat{R}\phi_e(r)] \tag{2.15.20}$$

In particular if $\psi_e = X_e = x$ and $\phi_e = Y_e = y$ we have

$$\hat{R}(xy) = (\hat{R}x)(\hat{R}y) = (xA_{11} + yA_{21} + zA_{31})(xA_{12} + yA_{22} + zA_{32}) \tag{2.15.21}$$

where A_{ij} are the elements of the matrix $\mathbf{A}$ which occurs in (2.15.18). If we take a rotation about the z axis by 45° we find

$$\hat{R}(xy) = \tfrac{1}{2}(x^2 - y^2)$$

which is the same as the result we obtained by considering xy as a single function of position.

References

Dirac, P. A. M. (1958), "The Principles of Quantum Mechanics", Oxford, Chapter 1.
Hamermesh, M. (1962), "Group Theory and Its Application to Physical Problems", Addison Wesley, Chapter 3.
Jacobson, N. (1953), "Lectures in Abstract Algebra", Vol. II, Van Nostrand, Chapter 2.
McWeeny, R. (1963), "Symmetry", Pergamon, Chapters 7 and 8.

CHAPTER 3

Group Representations

In Chapter 1 we introduced symmetry transformations in terms of geometrical operations. In this chapter we give algebraic significance to the symmetry transformations. The mathematical apparatus required is linear algebra which we discussed in Chapter 2.

3.1 Symmetry groups and linear spaces

Let $\mathscr{G} = \{A, B, \ldots, R, S, \ldots\}$ be a group of symmetry transformations. Let $\mathscr{L}$ be an n-dimensional linear space of vectors. We now associate with each element R of $\mathscr{G}$ a linear operator $\hat{R}$ which maps $\mathscr{L}$ onto itself. If $e_1, e_2, \ldots, e_n$ is a basis in $\mathscr{L}$ and $\hat{R}$ is a linear operator then as we have seen a new basis in $\mathscr{L}$ is given by

$$e_i' = \hat{R}e_i \qquad i = 1, 2, \ldots, n \tag{3.1.1}$$

which can be written in matrix notations as

$$\mathbf{e}' = \mathbf{eR} \tag{3.1.2}$$

The components α^i of an arbitrary *fixed* vector α in $\mathscr{L}$ undergo the contravariant transformations (see (2.2.9))

$$\boldsymbol{\alpha}' = \mathbf{R}^{-1}\boldsymbol{\alpha} \tag{3.1.3}$$

The vectors in $\mathscr{L}$ are regarded as invariant and (3.1.3) gives the relation between the components of α in the two bases e_i' and e_i.

We now define transformed vectors as follows. Consider transformations in $\mathscr{L}$ which carry the vectors along with the basis. Thus when

$$e_i' = \hat{R}e_i$$

we have

$$\bar{\alpha} = \hat{R}\alpha \tag{3.1.4}$$

and the transformed vector $\bar{\alpha}$ is a new vector whose components relative to the basis e_i are $\bar{\alpha}^i$ say. Since both vectors and basis have been changed simultaneously it is clear that the vector $\bar{\alpha}$ is required to have the same components in the transformed basis as the vector α had in the original basis. Thus in matrix notation

$$\bar{\alpha} = \mathbf{e}'\boldsymbol{\alpha} = \mathbf{eR}\boldsymbol{\alpha} = \mathbf{e}\bar{\boldsymbol{\alpha}} \tag{3.1.5}$$

and the components of the transformed vector relative to the fixed basis e are given by

$$\bar{\boldsymbol{\alpha}} = \mathbf{R}\boldsymbol{\alpha} \tag{3.1.6}$$

Note that $\bar{\boldsymbol{\alpha}}$ and $\boldsymbol{\alpha}'$ are quite distinct. Thus $\boldsymbol{\alpha}'$ and $\boldsymbol{\alpha}$ refer to the same vector α while $\bar{\boldsymbol{\alpha}}$ refers to a new vector $\bar{\alpha}$.

It is quite straightforward to show that the set of operators $\{\hat{A}, \hat{B}, \ldots, \hat{R}, \hat{S}, \ldots\}$ forms a group which has the same structure as $\mathscr{G}$. Thus if $RS = T$ we have $\hat{R}\hat{S} = \hat{T}$ and we say that the group of operators forms a *realization* of the group of symmetry transformations. When a basis e is defined in $\mathscr{L}$ the operators can be described by their matrix representatives. Now if $\hat{R}\hat{S} = \hat{T}$ we have for an arbitrary vector α in $\mathscr{L}$

$$\hat{R}\hat{S}\alpha = \hat{R}\hat{S}\mathbf{e}\boldsymbol{\alpha} = \hat{R}\,\mathbf{e}\mathbf{S}\boldsymbol{\alpha} = \mathbf{e}\mathbf{R}\mathbf{S}\boldsymbol{\alpha} = \hat{T}\alpha = \hat{T}\mathbf{e}\boldsymbol{\alpha} = \mathbf{e}\mathbf{T}\boldsymbol{\alpha}$$

and

$$\mathbf{RS} = \mathbf{T}$$

The group of matrices $\{\mathbf{A}, \mathbf{B}, \ldots, \mathbf{R}, \mathbf{S}, \ldots\}$ thus has the same structure as $\mathscr{G}$ and we say that the homomorphic mapping of $\mathscr{G}$ onto a group of $n \times n$ matrices gives us an *n-dimensional matrix representation* Γ of $\mathscr{G}$. The linear space $\mathscr{L}$ is called the *representation space.* When

$$\hat{R}\mathbf{e} = \mathbf{e}\mathbf{R} \equiv \mathbf{e}\boldsymbol{\Gamma}(R) \tag{3.1.7}$$

then $\boldsymbol{\Gamma}(R)$ is the matrix representative of R in the representation Γ. Whenever we have occasion to consider components it is clear that we must use (3.1.6) and not (3.1.3). Thus

$$\hat{R}\boldsymbol{\alpha} = \boldsymbol{\Gamma}(R)\boldsymbol{\alpha} \tag{3.1.8}$$

We say that the vectors $e_1, e_2, \ldots, e_n$ form a basis for Γ and that the components $\alpha^1, \alpha^2, \ldots, \alpha^n$ *carry* Γ.

The actual matrices of course depend upon the choice of basis in $\mathscr{L}$. A different basis gives an equivalent matrix representation and equivalent representations are not regarded as distinct (see (2.4.4)).

It is clearly desirable to fix the representation matrices once and for all. In the case of molecular symmetry groups this can be done by initially taking the representation space $\mathscr{L}$ to be E_3. We define a *standard* matrix representation by using a right handed Cartesian coordinate system where the principal axis is the Z-axis. In the case of cubic groups the Z-axis passes through a face of the cube.

Consider a molecule with a single principal axis. We embed in the molecule a set of orthonormal basis vectors e_1, e_2, e_3 such that e_3 is in the direction of the principal axis. Any symmetry transformation can be described by the effect of a linear operator acting on the basis vectors. When we rotate the base vectors by an angle θ about the Z-axis in a clockwise direction as viewed from the origin we readily find

$$\hat{R}e_1 = e_1' = e_1 \cos\theta + e_2 \sin\theta$$
$$\hat{R}e_2 = e_2' = -e_1 \sin\theta + e_2 \cos\theta$$
$$\hat{R}e_3 = e_3' = e_3$$

where $\hat{R}$ is the operator corresponding to the rotation. In matrix notation we have

$$\mathbf{e}' = \hat{R}\mathbf{e} = \mathbf{eR} \tag{3.1.9}$$

where

$$\mathbf{R} = \begin{bmatrix} \cos\theta & -\sin\theta & 0 \\ \sin\theta & \cos\theta & 0 \\ 0 & 0 & 1 \end{bmatrix} \tag{3.1.10}$$

For a general symmetry transformation R we write (3.1.9) as

$$\hat{R}\mathbf{e} = \mathbf{e}\boldsymbol{\Gamma}(R) \tag{3.1.11}$$

and we say that the vectors e_1, e_2, e_3 form a basis for a standard 3-dimensional representation Γ of R. The matrix which represents R is $\boldsymbol{\Gamma}(R)$. Any set of three linearly independent functions ψ_1, ψ_2, ψ_3 which transform according to

$$\hat{R}[\psi_1\psi_2\psi_3] = [\psi_1\psi_2\psi_3]\boldsymbol{\Gamma}(R) \tag{3.1.12}$$

are said to constitute a standard basis for the matrix representation Γ.

When R is the symmetry transformation $C_n(Z)$ we have $\theta = (2\pi/n)$ in (3.1.10) and the standard matrix representation of $C_n(Z)$ is given by

$$\boldsymbol{\Gamma}(C_n(Z)) = \begin{bmatrix} \cos\left(\frac{2\pi}{n}\right) & -\sin\left(\frac{2\pi}{n}\right) & 0 \\ \sin\left(\frac{2\pi}{n}\right) & \cos\left(\frac{2\pi}{n}\right) & 0 \\ 0 & 0 & 1 \end{bmatrix} \tag{3.1.13}$$

For $n = 3$ we find

$$\boldsymbol{\Gamma}(C_3(Z)) = \begin{bmatrix} -\frac{1}{2} & -\frac{\sqrt{3}}{2} & 0 \\ \frac{\sqrt{3}}{2} & -\frac{1}{2} & 0 \\ 0 & 0 & 1 \end{bmatrix} \tag{3.1.14}$$

The symmetry transformation $C_2(X)$ can readily be shown to have a matrix representation given by

$$\Gamma(C_2(X)) = \begin{bmatrix} 1 & 0 & 0 \\ 0 & -1 & 0 \\ 0 & 0 & -1 \end{bmatrix} \qquad (3.1.15)$$

in the standard basis e_1, e_2, e_3.

If we are interested in the group $\mathscr{D}_3$ then the standard matrix representations for the transformations C_3^2, C_2', C_2'' follow from the defining relations (1.2.2) together with (3.1.14) and (3.1.15). The vectors e_1, e_2, e_3 thus form a basis for a standard 3-dimensional matrix representation Γ of the group $\mathscr{D}_3$. This representation is given in full by

$$\Gamma: \quad \underset{E}{\begin{bmatrix} 1 & 0 & 0 \\ 0 & 1 & 0 \\ 0 & 0 & 1 \end{bmatrix}} \quad \underset{C_3}{\begin{bmatrix} -\frac{1}{2} & -\frac{\sqrt{3}}{2} & 0 \\ \frac{\sqrt{3}}{2} & -\frac{1}{2} & 0 \\ 0 & 0 & 1 \end{bmatrix}} \quad \underset{C_3^2}{\begin{bmatrix} -\frac{1}{2} & \frac{\sqrt{3}}{2} & 0 \\ -\frac{\sqrt{3}}{2} & -\frac{1}{2} & 0 \\ 0 & 0 & 1 \end{bmatrix}}$$

$$\underset{C_2}{\begin{bmatrix} 1 & 0 & 0 \\ 0 & -1 & 0 \\ 0 & 0 & -1 \end{bmatrix}} \quad \underset{C_2'}{\begin{bmatrix} -\frac{1}{2} & -\frac{\sqrt{3}}{2} & 0 \\ -\frac{\sqrt{3}}{2} & \frac{1}{2} & 0 \\ 0 & 0 & -1 \end{bmatrix}} \quad \underset{C_2''}{\begin{bmatrix} -\frac{1}{2} & \frac{\sqrt{3}}{2} & 0 \\ \frac{\sqrt{3}}{2} & \frac{1}{2} & 0 \\ 0 & 0 & -1 \end{bmatrix}} \qquad (3.1.16)$$

3.2 Character

If we use a different basis in the representation space $\mathscr{L}$ then according to (2.4.4) the matrix representation $\Gamma(\mathscr{G})$ is replaced by an equivalent matrix representation. If the transformation R is represented by the matrix $\boldsymbol{\Gamma}(R)$ then in an equivalent representation R is represented by the matrix $\boldsymbol{\Gamma}'(R) = \boldsymbol{\Delta}\boldsymbol{\Gamma}(R)\boldsymbol{\Delta}^{-1}$ where the matrix $\boldsymbol{\Delta}$ refers to the change of basis in $\mathscr{L}$.

The trace of the matrix $\boldsymbol{\Gamma}(R)$ is called the *character* of R in the representation Γ and is denoted by $\chi(R)$. Thus if $[\Gamma_{ij}(R)]$ is an $n \times n$ matrix we have

$$\chi(R) = \sum_{i=1}^{n} \Gamma_{ii}(R) \qquad (3.2.1)$$

It is not difficult to show that the characters of an element in equivalent representations are the same. Thus the character is invariant under a coordinate transformation and it characterizes the set of all equivalent representations. It is also quite straightforward to show that in a conjugate class all the elements of the class have the same character in a given representation. Unless we are concerned with the specific transformation properties of the basis vectors it follows that we can describe a matrix representation by listing the characters of one of the elements in each conjugate class. If the number of classes in the group is k then we have a set of characters $(\chi_1^{(i)}, \chi_2^{(i)}, \ldots, \chi_k^{(i)})$ for each matrix representation $\Gamma^{(i)}$. This set of characters can be regarded as a vector $\chi^{(i)}$ in a k-dimensional linear space.

For the representation (3.1.16) of $\mathscr{D}_3$ we have the characters

$$\begin{array}{cccc} & E & 2C_3 & 3C_2 \\ \Gamma & 3 & 0 & -1 \end{array} \tag{3.2.2}$$

The notation $2C_3$ denotes the class $\{C_3, C_3^2\}$ etc.

3.3 Irreducible representations

In the last section we constructed a 3-dimensional representation Γ of the group $\mathscr{D}_3$ in which the representation space $\mathscr{L}$ is simply E_3. If we examine the matrices (3.1.16) we see that, for all group elements, the vectors e_1 and e_2 are transformed amongst themselves and e_3 is left invariant. Thus, in accordance with Section 2.5, the space $\mathscr{L}_1$ spanned by (e_1, e_2) is an invariant subspace in the representation space. Likewise the space $\mathscr{L}_2$ spanned by e_3 is an invariant subspace in $\mathscr{L}$. Now

$$\mathscr{L} = \mathscr{L}_1 \oplus \mathscr{L}_2$$

and the group of matrices Γ is decomposable. We say that the representation Γ is *decomposable* and can be expressed as the *direct sum* of a 1-dimensional representation $\Gamma^{(1)}$ and a 2-dimensional representation $\Gamma^{(2)}$.

$$\Gamma = \Gamma^{(1)} \oplus \Gamma^{(2)}$$

From (3.1.16) the representations $\Gamma^{(1)}$ and $\Gamma^{(2)}$ are given by

$$\begin{array}{ccccccc} & E & C_3 & C_3^2 & C_2 & C_2' & C_2'' \\ \Gamma^{(1)}: & [1] & [1] & [1] & [-1] & [-1] & [-1] \end{array} \tag{3.3.1}$$

$$\Gamma^{(2)}: \begin{bmatrix} 1 & 0 \\ 0 & 1 \end{bmatrix} \begin{bmatrix} -\frac{1}{2} & -\frac{\sqrt{3}}{2} \\ \frac{\sqrt{3}}{2} & -\frac{1}{2} \end{bmatrix} \begin{bmatrix} -\frac{1}{2} & \frac{\sqrt{3}}{2} \\ -\frac{\sqrt{3}}{2} & -\frac{1}{2} \end{bmatrix} \begin{bmatrix} 1 & 0 \\ 0 & -1 \end{bmatrix} \begin{bmatrix} -\frac{1}{2} & -\frac{\sqrt{3}}{2} \\ -\frac{\sqrt{3}}{2} & \frac{1}{2} \end{bmatrix} \begin{bmatrix} -\frac{1}{2} & \frac{\sqrt{3}}{2} \\ \frac{\sqrt{3}}{2} & \frac{1}{2} \end{bmatrix} \tag{3.3.2}$$

In general let Γ be an n-dimensional matrix representation of a group $\mathcal{G}$ in which the representation space is $\mathcal{L}$. If there exists an m-dimensional invariant subspace in $\mathcal{L}$ then we say that Γ is a *reducible representation.* If on the other hand there is no proper invariant subspace in $\mathcal{L}$ then Γ is called an *irreducible representation* (I.R.). Suppose the representation space $\mathcal{L}$ can be written as the direct sum of two or more invariant subspaces $\mathcal{L}_1, \mathcal{L}_2, \ldots, \mathcal{L}_n$. Suppose further that each $\mathcal{L}_i$ is the representation space for an irreducible representation $\Gamma^{(i)}$ of $\mathcal{G}$. We say that the representation Γ induced by $\mathcal{L} = \mathcal{L}_1 \oplus \mathcal{L}_2 \oplus \ldots \oplus \mathcal{L}_n$ is *decomposable* into a direct sum of irreducible representations

$$\Gamma = \Gamma^{(1)} \oplus \Gamma^{(2)} \oplus \ldots \oplus \Gamma^{(n)}$$

The matrices $\mathbf{\Gamma}(R)$ are now expressible in block diagonal form (see (2.3.2)).

3.4 Representation theory of finite groups

In this section we collect together the most important results from the representation theory of finite groups (see, for example, Hamermesh, 1962).

I. The number of (non-equivalent) irreducible representations of a finite group is finite. In particular the number of classes of conjugate elements in the group is exactly equal to the number of irreducible representations.

Thus the group $\mathcal{D}_3$ has three conjugate classes and hence just three IR's.

II. From I we have a set of IR's $\Gamma^{(1)}$. $\Gamma^{(2)}, \ldots, \Gamma^{(k)}$ say. Let the dimensions of these IR's be $n_1, n_2, \ldots, n_k$ respectively. If g is the order of the group then

$$n_1^2 + n_2^2 + \ldots + n_k^2 = g \tag{3.4.1}$$

If g is not large this equation can be solved in positive integers and the possible dimensionalities of the IR's are thereby obtained.

Thus for $\mathcal{D}_3$ we have

$$n_1^2 + n_2^2 + n_3^2 = 6$$

with the solution $n_1 = 1$; $n_2 = 1$ and $n_3 = 2$. There are two 1-dimensional IR's and one 2-dimensional IR. One of the 1-dimensional IR's is the identity representation in which the unit 1×1 matrix is assigned to each group element. All groups possess the identity representation. The other 1-dimensional IR of $\mathcal{D}_3$ is clearly the $\Gamma^{(1)}$ of (3.3.1).

III. Every representation of a finite group is equivalent to a representation by unitary matrices.

IV. Every reducible representation of a finite group is decomposable.

V. Consider all the non-equivalent unitary IR's $\Gamma^{(\mu)}$ of a finite group $\mathcal{G}$. Let $\Gamma_{ij}^{(\mu)}(R)$ be the ijth element in the matrix representation $\Gamma^{(\mu)}(R)$ of the group element R. If $\mathcal{G}$ is of order g then

$$\sum_R \Gamma_{il}^{(\mu)}(R)\Gamma_{jm}^{(\nu)*}(R) = \left(\frac{g}{n_\mu}\right)\delta_{\mu\nu}\,\delta_{ij}\,\delta_{lm} \tag{3.4.2}$$

where n_μ is the dimension of $\Gamma^{(\mu)}$.

VI. If $\chi^{(\mu)}(R)$ is the character of R in the IR $\Gamma^{(\mu)}$ then

$$\sum_R \chi^{(\mu)}(R)\chi^{(\nu)*}(R) = g\,.\,\delta_{\mu\nu} \tag{3.4.3}$$

Further if $\mathcal{G}$ has k classes of conjugate elements and if g_i is the number of elements in the ith class then

$$\sum_{i=1}^{k} g_i\chi_i^{(\mu)}\chi_i^{(\nu)*} = g\,\delta_{\mu\nu} \tag{3.4.4}$$

where $\chi_i^{(\mu)}$ is the character of any element in the ith class in the μth IR. It might be helpful at this point to clarify the above results by some examples. We shall presently demonstrate that the $\Gamma^{(2)}$ given by (3.3.2) is the 2-dimensional IR of $\mathcal{D}_3$. Consider (3.4.2) and take $\mu=\nu=2$; $i=j=l=m=1$. We have

$$\sum_R \Gamma_{11}^{(2)}(R)\Gamma_{11}^{(2)}(R) = 1\cdot1+(-\tfrac{1}{2})(-\tfrac{1}{2})+(-\tfrac{1}{2})(-\tfrac{1}{2})+1\cdot1$$
$$+(-\tfrac{1}{2})(-\tfrac{1}{2})+(-\tfrac{1}{2})(-\tfrac{1}{2}) = 3 = \tfrac{6}{2}$$

and (3.4.2) is satisfied. For $\mu=1$, $\nu=2$ and $i=j=l=m=1$ we have

$$\sum_R \Gamma_{11}^{(1)}(R)\Gamma_{11}^{(2)}(R) = 1-\tfrac{1}{2}-\tfrac{1}{2}-1+\tfrac{1}{2}+\tfrac{1}{2}=0$$

and (3.4.2) is satisfied. The δ_{ij} and δ_{lm} properties can readily be tested by the reader for $\Gamma^{(2)}$. Consider next (3.4.3) for $\mathcal{D}_3$. With $\mu=\nu=2$ we have

$$\sum_R \chi^{(2)}(R)\chi^{(2)}(R) = 4+1+1+0+0+0=6$$

and (3.4.3) is satisfied. With $\mu=1$ and $\nu=2$ we have

$$\sum_R \chi^{(1)}(R)\chi^{(2)}(R) = 2-1-1+0+0+0=0$$

and (3.4.3) is satisfied. Finally to check (3.4.4) for $\mathcal{D}_3$ we first take $\mu=\nu=2$

and find

$$\sum_{i=1}^{3} g_i\chi_i^{(2)}\chi_i^{(2)} = 1\cdot 4 + 2\cdot 1 + 3\cdot 0 = 6$$

For $\mu = 1$ and $\nu = 2$ we have

$$\sum_{i=1}^{3} g_i\chi_i^{(1)}\chi_i^{(2)} = 1\cdot 2 + 2\,.\,(-1) + 3\cdot 0 = 0$$

and (3.4.4) is satisfied.

This last result can be used to derive an extremely useful formula. Let Γ be any representation of a finite group $\mathcal{G}$ of order g. According to IV this representation can be expressed as a direct sum of the IR's, $\Gamma^{(1)}, \Gamma^{(2)}, \ldots,$ $\Gamma^{(k)}$ of $\mathcal{G}$.

$$\Gamma = a_1\Gamma^{(1} \oplus a_2\Gamma^{(2)} \oplus \ldots \oplus a_k\Gamma^{(k)} \tag{3.4.5}$$

where a_i is the number of times $\Gamma^{(i)}$ appears in the decomposition. Now let $\sum^{\oplus}$ denote direct sum. We have

$$\Gamma(R) = \sum_{\nu}^{\oplus} a_\nu\Gamma^{(\nu)}(R)$$

If R is in the ith class then

$$\chi_i = \sum_{\nu} a_\nu\chi_i^{(\nu)} \tag{3.4.6}$$

and

$$\sum_i \chi_i^{(\mu)*}\chi_i g_i = \sum_{\nu} a_\nu \sum_i g_i\chi_i^{*(\mu)}\chi_i^{(\nu)} = g a_\mu$$

Thus

$$a_\mu = \frac{1}{g}\sum_i g_i\chi_i^{*(\mu)}\chi_i \tag{3.4.7}$$

This formula enables us to calculate the number of times a given IR occurs in the decomposition of an arbitrary representation.

From (3.4.6) we have

$$\chi_i^* = \sum_{\mu} a_\mu\chi_i^{*(\mu)}$$

and

$$\sum_i \chi_i\chi_i^* g_i = \sum_{\mu,\nu} a_\mu a_\nu \sum_i g_i\chi_i^{(\nu)}\chi_i^{*(\mu)} = \sum_{\mu,\nu} a_\mu a_\nu g\,\delta_{\mu\nu}$$

or

$$\sum_i g_i\,|\chi_i|^2 = g\sum_{\mu} a_\mu^2$$

If the original representation Γ is irreducible then all the coefficients a_μ are zero except one which is unity. Thus if we have

$$\sum_i g_i |\chi_i|^2 = g \tag{3.4.8}$$

then the representation with characters χ_i is irreducible. Consider the 2-dimensional representation $\Gamma^{(2)}$ of $\mathscr{D}_3$ given in (3.3.2). From (3.4.8) we have $1.4+2.1+3.0=6$ and $\Gamma^{(2)}$ is the 2-dimensional IR of $\mathscr{D}_3$. This is what we claimed earlier.

VII. Consider a finite group $\mathscr{G}$ of order g which can be expressed as the direct product of a subgroup $\mathscr{H}$ of order h and a subgroup $\mathscr{K}$ of order 2. Let the IR's of $\mathscr{H}$ be $\Gamma^{(1)}, \Gamma^{(2)}, \ldots, \Gamma^{(k)}$. From I and II it is clear that $\mathscr{K}$ has just two 1-dimensional IR's $\Delta^{(1)}$ and $\Delta^{(2)}$. It can be shown that the group $\mathscr{G}$ has twice as many IR's as $\mathscr{H}$. Thus the IR $\Gamma^{(i)}$ of $\mathscr{H}$ gives rise to the IR's $\Gamma^{(i)}\,\Delta^{(1)}$ and $\Gamma^{(i)}\,\Delta^{(2)}$ of $\mathscr{G}$. Since each element R of $\mathscr{G}$ can be expressed as $R = R_1R_2$ with R_1 in $\mathscr{H}$ and R_2 in $\mathscr{K}$ it follows that the characters of the IR's of $\mathscr{G}$ are the products of the characters for $\mathscr{H}$ and $\mathscr{K}$.

As an example consider the group $\mathscr{C}_{2h}$ which we introduced in Section 1.2. We have $\mathscr{C}_{2h} = \mathscr{C}_2 \otimes \mathscr{C}_i$. The IR's of $\mathscr{C}_2$ and $\mathscr{C}_i$ are given by

$\mathscr{C}_2$	E	C_2
$\Gamma^{(1)}$	[1]	[1]
$\Gamma^{(2)}$	[1]	[−1]

$\mathscr{C}_i$	E	I
$\Delta^{(1)}$	[1]	[1]
$\Delta^{(2)}$	[1]	[−1]

respectively. The IR's of $\mathscr{C}_{2h}$ are now found to be given by

$\mathscr{C}_{2h}$	E	C_2	I	$\sigma_h = C_2I$
$\Gamma^{(1)}\,\Delta^{(1)}$	[1]	[1]	[1]	[1]
$\Gamma^{(1)}\,\Delta^{(2)}$	[1]	[1]	[−1]	[−1]
$\Gamma^{(2)}\,\Delta^{(1)}$	[1]	[−1]	[1]	[−1]
$\Gamma^{(2)}\,\Delta^{(2)}$	[1]	[−1]	[−1]	[1]

VIII. Let $\{\mathscr{K}_i\}$ be the set of classes of a finite group. If for any element R in $\mathscr{K}_i$ the element R^{-1} is also in $\mathscr{K}_i$ (not in some other class), then we say that $\mathscr{K}_i$ is an *ambivalent* class. When all classes $\mathscr{K}_i$ are ambivalent then all characters must be real numbers.

3.5 Group representations and quantum mechanics

Let $\mathscr{G}$ be a group of symmetry transformations of a quantum mechanical system. An eigenfunction ψ of this system satsifies the Schrödinger equation

$$\hat{H}\psi = E\psi \tag{3.5.1}$$

and the totality $\{\psi\}$ of such functions forms a Hilbert space. If R is an element of $\mathcal{G}$ then application of the operator $\hat{R}$ on ψ will in general lead to a new function ψ'. Thus

$$\psi' = \hat{R}\psi \tag{3.5.2}$$

We now define the transformed Hamiltonian operator $\hat{H}'$ by

$$\hat{H}'\psi' = E\psi' \tag{3.5.3}$$

According to (3.5.1), (3.5.2) and (3.5.3) we have

$$E\psi' = \hat{R}E\psi = \hat{R}\hat{H}\psi = \hat{R}\hat{H}\hat{R}^{-1}\hat{R}\psi = (\hat{R}\hat{H}\hat{R}^{-1})\psi' = \hat{H}'\psi' \tag{3.5.4}$$

and it follows that

$$\hat{H}' = \hat{R}\hat{H}\hat{R}^{-1} \tag{3.5.5}$$

However since $\hat{R}$ is not just any operator but corresponds to a symmetry transformation R of the system we must have

$$\hat{H}' = \hat{H} = \hat{R}\hat{H}\hat{R}^{-1} \tag{3.5.6}$$

Thus

$$[\hat{R}, \hat{H}] = \hat{R}\hat{H} - \hat{H}\hat{R} = 0 \tag{3.5.7}$$

or

$$\hat{R}\hat{H} = \hat{H}\hat{R} \tag{3.5.8}$$

The linear operator $\hat{R}$ associated with a symmetry transformation R commutes with the Hamiltonian. Conversely if $\hat{R}$ is a linear operator which commutes with the Hamiltonian $\hat{H}$ of a quantum mechanical system then there exists a transformation R which is a symmetry transformation. The group of operators $\{\hat{R}_1, \hat{R}_2, \ldots\}$ together with $\hat{H}$ forms a set of mutually commuting operators and it follows that there exists a set of functions which are simultaneously eigenfunctions of each of the operators. The eigenvalues of $\hat{R}_1, \hat{R}_2, \ldots$ can thus be used to characterize the various stationary states of the system.

Now let E by an n-fold degenerate level so that there exists n linearly independent functions $\psi_1, \psi_2, \ldots, \psi_n$ all with energy E. These functions span an n-dimensional linear subspace of Hilbert space. The general vector of the subspace

$$\psi = \sum_{i=1}^{n} c_i\psi_i \tag{3.5.9}$$

where the c_i are scalars, has energy E. Now

$$\hat{R}\hat{H}\psi_j = E\hat{R}\psi_j = \hat{H}\hat{R}\psi_j$$

Thus $\hat{R}\psi_j$ is a vector in the subspace and we can write

$$\hat{R}\psi_j = \sum_{i=1}^{n} \psi_i \Gamma_{ij}(R) \tag{3.5.10}$$

where the coefficients c_i in (3.5.9) have been written in the more descriptive form as $\Gamma_{ij}(R)$. If S is a second symmetry transformation then

$$\hat{S}\psi_i = \sum_{k=1}^{n} \psi_k \Gamma_{ki}(S) \tag{3.5.11}$$

and

$$(\hat{S}\hat{R})\psi_j = \sum_{i,k=1}^{n} \psi_k \Gamma_{ki}(S)\Gamma_{ij}(R) \tag{3.5.12}$$

Since S and R are elements of a group the product $T = SR$ is a further symmetry transformation and we have

$$\hat{T}\psi_j = \sum_{k=1}^{n} \psi_k \Gamma_{kj}(T) \tag{3.5.13}$$

where $\hat{T} = \hat{S}\hat{R}$. It follows from (3.5.12) and (3.5.13) that

$$\Gamma_{kj}(T) = \sum_{i=1}^{n} \Gamma_{ki}(S)\Gamma_{ij}(R) \tag{3.5.14}$$

This can be written in matrix notation as

$$\mathbf{\Gamma}(T) = \mathbf{\Gamma}(S)\mathbf{\Gamma}(R)$$

It follows that the linear subspace of functions belonging to an n-fold degenerate energy level forms the representation space $\mathscr{L}$ for an n-dimensional matrix representation Γ of the symmetry group $\mathscr{G}$. Suppose the representation Γ is decomposable so that there exists at least two invariant subspaces $\mathscr{L}_1$ and $\mathscr{L}_2$ of $\mathscr{L}$ such that

$$\mathscr{L} = \mathscr{L}_1 \oplus \mathscr{L}_2$$

If this is the case then in general the eigenfunctions spanning $\mathscr{L}_1$ and $\mathscr{L}_2$ need not belong to the same eigenvalue. This contradicts the hypothesis that the eigenfunctions spanning $\mathscr{L}$ all belong to the same eigenvalue. Thus in general Γ is an irreducible representation of $\mathscr{G}$. When Γ is reducible we say that there is *accidental* degeneracy. This is a rare occurrence and is usually due to the very special nature of the potential energy of the system. Accidental degeneracy indicates the presence of hidden symmetries so that $\mathscr{G}$ is only a subgroup of the full symmetry group of the system.

If we know all the IR's of a symmetry group of a quantum mechanical system then we have at our disposal a means of classifying the stationary states of the system. This is so as we have seen how the eigenfunctions for these states span the IR's of the group. Severe limitations are thus placed on the possible types of eigenfunction that can occur.

References

Boerner, H. (1963), "Representations of Groups", North Holland, Chapter 3.

Hamermesh, M. (1962), "Group Theory and its application to Physical Problems", Addison-Wesley, Chapter 3.

Heine, V. (1960), "Group Theory in Quantum Mechanics", Pergamon Press.

McWeeny, R. (1963), "Symmetry", Pergamon.

CHAPTER 4

Irreducible Representations of Finite Molecular Symmetry Groups

In this chapter we show how the matrices for all standard IR's of a given finite molecular symmetry group are constructed. We shall not however display the results for every molecular symmetry group as an exhaustive treatment is already available (see McWeeny, 1963).

4.1 Spherical harmonics

The IR's of a given molecular symmetry group $\mathscr{G}$ can be obtained by starting from an n-dimensional linear space $\mathscr{L}$ of functions. Application of the linear operators, corresponding to the symmetry transformations, on the basis functions in $\mathscr{L}$ yields an n-dimensional representation of $\mathscr{G}$. This representation is then decomposed into IR's by looking for invariant subspaces of $\mathscr{L}$ If all the IR's are not obtained then we choose another linear space (often of higher dimension than $\mathscr{L}$) and repeat the process.

In quantum chemistry the spherical harmonics are of singular importance. These spherical harmonics are the functions $Y_{lm}(\theta, \phi)$ defined by

$$Y_{lm}(\theta, \phi) = (-1)^{(m+|m|)/2} T_{lm}(\theta) S_m(\phi) \tag{4.1.1}$$

with

$$S_m(\phi) = \frac{1}{\sqrt{2\pi}} e^{im\phi} \tag{4.1.2}$$

and

$$T_{lm}(\theta) = \left[\frac{(2l+1)}{2} \frac{(l-|m|)!}{(l+|m|)!}\right]^{\frac{1}{2}} P_l^{|m|}(\cos\theta) \tag{4.1.3}$$

where $P_l^{|m|}(\cos\theta)$ is the associated Legendre polynomial defined by

$$P_l^{|m|}(\cos\theta) = \frac{(1-\cos^2\theta)^{\frac{1}{2}|m|}}{2^l l!} \frac{d^{|m|+l}}{d(\cos\theta)^{|m|+l}} (\cos^2\theta - 1)^l \tag{4.1.4}$$

The variables θ, ϕ are angles in the system of spherical polar coordinates. For a given value of

$$l = 0, 1, 2, 3, \ldots$$

there are $(2l+1)$ spherical harmonics corresponding to

$$m = l, l-1, l-2, \ldots, 1, 0, -1, \ldots, (-l+1), -l$$

The spherical harmonics Y_{lm} form a basis for a $(2l+1)$-dimensional linear space.

For $m>0$ we define real spherical harmonics by

$$Y^{c}_{lm}=\frac{1}{\sqrt{2}}[(-1)^m Y_{lm}+Y_{l-m}]=\frac{1}{\sqrt{\pi}}T_{lm}(\theta)\cos m\phi \qquad (4.1.5)$$

$$Y^{s}_{lm}=\frac{-i}{\sqrt{2}}[(-1)^m Y_{lm}-Y_{l-m}]=\frac{1}{\sqrt{\pi}}T_{lm}(\theta)\sin m\phi$$

The first few real spherical harmonics are shown in Table 4. These real spherical harmonics are often referred to as *s*, *p*, *d*, *f*, . . . functions.

Since r is invariant it is clear that the functions Y^{c}_{11}, Y^{s}_{11}, Y_{10} transform in an identical fashion to the functions x, y, z under the operations of a molecular symmetry group. According to (2.15.18) the functions $X_e=x$, $Y_e=y$, $Z_e=z$ transform cogrediently to the standard base vectors e_1, e_2, e_3 and it follows that Y^{c}_{11}, Y^{s}_{11}, Y_{10} form the basis for a standard representation of the group.

We now consider the possibility of using real spherical harmonics with $l>1$ as the basis for standard representations of molecular symmetry groups. As an example we return to the group $\mathcal{D}_3$. The standard 2-dimensional IR of this group is given by (3.3.2) as

$$\hat{C}_3[e_1e_2]=[e_1e_2]\begin{bmatrix}-\frac{1}{2} & -\frac{\sqrt{3}}{2}\\ \frac{\sqrt{3}}{2} & -\frac{1}{2}\end{bmatrix}$$

$$\hat{C}_2[e_1e_2]=[e_1e_2]\begin{bmatrix}1 & 0\\ 0 & -1\end{bmatrix}$$

The functions $x=X_e$, $y=Y_e$, $z=Z_e$ transform cogrediently to the base vectors and it follows that

$$\hat{C}_3x=-\frac{1}{2}x+\frac{\sqrt{3}}{2}y \qquad \hat{C}_2x=x$$

$$\hat{C}_3y=-\frac{\sqrt{3}}{2}x-\frac{1}{2}y \qquad \hat{C}_2y=-y$$

Consider next the pair of functions

$$\psi_1=(x^2-y^2) \qquad \psi_2=2xy$$

which are proportional to Y^{c}_{22} and Y^{s}_{22} respectively. Using (2.15.17) and

TABLE 4. Real spherical harmonics

$Y_{00} = \dfrac{1}{\sqrt{4\pi}} = s$	s
$Y_{10} = \dfrac{\sqrt{3}}{\sqrt{4\pi}} \cos\theta = \dfrac{\sqrt{3}}{\sqrt{4\pi}}\left(\dfrac{z}{r}\right) = p_z$	p
$Y^c_{11} = \dfrac{\sqrt{3}}{\sqrt{4\pi}} \sin\theta\cos\phi = \dfrac{\sqrt{3}}{\sqrt{4\pi}}\left(\dfrac{x}{r}\right) = p_x$	
$Y^s_{11} = \dfrac{\sqrt{3}}{\sqrt{4\pi}} \sin\theta\sin\phi = \dfrac{\sqrt{3}}{\sqrt{4\pi}}\left(\dfrac{y}{r}\right) = p_y$	
$Y_{20} = \dfrac{\sqrt{5}}{2\sqrt{4\pi}} (3\cos^2\theta - 1) = \dfrac{\sqrt{5}}{2\sqrt{4\pi}}\left(\dfrac{2z^2 - x^2 - y^2}{r^2}\right) = d_{z^2}$	d
$Y^c_{21} = \dfrac{\sqrt{15}}{\sqrt{4\pi}} \sin\theta\cos\theta\cos\phi = \dfrac{\sqrt{15}}{\sqrt{4\pi}}\left(\dfrac{xz}{r^2}\right) = d_{xz}$	
$Y^s_{21} = \dfrac{\sqrt{15}}{\sqrt{4\pi}} \sin\theta\cos\theta\sin\phi = \dfrac{\sqrt{15}}{\sqrt{4\pi}}\left(\dfrac{yz}{r^2}\right) = d_{yz}$	
$Y^c_{22} = \dfrac{\sqrt{15}}{2\sqrt{4\pi}} \sin^2\theta\cos 2\phi = \dfrac{\sqrt{15}}{2\sqrt{4\pi}}\left(\dfrac{x^2 - y^2}{r^2}\right) = d_{x^2-y^2}$	
$Y^s_{22} = \dfrac{\sqrt{15}}{2\sqrt{4\pi}} \sin^2\theta\sin 2\phi = \dfrac{\sqrt{15}}{\sqrt{4\pi}}\left(\dfrac{xy}{r^2}\right) = d_{xy}$	
$Y_{30} = \dfrac{\sqrt{7}}{2\sqrt{4\pi}} (2\cos^3\theta - 3\sin^2\theta\cos\theta) = \dfrac{\sqrt{7}}{2\sqrt{4\pi}} \dfrac{z(2z^2 - 3x^2 - 3y^2)}{r^3} = f_0$	f
$Y^c_{31} = \dfrac{\sqrt{21}}{4\sqrt{2\pi}} \sin\theta(5\cos^2\theta - 1)\cos\phi = \dfrac{\sqrt{21}}{4\sqrt{2\pi}} \dfrac{x(4z^2 - x^2 - y^2)}{r^3} = f_1$	
$Y^s_{31} = \dfrac{\sqrt{21}}{4\sqrt{2\pi}} \sin\theta(5\cos^2\theta - 1)\sin\phi = \dfrac{\sqrt{21}}{4\sqrt{2\pi}} \dfrac{y(4z^2 - x^2 - y^2)}{r^3} = f_1'$	
$Y^c_{32} = \dfrac{\sqrt{105}}{4\sqrt{\pi}} \sin^2\theta\cos\theta\cos 2\phi = \dfrac{\sqrt{105}}{4\sqrt{\pi}} \dfrac{z(x^2 - y^2)}{r^3} = f_2$	f
$Y^s_{32} = \dfrac{\sqrt{105}}{4\sqrt{\pi}} \sin^2\theta\cos\theta\sin 2\phi = \dfrac{\sqrt{105}}{2\sqrt{\pi}}\left(\dfrac{xyz}{r^3}\right) = f_2'$	
$Y^c_{33} = \dfrac{\sqrt{35}}{4\sqrt{2\pi}} \sin^3\theta\cos 3\phi = \dfrac{\sqrt{35}}{4\sqrt{2\pi}} \dfrac{x(x^2 - 3y^2)}{r^3} = f_3$	
$Y^s_{33} = \dfrac{\sqrt{35}}{4\sqrt{2\pi}} \sin^3\theta\sin 3\phi = \dfrac{\sqrt{35}}{4\sqrt{2\pi}} \dfrac{y(3x^2 - y^2)}{r^3} = f_3'$	

(2.15.20) we readily find

$$\hat{C}_3[\psi_1\psi_2]=[\psi_1\psi_2]\begin{bmatrix} -\frac{1}{2} & \frac{\sqrt{3}}{2} \\ -\frac{\sqrt{3}}{2} & -\frac{1}{2} \end{bmatrix}$$

$$\hat{C}_2[\psi_1\psi_2]=[\psi_1\psi_2]\begin{bmatrix} 1 & 0 \\ 0 & -1 \end{bmatrix}$$

which is not the standard form of the 2-dimensional IR. However if we make a phase change and take as basis functions ψ_1 and $-\psi_2$ it can easily be verified that the standard form is obtained. It follows that the real spherical harmonics $(Y^c_{22}, -Y^s_{22})$ form a standard basis for the 2-dimensional IR of $\mathscr{D}_3$. This result can clearly be generalized. It is found that (see McWeeny, 1963) for molecular symmetry groups containing a single principal axis the real spherical harmonics form bases for the standard matrix representations provided due regard is paid to any possible phase changes. For cubic groups it is necessary to use cubic harmonics in order to obtain standard bases. The cubic harmonics coincide with the real spherical harmonics for $l=0, 1, 2$, while for $l=3$ they are given by

$$\begin{aligned} f_x &= -\tfrac{1}{4}(\sqrt{6}Y^c_{31}-\sqrt{10}Y^c_{33}) \\ f_y &= -\tfrac{1}{4}(\sqrt{6}Y^s_{31}+\sqrt{10}Y^s_{33}) \\ f_z &= Y_{30} \\ f'_x &= -\tfrac{1}{4}(\sqrt{10}Y^c_{31}+\sqrt{6}Y^c_{33}) \\ f'_y &= \tfrac{1}{4}(\sqrt{10}Y^s_{31}-\sqrt{6}Y^s_{33}) \\ f'_z &= Y^c_{32} \\ f_{xyz} &= Y^s_{32} \end{aligned} \qquad (4.1.6)$$

4.2 Notation for irreducible representations

In this book we shall adopt the Mulliken notation for the IR's of the finite molecular symmetry groups. This notation is shown in Table 5.

If there is a choice of secondary notation then g, *u* takes precedence over ′, ″ which in turn takes precedence over 1, 2, 3, ...

TABLE 5. Mulliken notation for irreducible representations

Property of basis functions	Behaviour of basis functions	Descriptive symbol
Dimension of IR	1	A or B
	2	E
	3	T
	4	U
	5	V
Symmetry of basis function with respect to rotation about principal axis for 1-dimensional IR's	Symmetric	A
	Antisymmetric	B
Symmetry of basis functions with respect to inversion in a centre of symmetry	Symmetric	g subscript
	Antisymmetric	u subscript
Symmetry of basis functions with respect to reflection in a σ_h plane	Symmetric	$'$ superscript
	Antisymmetric	$''$ superscript
Symmetry of basis function with respect to rotations C_2', C_2'', ... for 1-dimensional IR's	Most symmetric	1 subscript
	Other symmetries	2, 3 subscripts
Transformation properties of spherical harmonics Y_{lm} as basis functions for IR's of dimension two	$m = 1$ $m = 2$ ... $m = \nu$	subscripts $1, 2, \ldots, \nu$

4.3 Irreducible representations of the groups $\mathscr{C}_n$

The group $\mathscr{C}_n$ is given by $\mathscr{C}_n = \{E, C_n, C_n^2, \ldots, C_n^{n-1}\}$ and is a cyclic group of order n. Each element forms a class by itself and it follows from Section 3.4.I and Section 3.4.II that there are n IR's each of which is 1-dimensional. Since $C_n^n = E$ the character $\chi(C_n)$ of the element C_n satisfies

$$[\chi(C_n)]^n = 1 \tag{4.3.1}$$

From (4.3.1) we find the character of C_n in the m'th IR is given by

$$\chi(C_n) = e^{\frac{2\pi m i}{n}} \qquad m = 1, 2, 3, \ldots, n \tag{4.3.2}$$

A listing of the IR's of a group G together with their characters is called the *character table* for the group. It is customary to include in the character table simple functions like x, y, z which form bases for the various IR's.

For $\mathscr{C}_2$ we find from (4.3.2) the character table

$\mathscr{C}_2$	E	C_2
A; z	1	1
B; x, y	1	−1

The groups $\mathscr{C}_i$ and $\mathscr{C}_s$ are isomorphic with $\mathscr{C}_2$ and have character tables

$\mathscr{C}_i$	E	I
A_g	1	1
A_u; x, y, z	1	−1

$\mathscr{C}_s$	E	σ_h
A′; x, y	1	1
A″; z	1	−1

The character tables for isomorphic groups can clearly be written in the form of a single table. Thus for the isomorphic groups $\mathscr{C}_2$, $\mathscr{C}_i$ and $\mathscr{C}_s$ we have the table.

$\mathscr{C}_2$			E	C_2
	$\mathscr{C}_i$		E	I
		$\mathscr{C}_s$	E	σ_h
A; z	A_g	A′; x, y	1	1
B; x, y	A_u; x, y, z	A″; z	1	−1

We now turn to the group $\mathscr{C}_3$. If we use (4.3.2) and the relation

$$e^{4\pi i/3} = e^{-2\pi i/3}$$

we obtain the character table

$\mathscr{C}_3$	E	C_3	C_3^2
A	1	1	1
$\Gamma^{(1)}$	1	ε	ε^*
$\Gamma^{(2)}$	1	ε^*	ε

in which $\varepsilon = e^{2\pi i/3}$. We note that the IR's $\Gamma^{(1)}$ and $\Gamma^{(2)}$ are complex conjugate. From (4.3.2) it is not hard to see that for fixed l the spherical harmonics Y_{lm} form bases for the IR's of $\mathscr{C}_n$. Thus

$$\hat{C}_n Y_{lm}(\theta, \phi) = Y_{lm}\left(\theta, \phi - \frac{2\pi}{n}\right) = T_{lm}(\theta) e^{im[\phi - (2\pi/n)]}$$
$$= e^{-(2\pi i m/n)} Y_{lm}(\theta, \phi) \tag{4.3.3}$$

In particular for the group $\mathscr{C}_3$ if we take $l = 1$ we see that Y_{10} (or z) spans A. Also Y_{11} (or $x + iy$) spans $\Gamma^{(2)}$ and Y_{1-1} (or $x - iy$) spans $\Gamma^{(1)}$. Thus the basis functions for $\Gamma^{(1)}$ and $\Gamma^{(2)}$ are complex conjugate. In most applications to quantum chemistry the Hamiltonian $\hat{H}$ is real and it follows that the eigenfunctions ψ and ψ^* are degenerate. Because of this it is convenient to combine complex conjugate representations. Thus we write

$$E = \Gamma^{(1)} \oplus \Gamma^{(2)}$$

and the character table becomes

$\mathscr{C}_3$	E	C_3	C_3^2
$A;\ z$	1	1	1
$E;\ (x, y)$	2	-1	-1

Although E is strictly a reducible representation we usually treat it on the same footing as an IR.

4.4 Irreducible representations of the groups $\mathscr{D}_n$

As an example we shall work out the IR's for the group $\mathscr{D}_6$. This is sufficiently non-trivial to enable the general method to be clearly seen.

The group $\mathscr{D}_6$ is given by

$$\mathscr{D}_6 = \{E, C_6, C_6^2, C_6^3, C_6^4, C_6^5, C_2^{(1)}, C_2^{(2)}, C_2^{(3)}, C_2'^{(1)}, C_2'^{(2)}, C_2'^{(3)}\}$$

and is of order 12. The symmetry transformations are shown in Fig. 5. The classes of conjugate elements can readily be deduced from the figure below. They are

$$E = \{E\} \qquad 2C_6 = \{C_6, C_6^5\}; \qquad 2C_6^2 = \{C_6^2, C_6^4\}$$

$$C_6^3 = \{C_6^3\} \qquad 3C_2 = \{C_2^{(1)}, C_2^{(2)}, C_2^{(3)}\} \qquad 3C_2' = \{C_2'^{(1)}, C_2'^{(2)}, C_3'^{(3)}\}$$

Since there are six classes it follows that there are six IR's to be found. We write $C_2 = C_2^{(1)}$ and $C_2' = C_2'^{(3)}$. If we choose the elements C_6 and C_2 as generators then the only defining relation that is needed in order to

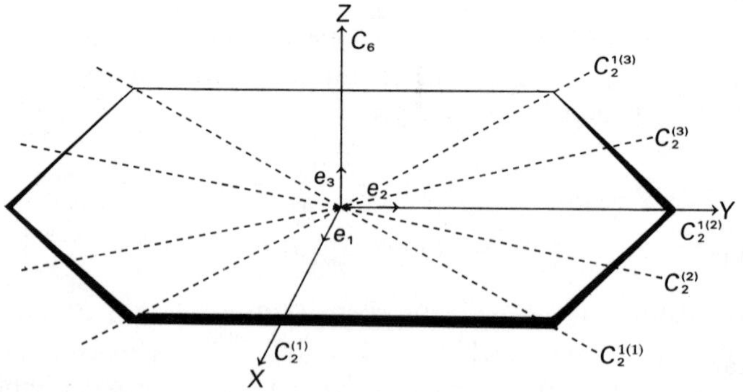

FIG. 5. The group $\mathscr{D}_6$.

construct the character table is

$$C_2' = C_2 C_6 \tag{4.4.1}$$

Since matrix representatives and group elements satisfy the same multiplication table it is sufficient to obtain the standard matrix representations for the generators in order to find those for all group elements.

Within the standard basis $[e_1\ e_2\ e_3]$ shown in Fig. 5 we have from (3.1.10) the matrix representation

$$\boldsymbol{\Gamma}(C_6) = \begin{bmatrix} \frac{1}{2} & -\frac{\sqrt{3}}{2} & 0 \\ \frac{\sqrt{3}}{2} & \frac{1}{2} & 0 \\ 0 & 0 & 1 \end{bmatrix}$$

It is also clear from Fig. 5 that in the same basis we have

$$\boldsymbol{\Gamma}(C_2) = \begin{bmatrix} 1 & 0 & 0 \\ 0 & -1 & 0 \\ 0 & 0 & -1 \end{bmatrix}$$

We have thus constructed a 1-dimensional representation A_2 with e_3 or z as basis together with a standard 2-dimensional representation which has basis (e_1, e_2) or (x, y). From (3.4.8) we see that the 2-dimensional representation is irreducible. Since the basis functions (x, y) come from spherical harmonics with $m = 1$ we denote this 2-dimensional IR by E_1. Of necessity we also have the identity representation A_1 which has basis Y_{00}.

At this stage we have obtained three of the six IR's. According to (3.4.1) the remaining three IR's are of dimensions 1, 1 and 2 respectively. We can readily obtain the remaining two 1-dimensional IR's by using the result (3.4.4). For the 1-dimensional IR's we have the system of characters

	E	$2C_6$	$2C_6^2$	C_6^3	$3C_2$	$3C_2'$
A_1	1	1	1	1	1	1
A_2	1	1	1	1	-1	-1
Γ	1	a	a^2	a^3	b	ab

where a and b are to be found for the representation Γ. Application of (3.4.4) using A_1 and Γ gives

$$1 + 2a + 2a^2 + a^3 + 3b + 3ab = 0 \tag{4.4.2}$$

Application of (3.4.4) using A_2 and Γ gives

$$1 + 2a + 2a^2 + a^3 - 3b - 3ab = 0 \tag{4.4.3}$$

If we add (4.4.2) and (4.4.3) we obtain

$$1+2a+2a^2+a^3=0 \tag{4.4.4}$$

For the group $\mathscr{D}_6$ it is evident that all the classes are ambivalent and it follows from Section 3.4.VIII that all characters are real numbers. The only real number which satisfies (4.4.4) is $a=-1$. Application of (3.4.4) using Γ with itself shows that $b=\pm 1$. Thus the other two 1-dimensional IR's of $\mathscr{D}_6$ are B_1 and B_2. If we examine the real spherical harmonics we see that bases for B_1 and B_2 are given by Y^c_{33} and Y^s_{33} respectively.

To obtain the other 2-dimensional IR of $\mathscr{D}_6$ we note that $\mathscr{D}_3$ is a subgroup of $\mathscr{D}_6$. We have already seen that the functions $(x^2-y^2, -2xy)$ form a standard basis for the 2-dimensional IR of $\mathscr{D}_3$. Application of the generators of $\mathscr{D}_6$ to these functions gives

$$\hat{C}_6[x^2-y^2 \quad -2xy]=[x^2-y^2 \quad -2xy]\begin{bmatrix} -\frac{1}{2} & \frac{\sqrt{3}}{2} \\ -\frac{\sqrt{3}}{2} & -\frac{1}{2} \end{bmatrix}$$

$$\hat{C}_2[x^2-y^2 \quad -2xy]=[x^2-y^2 \quad -2xy]\begin{bmatrix} 1 & 0 \\ 0 & -1 \end{bmatrix}$$

It follows that $(x^2-y^2, -2xy)$ form a basis for a 2-dimensional representation of $\mathscr{D}_6$. This representation can easily be shown to be irreducible and not equivalent to E_1. Since the basis functions come from spherical harmonics with $m=2$ we denote the IR by E_2. We shall also agree to call $(x^2-y^2, -2xy)$ a standard basis for E_2.

We have now constructed all the standard IR's of $\mathscr{D}_6$ and the character table is

$\mathscr{D}_6$	E	$2C_6$	$2C_6^2$	C_6^3	$3C_2$	$3C_2'$
A_1; d_{z^2}	1	1	1	1	1	1
A_2; p_z	1	1	1	1	−1	−1
B_1; f_3	1	−1	1	−1	1	−1
B_2; f_3'	1	−1	1	−1	−1	1
E_1; (p_x, p_y)	2	1	−1	−2	0	0
E_2; $(d_{x^2-y^2}, -d_{xy})$	2	−1	−1	2	0	0

4.5 Irreducible representations of the octahedral group $\mathscr{O}$

The octahedral group $\mathscr{O}$ consists of 24 elements. If we examine Figs. 4 and

6 the classes of conjugate elements are easily seen to be

$$8C_3 = \{C_3(\alpha), C_3^2(\alpha), C_3(\beta), C_3^2(\beta), C_3(\gamma), C_3^2(\gamma), C_3(\delta), C_3^2(\delta)\}$$
$$3C_2 = \{C_2(X), C_2(Y), C_2(Z)\}$$
$$6C_2' = \{C_2(a), C_2(b), C_2(c), C_2(d), C_2(e), C_2(f)\}$$
$$6C_4 = \{C_4(X), C_4^3(X), C_4(Y), C_4^3(Y), C_4(Z), C_4^3(Z)\}$$
$$E = \{E\}$$

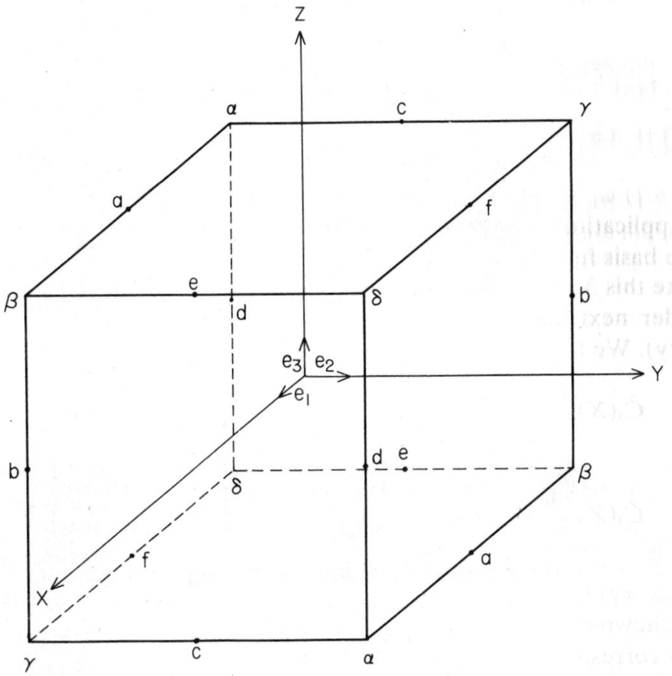

FIG. 6. Labelling of axes in $\mathcal{O}$.

If we choose $C_4(X)$ and $C_4(Z)$ as generators then we have defining relations such as

$$E = C_4^4(Z) \qquad C_2(Z) = C_4^2(Z)$$
$$C_3(\alpha) = C_4(X)C_4^3(Z) \qquad C_2'(a) = C_4(X)C_4^2(Z) \tag{4.5.1}$$

for elements in the other four classes.

Since there are five conjugate classes it follows that there are five IR's to be found. One of these is the identity representation A_1. Within the standard basis (e_1, e_2, e_3) we readily find

$$\hat{C}_4(X)[e_1 \quad e_2 \quad e_3] = [e_1 \quad e_2 \quad e_3]\begin{bmatrix} 1 & 0 & 0 \\ 0 & 0 & -1 \\ 0 & 1 & 0 \end{bmatrix}$$

$$\hat{C}_4(Z)[e_1 \quad e_2 \quad e_3] = [e_1 \quad e_2 \quad e_3]\begin{bmatrix} 0 & -1 & 0 \\ 1 & 0 & 0 \\ 0 & 0 & 1 \end{bmatrix}$$

for the group generators. It follows that the functions (x, y, z) form a basis for a standard 3-dimensional representation Γ of $\mathcal{O}$ given by

$$\Gamma(C_4(X)) = \begin{bmatrix} 1 & 0 & 0 \\ 0 & 0 & -1 \\ 0 & 1 & 0 \end{bmatrix} \qquad \Gamma(C_4(Z)) = \begin{bmatrix} 0 & -1 & 0 \\ 1 & 0 & 0 \\ 0 & 0 & 1 \end{bmatrix}$$

From (4.5.1) we obtain the matrix representatives of E, $C_2(Z)$, $C_3(\alpha)$ and $C_2'(a)$. Application of (3.4.8) shows that the representation Γ is irreducible. Since the basis functions correspond to real spherical harmonics with $l = 1$ we denote this 3-dimensional IR by T_1.

Consider next the effect of the generators on the product functions (yz, zx, xy). We find

$$\hat{C}_4(X)[yz \quad zx \quad xy] = [yz \quad zx \quad xy]\begin{bmatrix} -1 & 0 & 0 \\ 0 & 0 & 1 \\ 0 & -1 & 0 \end{bmatrix}$$

$$\hat{C}_4(Z)[yz \quad zx \quad xy] = [yz \quad zx \quad xy]\begin{bmatrix} 0 & 1 & 0 \\ -1 & 0 & 0 \\ 0 & 0 & -1 \end{bmatrix}$$

and (yz, zx, xy) form a basis for a 3-dimensional representation of $\mathcal{O}$. It can easily be shown that this is an IR which is not equivalent to T_1. Since the basis functions correspond to real spherical harmonics with $l = 2$ we denote this 3-dimensional IR by T_2. We shall agree to call (yz, zx, xy) a standard basis for T_2. For the group $\mathcal{O}$ the functions yz, zx and xy span a 3-dimensional subspace of the 5-dimensional space of d functions. The remaining d functions are $\sqrt{3}(x^2 - y^2)$ and $(2z^2 - x^2 - y^2)$ where the $\sqrt{3}$ appears with $(x^2 - y^2)$ due to the difference of normalization factors (see Table 4). Now

$$\hat{C}_4(X)[(2z^2 - x^2 - y^2) \quad \sqrt{3}(x^2 - y^2)] = [(2z^2 - x^2 - y^2) \quad \sqrt{3}(x^2 - y^2)]\begin{bmatrix} -\frac{1}{2} & -\frac{\sqrt{3}}{2} \\ \frac{\sqrt{3}}{2} & \frac{1}{2} \end{bmatrix}$$

$$\hat{C}_4(Z)[(2z^2 - x^2 - y^2 \quad \sqrt{3}(x^2 - y^2)] = [(2z^2 - x^2 - y^2) \quad \sqrt{3}(x^2 - y^2)]\begin{bmatrix} 1 & 0 \\ 0 & -1 \end{bmatrix}$$

and we have constructed a 2-dimensional representation of $\mathcal{O}$ which can readily be shown to be irreducible. We shall agree to regard $((2z^2-x^2-y^2), \sqrt{3}(x^2-y^2))$ as forming a standard basis for the 2-dimensional IR E of the group $\mathcal{O}$. There is only one 2-dimensional IR of $\mathcal{O}$ as we have already obtained two 3-dimensional IR's and it follows from (3.4.1) that there remains a single new 1-dimensional IR to be found. Let this 1-dimensional IR be denoted by Γ. Consider the system of characters

	E	$8C_3$	$3C_2$	$6C_2'$	$6C_4$
A_1	1	1	1	1	1
Γ	1	a^4	a^2	a^3	a

According to (3.4.4) we have

$$1+8a^4+3a^2+6a^3+6a=0$$

which gives $a=-1$. The other roots can be discarded since they are complex and the classes of $\mathcal{O}$ are all ambivalent. The new representation is A_2. By examining Table 4 we see that $xyz = Y_{32}^s = f_{xyz}$ is a basis function for A_2.

The character table for the octahedral group now follows.

$\mathcal{O}$	E	$8C_3$	$3C_2$	$6C_2'$	$6C_4$
A_1; s	1	1	1	1	1
A_2; f_{xyz}	1	1	1	-1	-1
E; $(d_{z^2}, d_{x^2-y^2})$	2	-1	2	0	0
T_1; (p_x, p_y, p_z)	3	0	-1	-1	1
T_2; (d_{yz}, d_{zx}, d_{xy})	3	0	-1	1	-1

References

Hamermesh, M. (1962) "Group Theory and its application to physical problems", Addison-Wesley, Chapter 4 & 5.

McWeeny, R. (1963) "Symmetry", Pergamon Press, Chapters 4, 5, and Appendix I.

CHAPTER 5

Applications of Irreducible Representations

5.1 Projection operators and molecular orbitals

In the previous chapter we saw how to obtain the standard IR's of the molecular symmetry groups. This was done by considering the effect of the operators corresponding to symmetry transformations acting on basis functions for some linear space. Any set of basis functions which transform in a standard way are called *standard basis functions.* A standard basis function which transforms like the ith basis vector of the IR Γ is called a *symmetry species* and is denoted by $\{\Gamma, i\}$. In quantum chemistry we are often faced with the situation in which we have a function of arbitrary symmetry and we wish to obtain from it a symmetry species $\{\Gamma, i\}$. One way of doing this is by the use of a projection operator. Before defining projection operators however it is instructive to consider a definite example.

We take the planar AB_3 molecule shown in Fig. 2. The full symmetry is $\mathscr{D}_{3h}$ but for many purposes it is sufficient to use the subgroup $\mathscr{D}_3$. From (3.3.1) and (3.3.2) we obtain the behaviour of standard basis functions under the group generators. This is shown in Table 6. Let $\phi_1 = \psi(x, y, z)$ be an arbitrary even function of x, y, z such as an s orbital centered on atom B_1 as shown in Fig. 7.

TABLE 6. Standard IR's for $\mathscr{D}_3$

Standard basis function	Example	Effect of $C_3(Z)$	Effect of $C_2(X)$
a_1	z^2	a_1	a_1
a_2	z	a_2	$-a_2$
e_x	x	$-\frac{1}{2}e_x - \frac{\sqrt{3}}{2}e_y$	e_x
e_y	y	$\frac{\sqrt{3}}{2}e_x - \frac{1}{2}e_y$	$-e_y$

We have

$$\hat{C}_3\phi_1 = \phi_2 \qquad \hat{C}_3\phi_2 = \phi_3 \qquad \hat{C}_3\phi_3 = \phi_1$$

where ϕ_2 and ϕ_3 are the corresponding s orbitals centered on atoms B_2 and

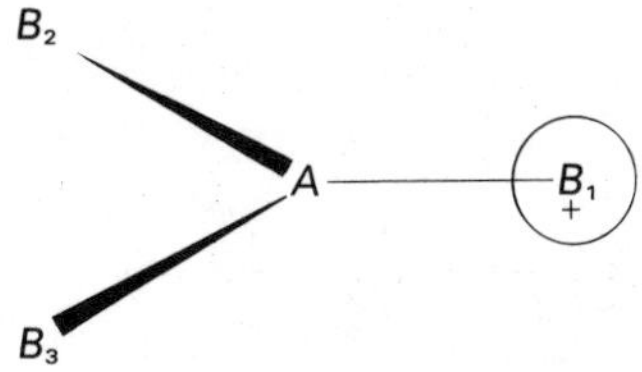

FIG. 7. s orbital centered on B_1.

B_3 respectively. We also have

$$\hat{C}_2\phi_1 = \phi_1 \qquad \hat{C}_2\phi_2 = \phi_3 \qquad \hat{C}_2\phi_3 = \phi_2$$

where C_2 refers to the axis AB_1. It follows that (ϕ_1, ϕ_2, ϕ_3) forms a basis for the representation Γ of $\mathscr{D}_3$ given by

$$\boldsymbol{\Gamma}(C_3) = \begin{bmatrix} 0 & 0 & 1 \\ 1 & 0 & 0 \\ 0 & 1 & 0 \end{bmatrix} \qquad \boldsymbol{\Gamma}(C_2) = \begin{bmatrix} 1 & 0 & 0 \\ 0 & 0 & 1 \\ 0 & 1 & 0 \end{bmatrix}$$

If we consider (1.2.2), (3.4.7) and the character table for $\mathscr{D}_3$ we find

$$\Gamma = A_1 \oplus E$$

It follows that the space spanned by (ϕ_1, ϕ_2, ϕ_3) can be expressed as the direct sum of two subspaces whose basis functions χ_1 and (χ_2, χ_3) respectively span the IR's A_1 and E. The new basis functions χ_i are given in matrix notation by

$$\boldsymbol{\chi} = \boldsymbol{\phi}\mathbf{A}$$

where the matrix $\mathbf{A} = [a_{ij}]$ is to be determined. If χ_1 is to transform according to A_1 we must have

$$\hat{C}_3\chi_1 = \chi_1$$

and it follows that

$$a_{11} = a_{21} = a_{31} = \alpha$$

where α is a normalization factor. The required basis function is thus

$$\chi_1 = \chi_{a_1}^{(A_1)} = \alpha(\phi_1 + \phi_2 + \phi_3) \tag{5.1.1}$$

If χ_2 and χ_3 are to transform according to E we require

$$\hat{C}_3[\chi_2 \quad \chi_3] = [\chi_2 \quad \chi_3]\begin{bmatrix} -\frac{1}{2} & -\frac{\sqrt{3}}{2} \\ \frac{\sqrt{3}}{2} & -\frac{1}{2} \end{bmatrix}$$

$$\hat{C}_2[\chi_2 \quad \chi_3] = [\chi_2 \quad \chi_3]\begin{bmatrix} 1 & 0 \\ 0 & -1 \end{bmatrix}$$

from which it follows that

$$a_{13}=0 \qquad a_{33}=-a_{23} \qquad a_{22}=a_{32}$$

$$a_{12}=-\frac{1}{2}a_{22}+\frac{\sqrt{3}}{2}a_{23}$$

and the required symmetry species are

$$\begin{aligned}\chi_2&=\chi_x^{(E)}=\beta(\phi_1-\tfrac{1}{2}\phi_2-\tfrac{1}{2}\phi_3)\\ \chi_3&=\chi_y^{(E)}=\gamma(\phi_2-\phi_3)\end{aligned} \tag{5.1.2}$$

where β and γ are normalization factors. The equations (5.1.1) and (5.1.2) can be inverted and we find for example

$$\phi_1=\left(\frac{1}{3\alpha}\right)\chi_{a_1}^{(A_1)}+\left(\frac{2}{3\beta}\right)\chi_x^{(E)} \tag{5.1.3}$$

This shows that a function of arbitrary symmetry can be expressed as a linear combination of symmetry species.

In general for a group $\mathcal{G}$ of order g we can write (5.1.3) as

$$\phi=\sum_{\nu}\sum_{i=1}^{n_\nu}\chi_i^{(\nu)} \tag{5.1.4}$$

which expresses a function ϕ of arbitrary symmetry as a linear combination of symmetry species $\chi_i^{(\nu)}$ which transform like the ith basis vector of the IR $\Gamma^{(\nu)}$ of dimension n_ν. We have

$$\hat{R}\phi=\sum_{\nu}\sum_{i=1}^{n_\nu}\hat{R}\chi_i^{(\nu)} \tag{5.1.5}$$

and since $\chi_i^{(\nu)}$ are basis functions for $\Gamma^{(\nu)}$ we also have

$$\hat{R}\chi_i^{(\nu)}=\sum_{j=1}^{n_\nu}\chi_j^{(\nu)}\Gamma_{ji}^{(\nu)}(R) \tag{5.1.6}$$

where $\Gamma_{ji}^{(\nu)}(R)$ is the jith element in the matrix representation of R. Substitution of (5.1.6) into (5.1.5) leads to

$$\hat{R}\phi=\sum_{\nu}\sum_{i,j=1}^{n_\nu}\chi_j^{(\nu)}\Gamma_{ji}^{(\nu)}(R)$$

We now multiply this equation by $\Gamma_{kl}^{*(\nu)}(\mathrm{R})$ and sum over R to obtain

$$\sum_{R}\Gamma_{kl}^{*(\mu)}\hat{R}\phi=\left(\frac{g}{n_\mu}\right)\delta_{\mu\nu}\,\delta_{kj}\,\delta_{li}\chi_k^{(\mu)}$$

where we have used (3.4.2). This result enables us to obtain a symmetry species $\{\mu, k\}$ from a function of arbitrary symmetry. We define a projection operator $\hat{P}^{(\mu)}_{kl}$ by

$$\hat{P}^{(\mu)}_{kl} = \left(\frac{n_\mu}{g}\right) \sum_R \Gamma^{*(\mu)}_{kl}(R)\hat{R} \tag{5.1.7}$$

and we have

$$\chi^{(\mu)}_k = \hat{P}^{(\mu)}_{kl}\phi \tag{5.1.8}$$

The results (5.1.1) and (5.1.2) can now be obtained directly by use of the projection operators $\hat{P}^{(A_1)}_{a_1a_1}$, $\hat{P}^{(E)}_{xx}$ and $\hat{P}^{(E)}_{yy}$.

As we have seen a set of orbitals $(\phi_1, \phi_2, \ldots, \phi_n)$ where ϕ_i is centered on atom i of a molecule, forms a basis for a representation Γ of the symmetry group $\mathcal{G}$. In such a representation a symmetry transformation R of $\mathcal{G}$ can affect the basis function ϕ_i in three different ways: (i) it can leave ϕ_i unchanged in which case $+1$ will appear as the iith element in the representation matrix, (ii) it can change the sign of ϕ_i in which case -1 will appear as the iith element, and (iii) it can change ϕ_i into some other function centered on atom $j \neq i$ in which case 0 will appear as the iith element. It follows that the characters in the representation Γ can readily be obtained and hence the various IR's which occur in the decomposition of Γ can be found.

Consider the set of three p_z functions (z_1, z_2, z_3) on the atoms B_1, B_2 and B_3 respectively. These functions clearly span the representation $A_2 \oplus E$ of $\mathcal{D}_3$. In order to discuss the transformation properties of the p_x functions (x_1, x_2, x_3) and the p_y functions (y_1, y_2, y_3) it is convenient to introduce the coordinate axes shown in Fig. 8.

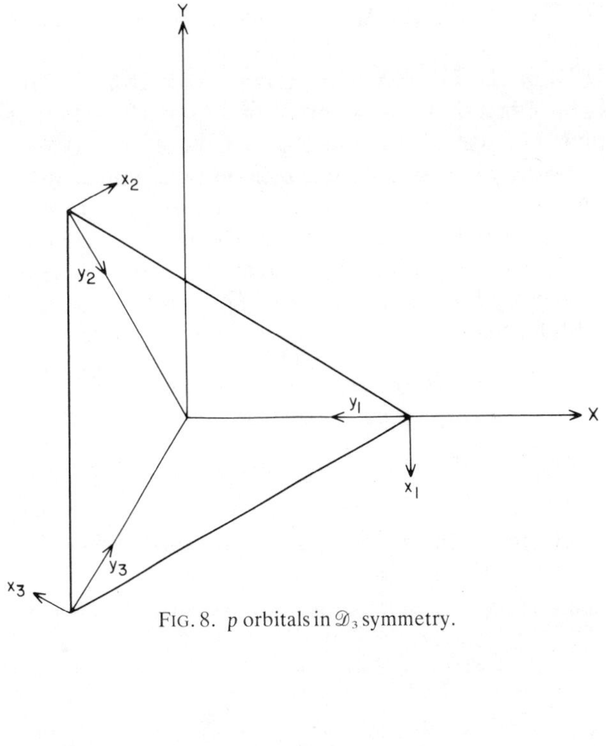

FIG. 8. p orbitals in $\mathcal{D}_3$ symmetry.

Within the basis (x_1, x_2, x_3) we have

$$\Gamma(C_3(Z)) = \begin{bmatrix} 0 & 0 & 1 \\ 1 & 0 & 0 \\ 0 & 1 & 0 \end{bmatrix} \qquad \Gamma(C_2(X)) = \begin{bmatrix} -1 & 0 & 0 \\ 0 & 0 & -1 \\ 0 & -1 & 0 \end{bmatrix}$$

and it follows that the three p_x functions span the representation $A_2 \oplus E$. Similarly in the basis (y_1, y_2, y_3) we find

$$\Gamma(C_3(Z)) = \begin{bmatrix} 0 & 0 & 1 \\ 1 & 0 & 0 \\ 0 & 1 & 0 \end{bmatrix} \qquad \Gamma(C_2(X)) = \begin{bmatrix} 1 & 0 & 0 \\ 0 & 0 & 1 \\ 0 & 1 & 0 \end{bmatrix}$$

and the three p_y functions span the representation $A_1 \oplus E$. We now use the projection operators

$$\hat{P}^{(A_1)}_{a_1 a_1} = \tfrac{1}{6}(\hat{E} + \hat{C}_3 + \hat{C}_3^2 + \hat{C}_2 + \hat{C}_2' + \hat{C}_2'')$$
$$\hat{P}^{(A_2)}_{a_2 a_2} = \tfrac{1}{6}(\hat{E} + \hat{C}_3 + \hat{C}_3^2 - \hat{C}_2 - \hat{C}_2' - \hat{C}_2'')$$
$$\hat{P}^{(E)}_{xx} = \tfrac{1}{3}(\hat{E} - \tfrac{1}{2}\hat{C}_3 - \tfrac{1}{2}\hat{C}_3^2 + \hat{C}_2 - \tfrac{1}{2}\hat{C}_2' - \tfrac{1}{2}\hat{C}_2'')$$
$$\hat{P}^{(E)}_{yy} = \tfrac{1}{3}(\hat{E} - \tfrac{1}{2}\hat{C}_3 - \tfrac{1}{2}\hat{C}_3^2 - \hat{C}_2 + \tfrac{1}{2}\hat{C}_2' + \tfrac{1}{2}\hat{C}_2'')$$

on the functions x_1, y_1 etc. in order to obtain the appropriate symmetry species. The linear combinations of atomic orbitals which belong to such symmetry species are called *symmetry orbitals* or group orbitals. If we denote respectively by h_i and k_i the $1s$ and $2s$ orbitals on atom B_i then we obtain the set of symmetry orbitals shown in Table 7. In Table 7 we have used the IR's for the full symmetry group $\mathscr{D}_{3h} = \mathscr{D}_3 \otimes \mathscr{C}_s$. This is done by considering the effect of the operator $\hat{\sigma}_h$ on the basis functions as can be seen by inspection. The use of $\mathscr{D}_{3h}$ and not $\mathscr{D}_3$ is necessary now since we have functions of both primed and double primed types appearing.

Suppose we consider a molecule like BF_3 and we agree to set up molecular orbitals as a linear combination of atomic orbitals, using the limited basis set of $1s$, $2s$ and $2p$ functions on each of the atoms. From Table 6 we have the following symmetry species for functions centered on boron. (See Table 7)

Since the MO's must be symmetry species we use Table 7 and set them up as

$$\psi(a_1') = c_1(1s) + c_2(2s) + c_3(h_1 + h_2 + h_3) + c_4(k_1 + k_2 + k_3) + c_5(y_1 + y_2 + y_3)$$

$$\psi(a_2') = x_1 + x_2 + x_3$$

$$\psi(a_2'') = c_1(2p_z) + c_2(z_1 + z_2 + z_3)$$

$$\psi(e_x') = c_1(2p_x) + c_2(h_1 - \tfrac{1}{2}h_2 - \tfrac{1}{2}h_3) + c_3(k_1 - \tfrac{1}{2}k_2 - \tfrac{1}{2}k_3) + c_4(x_2 - x_3) + c_5(2y_1 - y_2 - y_3)$$

$$\psi(e_x'') = z_2 - z_3$$

TABLE 7. Symmetry orbitals from 1s, 2s and 2p AO's in $\mathcal{D}_{3h}$

Symmetry orbital	Type
$h_1+h_2+h_3$	a_1'
$h_1-\frac{1}{2}h_2-\frac{1}{2}h_3$	e_x'
h_2-h_3	e_y'
$k_1+k_2+k_3$	a_1'
$k_1-\frac{1}{2}k_2-\frac{1}{2}k_3$	e_x'
k_2-k_3	e_y'
$z_1+z_2+z_3$	a_2''
z_2-z_3	e_x''
$2z_1-z_2-z_3$	e_y''
$x_1+x_2+x_3$	a_2'
x_2-x_3	e_x'
$-x_1+\frac{1}{2}x_2+\frac{1}{2}x_3$	e_y'
$y_1+y_2+y_3$	a_1'
$2y_1-y_2-y_3$	e_x'
y_2-y_3	e_y'
Boron orbital	Type
$1s$	a_1'
$2s$	a_1'
$2p_z$	a_2''
$2p_x$	e_x'
$2p_y$	e_y'

We note that within this limited set of basis functions the MO's $\psi(a_2')$ and $\psi(e_x'')$ are determined by symmetry alone. The other MO's can be found by setting up and solving secular equations for the coefficients c_i (see McWeeny & Sutcliffe, 1969).

In simple cases it is possible to obtain useful qualitative information without actually solving the secular equations. Consider the planar species CH_3. In an initial approximation we can neglect the $2s$ and $2p$ orbitals on the H atoms and the MO's are

$$\psi(a_1') = c_1(1s) + c_2(2s) + c_3(h_1+h_2+h_3)$$

$$\psi(a_2'') = (2p_z)$$

$$\psi(e_x') = c_1(2p_x) + c_2(h_1 - \tfrac{1}{2}h_2 - \tfrac{1}{2}h_3)$$

The carbon $1s$ orbital clearly has by far the lowest energy and this will dominate the $1a_1'$ MO. If we take simple bonding and antibonding approximations to the solutions of the secular equations for the other MO's we have

$$\psi(1a_1') \sim (1s)$$
$$\psi(2a_1') \sim (2s) + (h_1 + h_2 + h_3)$$
$$\psi(3a_1') \sim (2s) - (h_1 + h_2 + h_3)$$
$$\psi(1a_2'') = (2p_z)$$
$$\psi(1e_x') \sim (2p_x) + (h_1 - \tfrac{1}{2}h_2 - \tfrac{1}{2}h_3)$$
$$\psi(2e_x') \sim (2p_x) - (h_1 - \tfrac{1}{2}h_2 - \tfrac{1}{2}h_3)$$

These MO's are illustrated in Fig. 9 and the schematic energy level diagram shown in Fig. 10 is now obtained. Diagrams such as Fig. 10 are extremely useful for a qualitative interpretation of the electronic spectra of polyatomic molecules. Not only do the diagrams show the effect of mixing of symmetry orbitals of indicated symmetry species but they also show what types of electronic transition might be expected to occur. We shall return to this later.

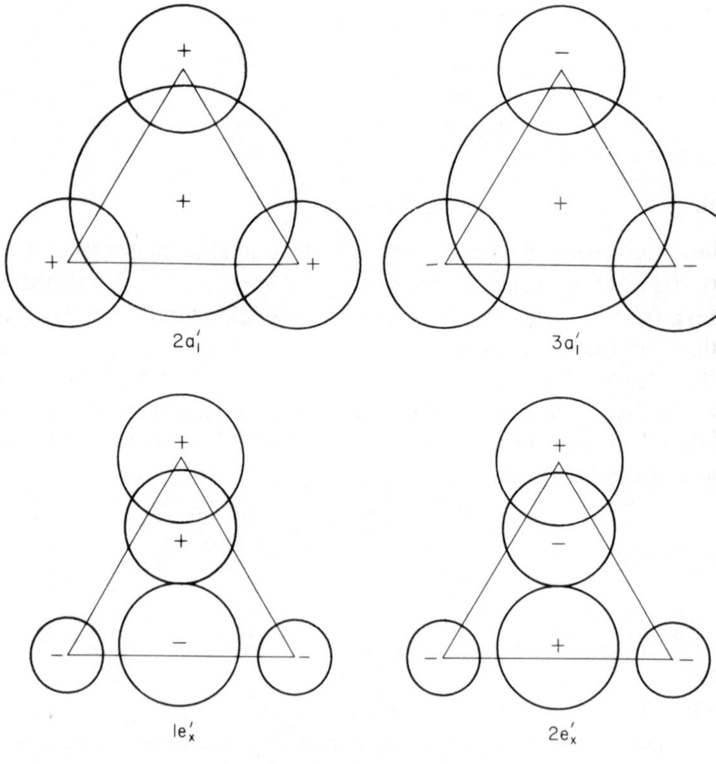

FIG. 9

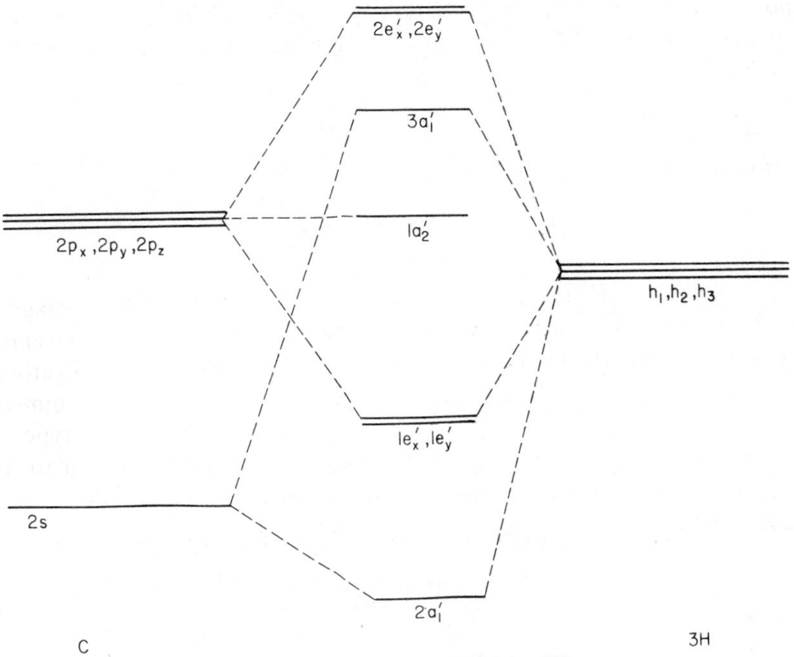

FIG. 10. Schematic energy levels for CH_3.

5.2 Direct products

In a large part of quantum chemistry we are concerned with the evaluation of matrix elements or integrals. Many of these integrals turn out to vanish. It is clear that a great simplification occurs if we know in advance which integrals are vanishing. The integrals with which we are concerned involve products of functions which transform according to the IR's of a given molecular symmetry group. Let $\phi_i^{(\mu)}$ and $\psi_j^{(\nu)}$ be the basis functions respectively for the IR's $\Gamma^{(\mu)}$ and $\Gamma^{(\nu)}$ of a group $\mathcal{G}$. Let the dimensions of $\Gamma^{(\mu)}$ and $\Gamma^{(\nu)}$ be n_μ and n_ν respectively and suppose that $\mu \neq \nu$. We have

$$\hat{R}\phi_i^{(\mu)} = \sum_{j=1}^{n_\mu} \phi_j^{(\mu)} \Gamma_{ji}^{(\mu)}(R)$$

and

$$\hat{R}\psi_k^{(\nu)} = \sum_{l=1}^{n_\nu} \psi_l^{(\nu)} \Gamma_{lk}^{(\nu)}(R)$$

It follows that

$$\begin{aligned} \hat{R}(\phi_i^{(\mu)}\psi_k^{(\nu)}) &= \hat{R}\phi_i^{(\mu)}\hat{R}\psi_k^{(\nu)} \\ &= \sum_{j,l} \phi_j^{(\mu)}\psi_l^{(\nu)}\Gamma_{ji}^{(\mu)}(R)\Gamma_{lk}^{(\nu)}(R) \end{aligned} \tag{5.2.1}$$

and the $n_\mu n_\nu$ functions $\phi_i^{(\mu)}\psi_j^{(\nu)}$ form the basis for a $n_\mu n_\nu$-dimensional representation of $\mathcal{G}$. This representation is called the direct product of $\Gamma^{(\mu)}$ and $\Gamma^{(\nu)}$ and we write

$$\Gamma^{(\mu\otimes\nu)} = \Gamma^{(\mu)} \otimes \Gamma^{(\nu)}$$

The equation (5.2.1) is written

$$\hat{R}(\phi_i^{(\mu)}\psi_k^{(\nu)}) = \sum_{jl} \phi_j^{(\mu)}\psi_l^{(\nu)}\Gamma_{jl;ik}^{(\mu\otimes\nu)}(R)$$

The matrix $\mathbf{\Gamma}^{(\mu\otimes\nu)}(R)$ is the direct product of $\mathbf{\Gamma}^{(\mu)}(R)$ and $\mathbf{\Gamma}^{(\nu)}(R)$ and is given by (2.3.4). From the definition it is clear that the character of an element in the direct product is the product of the characters in the constituent IR's.

$$\chi^{(\mu\otimes\nu)}(R) = \chi^{(\mu)}(R)\chi^{(\nu)}(R) \qquad (5.2.2)$$

This result is conjunction with (3.4.7) is used to find the IR's of $\mathcal{G}$ which are contained in the direct product representations. As an example we have Table 8 which lists the various direct products for the group $\mathcal{D}_3$.

TABLE 8. Direct products in $\mathcal{D}_3$

$\otimes$	A_1	A_2	E
A_1	A_1	A_2	E
A_2	A_2	A_1	E
E	E	E	$A_1 \oplus A_2 \oplus E$

A further important development occurs in the special case $\mu = \nu$ and $n_\mu > 1$. From (5.2.1) we have

$$\hat{R}\phi_i^{(\mu)}\psi_k^{(\mu)} = \sum_{j,l} \phi_j^{(\mu)}\psi_l^{(\mu)}\Gamma_{ji}^{(\mu)}(R)\Gamma_{lk}^{(\mu)}(R)$$

and

$$\hat{R}\phi_k^{(\mu)}\psi_i^{(\mu)} = \sum_{j,l} \phi_j^{(\mu)}\psi_l^{(\mu)}\Gamma_{jk}^{(\mu)}(R)\Gamma_{li}^{(\mu)}(R)$$

which is obtained on interchange of i and k. It follows that

$$\hat{R}\{\phi_i^{(\mu)}\psi_k^{(\mu)} \pm \phi_k^{(\mu)}\psi_i^{(\mu)}\} = \sum_{j,l} \phi_j^{(\mu)}\psi_l^{(\mu)}\{\Gamma_{ji}^{(\mu)}(R)\Gamma_{lk}^{(\mu)}(R) \pm \Gamma_{jk}^{(\mu)}(R)\Gamma_{li}^{(\mu)}(R)\} \qquad (5.2.3)$$

and

$$\hat{R}\{\phi_i^{(\mu)}\psi_k^{(\mu)} \pm \phi_k^{(\mu)}\psi_i^{(\mu)}\} = \sum_{j,l} \phi_l^{(\mu)}\psi_j^{(\mu)}\{\Gamma_{li}^{(\mu)}(R)\Gamma_{jk}^{(\mu)}(R) \pm \Gamma_{lk}^{(\mu)}(R)\Gamma_{ji}^{(\mu)}(R)\} \qquad (5.2.4)$$

The last equation comes by interchanging j and l. We now add (5.2.3) and

(5.2.4) to obtain

$$\hat{R}\{\phi_i^{(\mu)}\psi_k^{(\mu)} \pm \phi_k^{(\mu)}\psi_i^{(\mu)}\} = \tfrac{1}{2}\sum_{j,l} (\phi_j^{(\mu)}\psi_l^{(\mu)} \pm \phi_l^{(\mu)}\psi_j^{(\mu)})$$
$$\times (\Gamma_{ji}^{(\mu)}(R)\Gamma_{lk}^{(\mu)}(R) \pm \Gamma_{jk}^{(\mu)}(R)\Gamma_{li}^{(\mu)}(R)) \quad (5.2.5)$$

It follows that the functions $\phi_i^{(\mu)}\psi_k^{(\mu)} \pm \phi_k^{(\mu)}\psi_i^{(\mu)}$ form the basis for a $\frac{1}{2}n_\mu(n_\mu \pm 1)$ representation of $\mathscr{G}$ which is called the $\begin{pmatrix}\text{symmetric}\\ \text{antisymmetric}\end{pmatrix}$ product representation and is denoted by $(\Gamma^{(\mu)} \otimes \Gamma^{(\nu)})^{\pm}$. From the method of construction it is clear that

$$\Gamma^{(\mu)} \otimes \Gamma^{(\mu)} = (\Gamma^{(\mu)} \otimes \Gamma^{(\mu)})^{+} \oplus (\Gamma^{(\mu)} \otimes \Gamma^{(\mu)})^{-} \quad (5.2.6)$$

If we return to (5.2.5) it is not hard to show that for the characters in $(\Gamma^{(\mu)} \otimes \Gamma^{(\mu)})^{\pm}$ we have

$$\chi^{(\mu\otimes\mu)^{\pm}}(R) = \tfrac{1}{2}\{(\chi^{(\mu)}(R))^2 \pm \chi^{(\mu)}(R^2)\} \quad (5.2.7)$$

This result enables us to find which IR's are contained in $(\Gamma^{(\mu)} \otimes \Gamma^{(\mu)})^{\pm}$. Thus if we use for example the character table for $\mathscr{D}_3$ we find

$$\begin{aligned}(E \otimes E)^{+} &= A_1 \oplus E\\ (E \otimes E)^{-} &= A_2\end{aligned} \quad (5.2.8)$$

If we happen to use identical basis functions so that $\phi_i^{(\mu)} = \psi_i^{(\mu)}$ then it is clear from (5.2.5) that only the symmetric product representation can occur.

Before going on to consider applications of direct products we derive one further result of interest. Suppose the IR $\Gamma^{(\nu)}$ is contained in the direct product $\Gamma^{(\lambda)} \otimes \Gamma^{(\mu)}$ a total of a_ν times. We have from (5.2.3)

$$a_\nu = \frac{1}{g}\sum_R \chi^{(\lambda\otimes\mu)}(R)\chi^{(\nu)}(R) = \frac{1}{g}\sum_R \chi^{(\lambda)}(R)\chi^{(\mu)}(R)\chi^{(\nu)}(R)$$

Now if a_μ is the number of times $\Gamma^{(\mu)}$ occurs in $\Gamma^{(\lambda)} \otimes \Gamma^{(\nu)}$ we have

$$a_\mu = \frac{1}{g}\sum_R \chi^{(\lambda\otimes\nu)}(R)\chi^{(\mu)}(R) = \frac{1}{g}\sum_R \chi^{(\lambda)}(R)\chi^{(\nu)}(R)\chi^{(\mu)}(R) = a_\nu$$

Thus if $\Gamma^{(\lambda)} \otimes \Gamma^{(\mu)}$ contains $\Gamma^{(\nu)}$ a total of k times then $\Gamma^{(\lambda)} \otimes \Gamma^{(\nu)}$ contains $\Gamma^{(\mu)}$ a total of k times. If $a_\mu = 0$ then $\Gamma^{(\mu)}$ does not appear in the decomposition of $\Gamma^{(\lambda)} \otimes \Gamma^{(\nu)}$.

5.3 Selection rules

We have seen how a set of functions can form the basis vectors for a representation of a group $\mathscr{G}$. In a similar way a set of operators can often be regarded as basis vectors for a representation of $\mathcal{G}$. Consider for example the group $\mathcal{D}_3$ and the orbital angular momentum operators

$$\hat{l}_x = i\left(z\frac{\partial}{\partial y} - y\frac{\partial}{\partial z}\right)$$

$$\hat{l}_y = i\left(x\frac{\partial}{\partial z} - z\frac{\partial}{\partial x}\right)$$

$$\hat{l}_y = i\left(y\frac{\partial}{\partial x} - x\frac{\partial}{\partial y}\right)$$

The quantities x, y, z are vector components. Within the standard basis (e_1, e_2, e_3) we have seen that

$$\hat{C}_3[e_1\ e_2\ e_3] = [e_1\ e_2\ e_3]\begin{bmatrix} -\frac{1}{2} & -\frac{\sqrt{3}}{2} & 0 \\ \frac{\sqrt{3}}{2} & -\frac{1}{2} & 0 \\ 0 & 0 & 1 \end{bmatrix}$$

This can be regarded as a change of basis and it follows that x, y, z undergo the contravariant transformation

$$\begin{bmatrix} x' \\ y' \\ z' \end{bmatrix} = \hat{C}_3\begin{bmatrix} x \\ y \\ z \end{bmatrix} = \begin{bmatrix} -\frac{1}{2} & -\frac{\sqrt{3}}{2} & 0 \\ \frac{\sqrt{3}}{2} & -\frac{1}{2} & 0 \\ 0 & 0 & 1 \end{bmatrix}^{-1}\begin{bmatrix} x \\ y \\ z \end{bmatrix} = \begin{bmatrix} -\frac{1}{2} & \frac{\sqrt{3}}{2} & 0 \\ -\frac{\sqrt{3}}{2} & -\frac{1}{2} & 0 \\ 0 & 0 & 1 \end{bmatrix}\begin{bmatrix} x \\ y \\ z \end{bmatrix} \tag{5.3.1}$$

from which we find

$$x = -\frac{1}{2}x' - \frac{\sqrt{3}}{2}y'$$

$$y = \frac{\sqrt{3}}{2}x' - \frac{1}{2}y'$$

$$z = z'$$

Now

$$\begin{aligned} \frac{\partial}{\partial x'} &= \frac{\partial x}{\partial x'}\frac{\partial}{\partial x} + \frac{\partial y}{\partial x'}\frac{\partial}{\partial y} + \frac{\partial z}{\partial x'}\frac{\partial}{\partial z} = -\frac{1}{2}\frac{\partial}{\partial x} + \frac{\sqrt{3}}{2}\frac{\partial}{\partial y} \\ \frac{\partial}{\partial y'} &= -\frac{\sqrt{3}}{2}\frac{\partial}{\partial x} - \frac{1}{2}\frac{\partial}{\partial y} \\ \frac{\partial}{\partial z'} &= \frac{\partial}{\partial z} \end{aligned} \tag{5.3.2}$$

If we define the transformed operator $\hat{l}'_x$ by

$$\hat{l}'_x = \hat{C}_3\hat{l}_x = \hat{l}_{x'} \qquad (5.3.3)$$

then we have

$$\hat{l}'_x = i\left(z'\frac{\partial}{\partial y'} - y'\frac{\partial}{\partial z'}\right) \qquad (5.3.4)$$

Upon using (5.3.1) and (5.3.2) in (5.3.4) we find

$$\hat{l}'_x = \hat{C}_3\hat{l}_x = -\frac{1}{2}\hat{l}_x + \frac{\sqrt{3}}{2}\hat{l}_y \qquad (5.3.5)$$

Similarly

$$\hat{l}'_y = \hat{C}_3\hat{l}_y = \hat{l}_{y'} = -\frac{\sqrt{3}}{2}\hat{l}_x - \frac{1}{2}\hat{l}_y \qquad (5.3.6)$$

$$\hat{l}'_z = \hat{C}_3\hat{l}_z = \hat{l}_z \qquad (5.3.7)$$

For the other generator C_2 we have seen that

$$\hat{C}_2[e_1 \quad e_2 \quad e_3] = [e_1 \quad e_2 \quad e_3]\begin{bmatrix} 1 & 0 & 0 \\ 0 & -1 & 0 \\ 0 & 0 & -1 \end{bmatrix}$$

which is a further basis change and implies

$$\hat{C}_2\begin{bmatrix} x \\ y \\ z \end{bmatrix} = \begin{bmatrix} x'' \\ y'' \\ z'' \end{bmatrix} = \begin{bmatrix} 1 & 0 & 0 \\ 0 & -1 & 0 \\ 0 & 0 & -1 \end{bmatrix}\begin{bmatrix} x \\ y \\ z \end{bmatrix}$$

If we write

$$\hat{C}_2\hat{l}_x = \hat{l}'_x = \hat{l}_{x''} \quad \text{etc.}$$

then we find

$$\hat{C}_2\hat{l}_x = \hat{l}_x \qquad \hat{C}_2\hat{l}_y = -\hat{l}_y \qquad \hat{C}_2\hat{l}_z = -\hat{l}_z \qquad (5.3.8)$$

From (5.3.5), (5.3.6), (5.3.7) and (5.3.8) we have

$$\hat{C}_3[\hat{l}_x \quad \hat{l}_y] = [\hat{l}_x \quad \hat{l}_y]\begin{bmatrix} -\frac{1}{2} & -\frac{\sqrt{3}}{2} \\ \frac{\sqrt{3}}{2} & -\frac{1}{2} \end{bmatrix}$$

$$\hat{C}_2[\hat{l}_x \quad \hat{l}_y] = [\hat{l}_x \quad \hat{l}_y]\begin{bmatrix} 1 & 0 \\ 0 & -1 \end{bmatrix}$$

$$\hat{C}_3\hat{l}_z = \hat{l}_z \qquad \hat{C}_2\hat{l}_z = -\hat{l}_z$$

It follows that $(\hat{l}_x, \hat{l}_y)$ can be taken as a basis for the standard IR E of $\mathscr{D}_3$. Similarly $\hat{l}_z$ spans the IR A_2 of $\mathscr{D}_3$.

Note that we have regarded $(\hat{l}_x, \hat{l}_y, \hat{l}_z)$ as basis vectors. In elementary accounts of angular momentum $(\hat{l}_x, \hat{l}_y, \hat{l}_z)$ are taken as the components of a vector operator $\hat{\mathbf{l}}$. Strictly speaking $\hat{\mathbf{l}}$ is only a vector operator if we exclude improper rotations. When improper rotations are taken into consideration $\hat{\mathbf{l}}$ is not a vector operator but only a pseudo vector operator (see McWeeny, 1963).

Before considering matrix elements such as

$$M = \langle \phi_i^{(\mu)} | \hat{A}_k^{(\lambda)} | \psi_j^{(\nu)} \rangle \equiv \int \phi_i^{(\mu)*} \hat{A}_k^{(\lambda)} \psi_j^{(\nu)} \, d\tau$$

which involve operators $\hat{A}_k^{(\lambda)}$ we investigate the special case of the scalar product

$$S = \langle \phi_i^{(\mu)} | \psi_j^{(\nu)} \rangle \equiv \int \phi_i^{(\mu)*} \psi_j^{(\nu)} \, d\tau$$

The function $\phi_i^{(\mu)*}$ transforms like the ith basis vector of the IR $\Gamma^{(\mu)*}$ of a group $\mathcal{G}$. Similarly the function $\psi_j^{(\nu)}$ transforms like the jth basis vector of the IR $\Gamma^{(\nu)}$ of $\mathcal{G}$.

Integrals of the form M and S occur frequently in quantum chemistry when we are calculating the properties of atomic and molecular systems. Integrals like M also occur in the theory of atomic and molecular spectra. Now it is found that the integrals are very often zero. It is clearly advantageous to know in advance which integrals are going to vanish. When the functions and operators which occur in the integrand have group theoretical significance as indicated above then we can establish criterion which tell us when a given integral vanishes. In order to establish these important criteria for the vanishing of integrals we proceed as follows.

From III of Section (3.4) the operators $\hat{R}$ corresponding to the group elements R can be taken as unitary operators. It follows from (2.8.14) that

$$\langle \hat{R}\phi_i^{(\mu)} | \hat{R}\psi_j^{(\nu)} \rangle = \langle \phi_i^{(\mu)} | \psi_j^{(\nu)} \rangle$$

for all R in $\mathcal{G}$. If $\mathcal{G}$ is of order g the last equation can be written as

$$\begin{aligned}\langle \phi_i^{(\mu)} | \psi_j^{(\nu)} \rangle &= \frac{1}{g} \sum_R \langle \hat{R}\phi_i^{(\mu)} | \hat{R}\psi_j^{(\nu)} \rangle \\ &= \frac{1}{g} \sum_R \sum_{k,l} \Gamma_{ki}^{(\mu)*}(R) \Gamma_{lj}^{(\nu)}(R) \langle \phi_k^{(\mu)} | \psi_l^{(\nu)} \rangle \\ &= \sum_{k,l} \frac{1}{n_\mu} \delta_{\mu\nu} \, \delta_{kl} \, \delta_{ij} \langle \phi_k^{(\mu)} | \psi_l^{(\nu)} \rangle \end{aligned} \tag{5.3.9}$$

where we have used (3.4.2). From this result we see that $\langle \phi_i^{(\mu)} | \psi_j^{(\nu)} \rangle$ vanishes unless $\mu = \nu$ and $i = j$. When this is so we have

$$\langle \phi_i^{(\mu)} | \psi_i^{(\mu)} \rangle = \frac{1}{n_\mu} \sum_k \langle \phi_k^{(\mu)} | \psi_k^{(\nu)} \rangle \tag{5.3.10}$$

and we see that the value of S is independent of i. Now the function $\phi_i^{(\mu)*}\psi_j^{(\nu)}$ is a basis function for the direct product representation $\Gamma^{(\mu)*} \otimes \Gamma^{(\nu)}$. This direct product can be decomposed into IR's as

$$\Gamma^{(\mu)*} \otimes \Gamma^{(\nu)} = \sum_{\sigma}^{\oplus} a_\sigma \Gamma^{(\sigma)} \tag{5.3.11}$$

where a_σ is the number of times the IR $\Gamma^{(\sigma)}$ occurs. According to (5.1.4) we can write

$$\phi_i^{(\mu)*}\psi_j^{(\nu)} = \sum_{\sigma} \sum_{s} \xi_s^{(\sigma)} \tag{5.3.12}$$

where $\xi_s^{(\sigma)}$ is a function transforming like the sth basis vector of the IR $\Gamma^{(\sigma)}$. We now have

$$\langle \phi_i^{(\mu)} \mid \psi_j^{(\nu)} \rangle = \sum_{\sigma,s} \int \xi_s^{(\sigma)}\, d\tau \equiv \sum_{\sigma,s} \langle 1 \mid \xi_s^{(\sigma)} \rangle \tag{5.3.13}$$

A constant such as unity transforms according to the identity representation A_1 and so from (5.3.9) we see that $\langle \phi_i^{(\mu)} \mid \psi_j^{(\nu)} \rangle = 0$ unless $\Gamma^{(\sigma)} = A_1$ occurs in the decomposition (5.3.11).

We can generalize our findings to matrix elements of the form

$$M = \langle \phi_i^{(\mu)} |\, \hat{A}_k^{(\lambda)}\, | \psi_j^{(\nu)} \rangle$$

where the operator $\hat{A}_k^{(\lambda)}$ transforms like the kth basis vector of the IR $\Gamma^{(\lambda)}$. The function $\hat{A}_k^{(\lambda)}\psi_j^{(\nu)}$ is a basis function for $\Gamma^{(\lambda)} \otimes \Gamma^{(\nu)}$. If

$$\Gamma^{(\lambda)} \otimes \Gamma^{(\nu)} = \sum_{\alpha}^{\oplus} a_\alpha \Gamma^{(\alpha)} \tag{5.3.14}$$

then, as in (5.3.12), we can write

$$\hat{A}_k^{(\lambda)}\psi_j^{(\nu)} = \sum_{\alpha} \sum_{p} \eta_p^{(\alpha)}$$

where $\eta_p^{(\alpha)}$ transforms like the pth basis vector of the IR $\Gamma^{(\alpha)}$. We now have

$$M = \sum_{\alpha,p} \langle \phi_i^{(\mu)} \mid \eta_p^{(\alpha)} \rangle$$

and it follows from (5.3.9) that $M = 0$ unless $\Gamma^{(\alpha)} = \Gamma^{(\mu)*}$ is contained in (5.3.14). This result is basic to the determination of selection rules in atomic and molecular spectroscopy.

As an example let us consider the selection rules for electric dipole transitions between energy levels for a molecule with $\mathcal{D}_3$ symmetry. For light polarized in the z-direction the relevant transition integral is

$$\langle \phi_i^{(\mu)} |\, z\, | \phi_j^{(\nu)} \rangle$$

Since z transforms according to A_2 we immediately see from Table 8 that the only allowed transitions are

$$A_1 \leftrightarrow A_2 \qquad E \leftrightarrow E$$

For light polarized in the x (or y) direction the relevant transition integral is

$$\langle \phi_i^{(\mu)} | \, x \, | \phi_j^{(\nu)} \rangle$$

Since x (or y) transforms according to E the only allowed transitions are

$$A_1 \leftrightarrow E \qquad A_2 \leftrightarrow E \qquad E \leftrightarrow E$$

5.4 Branching rules

Let Γ be an IR of a molecular symmetry group $\mathscr{G}$. If we consider a subgroup $\mathscr{H}$ of $\mathscr{G}$ then in genral Γ will be a reducible representation of $\mathscr{H}$. If we denote the IR's of $\mathscr{H}$ by $\Delta^{(\alpha)}$ then we can write

$$\Gamma = \sum_{\alpha}^{\oplus} a_\alpha \Delta^{(\alpha)} \tag{5.4.1}$$

A relation such as (5.4.1) is called a *branching rule* for the reduction of symmetry $\mathscr{G} \to \mathscr{H}$.

Consider a quantum mechanical system whose Hamiltonian $\hat{H}_0$ is invariant under a group of transformations $\mathscr{G}$. The energy levels are classified according to the IR's of $\mathscr{G}$. If we now apply a perturbation $\lambda \hat{H}_1$ to the system

TABLE 9. Corresponding symmetry transformations in $\mathscr{D}_4$ and $\mathscr{O}$

Symmetry transformations in $\mathscr{D}_4$	Symmetry transformations in $\mathscr{O}$
E	E
$C_4(Z)$	$C_4(Z)$
$C_4^2(Z)$	$C_2(Z)$
$C_2(X)$	$C_2(X)$
C_2'	C_2'

TABLE 10. *IR's of* $\mathscr{O}$ as representations of $\mathscr{D}_4$

$\mathscr{D}_4$	E	C_4^2	$2C_4$	$2C_2$	$2C_2'$
A_1	1	1	1	1	1
A_2	1	1	−1	1	−1
E	2	2	0	2	0
T_1	3	−1	1	−1	−1
T_2	3	−1	−1	−1	1

TABLE 11. Branching rules for $\mathcal{O} \rightarrow \mathcal{D}_4$

$A_1 = A_1$
$A_2 = B_1$
$E = A_1 \oplus B_1$
$T_1 = A_2 \oplus E$
$T_2 = B_2 \oplus E$

which is only invariant under the transformations of a subgroup $\mathcal{H}$ of $\mathcal{G}$ then the energy levels of the perturbed system with Hamiltonian $\hat{H}_0 + \lambda \hat{H}_1$ must be classified according to the IR's of the subgroup $\mathcal{H}$. If a given energy level of the unperturbed system corresponds to the IR Γ of $\mathcal{G}$ and if Γ is reducible as a representation of $\mathcal{H}$ then it is clear that the effect of the perturbation is to remove the degeneracy of the unperturbed level either partially or completely. The branching rules enable us to predict the degeneracies of the energy levels upon the application of a perturbation.

As an example consider the application of a tetragonal perturbation to an octahedral complex. The symmetry is reduced from $\mathcal{O}$ to $\mathcal{D}_4$. The branching rules for the reduction $\mathcal{O} \rightarrow \mathcal{D}_4$ can be found as follows. Firstly we set up the correspondence shown in Table 9. By using the character table for $\mathcal{O}$ we can now write down the characters of the IR's of $\mathcal{O}$ as representation of $\mathcal{D}_4$. This is shown in Table 10. Finally we use the character table for $\mathcal{D}_4$ and the result (3.4.7) to obtain the branching rules as shown in Table 11.

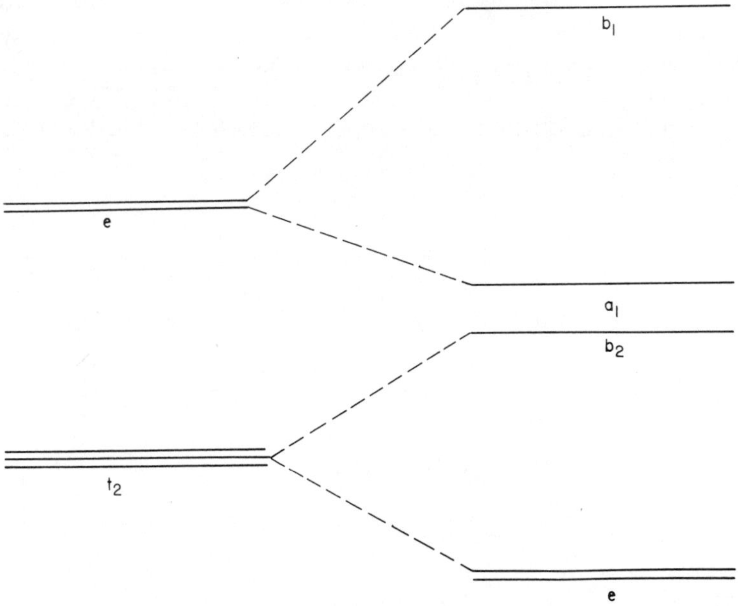

FIG. 11. Tetragonal distortion.

We can use Table 11 to obtain a schematic energy level diagram for the splitting of the d-orbital energy levels in octahedral symmetry when a tetragonal distortion is applied. This is shown in Fig. 11.

References

Hamermesh, M. (1962) "Group theory and its application to physical problems" Addison-Wesley, Chapters 5 & 6.

McWeeny, R. (1963) "Symmetry" Pergamon Press, Chapters 5, 6 and 7.

McWeeny, R. & Sutcliffe, B. T. (1969) "Methods of Molecular Quantum Mechanics" Academic Press, Chapter 2.

CHAPTER 6

The Symmetric Group

In previous chapters we have been concerned with molecular symmetry groups. These groups consist of transformations which can be regarded as permutations of identical nuclei. Since the electrons in a molecule are indistinguishable we are led to consider further symmetry transformations of the system. Within the Born-Oppenhiemer approximation in which we are working the nuclei are stationary. The electrons however are in motion and we cannot therefore consider rotations that send one electron into another. We must use instead permutations of the electron coordinates. We shall use the label i to denote the coordinates of the ith electron. The set of all permutations of the labels, $1, 2, \ldots, n$ forms a finite group of order $n!$ called the *symmetric group of degree n* and denoted by $\mathscr{S}_n$. Since $\mathscr{S}_n$ is a finite group it is possible to use the theory developed in Chapter 3 in order to obtain and discuss the representations of the symmetric group. This becomes very tedious as n increases and is unnecessary since there exists a powerful and elegant method for studying the representations of the general symmetric group.

6.1 Permutations

Consider the labels

$$1, 2, \ldots, n \tag{6.1.1}$$

and the permutation P of $\mathscr{S}_n$ which replaces 1 by p_1, 2 by $p_2, \ldots, n$ by p_n where

$$p_1, p_2, \ldots, p_n$$

are the labels (6.1.1) in some order. We denote this permutation P by

$$P = \begin{pmatrix} 1 & 2 & \ldots & n \\ p_1 & p_2 & \ldots & p_n \end{pmatrix}$$

If Q is a second permutation of $\mathscr{S}_n$ given by

$$Q = \begin{pmatrix} 1 & 2 & \ldots & n \\ q_1 & q_2 & \ldots & q_n \end{pmatrix}$$

then the product PQ is defined by

$$PQ = \begin{pmatrix} 1 & 2 & \dots & n \\ p_1 & p_2 & \dots & p_n \end{pmatrix} \begin{pmatrix} 1 & 2 & \dots & n \\ q_1 & q_2 & \dots & q_n \end{pmatrix}$$

$$= \begin{pmatrix} q_1 & q_2 & \dots & q_n \\ r_1 & r_2 & \dots & r_n \end{pmatrix} \begin{pmatrix} 1 & 2 & \dots & n \\ q_1 & q_2 & \dots & q_n \end{pmatrix}$$

$$= \begin{pmatrix} 1 & 2 & \dots & n \\ r_1 & r_2 & \dots & r_n \end{pmatrix}$$

where we have rearranged P as $\begin{pmatrix} q_1 \dots q_n \\ r_1 \dots r_n \end{pmatrix}$.

A permutation which interchanges m labels cyclically is called an m-*cycle* and is written as

$$\begin{pmatrix} 1 & 2 & \dots & m-1 & m \\ 2 & 3 & \dots & m & 1 \end{pmatrix} \equiv (1 \quad 2 \quad \dots \quad m)$$

A 2-cycle is called a *transposition.* It is not hard to show that every permutation can be written as a product of cycles which operate on mutually exclusive sets of labels. Thus for example

$$\begin{pmatrix} 1 & 2 & 3 & 4 & 5 & 6 \\ 2 & 4 & 5 & 1 & 3 & 6 \end{pmatrix} = (1 \quad 2 \quad 4)(3 \quad 5)(6)$$

Now let $x_1, x_2, \dots, x_n$ be independent variables and consider the function

$$\begin{aligned} \Delta(x_1, x_2, \dots, x_n) = (x_1 - x_2)(x_1 - x_3) \dots (x_1 - x_n) \\ \times (x_2 - x_3) \dots (x_2 - x_n) \dots \\ \times (x_{n-1} - x_n) \end{aligned} \qquad (6.1.2)$$

If P is an element of $\mathscr{S}_n$ it is clear that

$$P\,\Delta = (-1)^p\,\Delta$$

where p is a non-negative integer called the *parity* of P. If p is even we say that the permutation P is *even* while if p is odd we say that the permutation P is *odd.*

It is not difficult to show that every permutation can be expressed as a product of transpositions. Thus for example $(123) = (13)(12)$. In particular it can be shown that the $(n-1)$ transpositions

$$(12), (13), \dots, (1n) \qquad (6.1.3)$$

may be taken as generators for the group $\mathscr{S}_n$, (see, for example, Ledermann, 1957).

6.2 Partitions and conjugate classes

Let P be an element of $\mathcal{S}_n$. According to Section 1.2 an element Q of $\mathcal{S}_n$ is in the same class as P if there exists an element T of $\mathcal{S}_n$ for which

$$Q = TPT^{-1}$$

Now we can decompose P into cycles. Suppose that in the decomposition there occurs ν_1 1-cycles, . . . , ν_2 2-cycles, . . . , ν_n n-cycles. We say that P has the *cycle structure*

$$(\nu) \equiv (1^{\nu_1} \quad 2^{\nu_2} \quad \dots \quad n^{\nu_n}) \tag{6.2.1}$$

If we denote the cycles by C_i then

$$P = C_1 C_2 \dots C_p \tag{6.2.2}$$

where $p = \nu_1 + \nu_2 + \dots + \nu_n$. Since the total number of labels is n it follows that

$$\nu_1 + 2\nu_2 + \dots + n\nu_n = n \tag{6.2.3}$$

Now the conjugate element Q is given by

$$Q = (TC_1T^{-1})(TC_2T^{-1}) \dots (TC_pT^{-1})$$

and it is not difficult to show from this that Q must have the same cycle structure as P. Thus all elements in a given conjugate class (see Ledermann (1957)) have the same cycle structure. It follows that each solution of (6.2.3) in non-negative integers $\nu_1, \nu_2, \dots, \nu_n$ determines a cycle structure and hence a conjugate class. Thus the number of classes in $\mathcal{S}_n$ is given by the number of solutions of (6.2.3). If we write

$$\begin{aligned} \nu_1 + \nu_2 + \dots + \nu_n &= \lambda_1 \\ \nu_2 + \dots + \nu_n &= \lambda_2 \\ \dots\dots \nu_n &= \lambda_n \end{aligned} \tag{6.2.4}$$

then we have

$$\lambda_1 + \lambda_2 + \dots + \lambda_n = n \tag{6.2.5}$$

with

$$\lambda_1 \geqslant \lambda_2 \geqslant \dots \geqslant \lambda_n \geqslant 0 \tag{6.2.6}$$

We say that (6.2.5) is a *partition* of n and we denote this partition by $[\lambda] \equiv [\lambda_1\lambda_2 \dots \lambda_n]$. It is clear that there is a one to one correspondence between partitions of n and solutions of (6.2.3) since from (6.2.4) we have

$$\begin{aligned} \nu_1 &= \lambda_1 - \lambda_2 \\ \nu_2 &= \lambda_2 - \lambda_3 \\ &\dots\dots\dots \\ \nu_n &= \lambda_n \end{aligned} \tag{6.2.7}$$

Thus the number of classes in $\mathscr{S}_n$ is given by the number of partitions of n.

It can be shown that the number of elements in the conjugate class with cycle structure

$$(\nu)=(1^{\nu_1}2^{\nu_2}\ldots n^{\nu_n})$$

is given by

$$n_{(\nu)}=\frac{n!}{1^{\nu_1}\nu_1!\,2^{\nu_2}\nu_2!\,3^{\nu_3}\nu_3!\ldots n^{\nu_n}\nu_n!}$$

As an illustration consider $\mathscr{S}_4$ (see Ledermann, 1957). The partitions of 4 are [4], [3 1], [2 2]≡[2^2], [2 1 1]≡[2 1^2] and [1 1 1 1]≡[1^4]. Thus there are 5 conjugate classes in $\mathscr{S}_4$. Using (6.2.7) and (6.2.8) we obtain Table 12.

TABLE 12. Partitions of four

Partition	Cycle structure	Number of elements in class	Example
[4]	(1^4)	1	E
[1^4]	(4^1)	6	(1432)
[2^2]	(2^2)	3	(14)(23)
[21^2]	$(1^1 3^1)$	8	(132)
[31]	$(1^2 2^1)$	6	(12)

6.3 Young tableaux

Since the number of conjugate classes in $\mathscr{S}_n$ is given by the number of partitions of n it follows that the number of IR's of $\mathscr{S}_n$ is given by the number of partitions of n. Thus corresponding to each partition of n there is an IR of $\mathscr{S}_n$. In this way we have at our disposal a very convenient method for labelling the various IR's of $\mathscr{S}_n$.

Corresponding to each partition $[\lambda]=[\lambda_1\lambda_2\ldots\lambda_p]$ we can draw a *shape* $S^{[\lambda]}$ consisting of λ_1 cells in the first row, λ_2 cells in the second row and so on down to λ_p cells in the last row.

$S^{[\lambda]}=$ 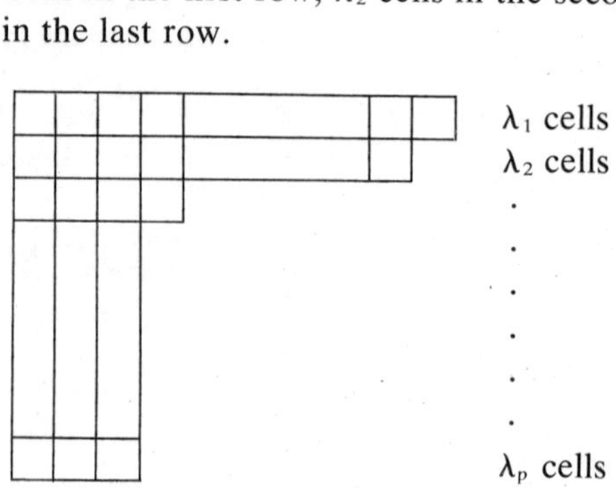

If the numbers $1, 2, \ldots, n$ are now inserted into the cells we obtain a *Young tableau*. If the numbers are inserted into the cells in such a way that they increase on going down a column and increase on going along a row from left to right then we have a *standard* Young tableau $T^{[\lambda]}$.

We now state without proof a theorem of fundamental importance (see Hamermesh, 1962).

THEOREM (YOUNG). The dimension n_λ of the IR denoted by the partition $[\lambda]$ is given by the number of standard Young tableaux $T_1^{[\lambda]}, \ldots, T_{n_\lambda}^{[\lambda]}$ which can be constructed from the shape $S^{[\lambda]}$.

The result of applying this theorem to $\mathscr{S}_4$ is shown in Table 13. From Table 13 we note that corresponding to each IR $[\lambda]$ there exists an IR $[\tilde{\lambda}]$ in which the rows and columns of the shape have been interchanged. The IR $[\tilde{\lambda}]$ is called the *dual* of $[\lambda]$. We see also that the IR $[2^2]$ is self-dual and that dual IR's have the same dimension. These results can be shown to be quite general (Rutherford, 1948).

TABLE 13. IR's of $\mathscr{S}_4$

IR	Standard Young tableaux	Dimension of IR
[4]	1 2 3 4	1
$[1^4]$	1 / 2 / 3 / 4	1
$[2^2]$	1 2 / 3 4; 1 3 / 2 4	2
[31]	1 2 3 / 4; 1 2 4 / 3; 1 3 4 / 2	3
$[21^2]$	1 4 / 2 / 3; 1 3 / 2 / 4; 1 2 / 3 / 4	3

It is evident from the mode of construction that the group $\mathscr{S}_n$ possesses two 1-dimensional IR's namely $[n]$ and its dual $[1^n]$. Furthermore all other IR's must be of dimension greater than one. If we consider the generators $(1i)$; $i = 2, 3, \ldots n$ and a basis function $\psi(1, 2, \ldots, n)$ then, in the identity representation we have

$$(1i)\psi(1, 2, \ldots, n) = \psi(1, 2, \ldots, n)$$

It is conventional to choose for the identity representation the label $[n]$. It follows that the IR $[n]$ is spanned by any function which is totally symmetric. The dual IR $[1^n]$ can be obtained by assigning $+1$ to all even permutations and -1 to all odd permutations. Since the transpositions $(1i)$ are all odd it follows that $[1^n]$ is spanned by a function $\psi(1, 2, \ldots, n)$ which is totally antisymmetric, i.e.

$$(1i)\psi(1, 2, \ldots, n) = -\psi(1, 2, \ldots, n)$$

We now associate with the standard Young tableau

1	2		n

the symmetrizing operator

$$\hat{S} = \sum_{P} \hat{P} \tag{6.3.1}$$

where the sum runs over all permutations. Now if $\Phi(1, 2, \ldots, n)$ is an arbitrary function then the function $\hat{S}\Phi$ is a basis function for the IR $[n]$. Similarly we associate with the standard Young tableau

1
2
n

the antisymmetrizing operator

$$\hat{A} = \sum_{P} (-1)^{p_P} P \tag{6.3.2}$$

where p_P is the parity of the permutation P. The function $\hat{A}\Phi$ is then a basis function for the IR $[1^n]$. The operators $\hat{S}$ and $\hat{A}$ are projection operators of the type introduced in Section 5.1. It is customary to call such operators *Young operators*.

The ideas introduced above can be generalized to IR's of dimension greater than one. We begin by defining two types of permutation. Horizontal permutations are those which interchange only symbols in the same row of a standard tableau. Vertical permutations are those which interchange only symbols in the same column of a standard tableau. We now associate with a given standard tableau $T_i^{[\lambda]}$ a Young operator $\hat{Y}_i^{[\lambda]}$. This is defined by

$$\hat{Y}_i^{[\lambda]} = \hat{A}\hat{S} \tag{6.3.4}$$

Here $\hat{S} = \sum_P \hat{P}$ and the sum runs over all horizontal permutations in $T_i^{[\lambda]}$ while $\hat{A} = \sum_P (-1)^{p_P} \hat{P}$ with the sum running over all vertical permutations in $T_i^{[\lambda]}$.

As an example let us consider the 2-dimensional IR [21] of $\mathscr{S}_3$. Corresponding to this IR we have the standard tableaux

$$T_1 = \begin{array}{|c|c|}\hline 1 & 2 \\ \hline 3 \\ \cline{1-1}\end{array} \qquad T_2 = \begin{array}{|c|c|}\hline 1 & 3 \\ \hline 2 \\ \cline{1-1}\end{array} \tag{6.3.5}$$

The Young operators which are associated with T_1 and T_2 are

$$\left.\begin{aligned}\hat{Y}_1 &= [E-(13)][E+(12)] \\ &\text{and} \\ \hat{Y}_2 &= [E-(12)][E+(13)]\end{aligned}\right\} \tag{6.3.6}$$

respectively.

Now let $T_1^{[\lambda]}, T_2^{[\lambda]}, \ldots, T_{n_\lambda}^{[\lambda]}$ be the standard Young tableaux for the n_λ-dimensional IR $[\lambda]$ of $\mathscr{S}_n$. These standard tableaux are said to be in *dictionary order* when the following criterion is satisfied. Take any two standard tableaux $T_i^{[\lambda]}$ and $T_j^{[\lambda]}$. Let the symbols occurring in $T_i^{[\lambda]}$ be r and the symbols occurring in $T_j^{[\lambda]}$ be s. Read the symbols r and s from the tableaux $T_i^{[\lambda]}$ and $T_j^{[\lambda]}$ as if reading the page of a book. If the first non-zero difference $(s-r)$ between symbols placed in the same cell in the two tableaux is positive then $T_i^{[\lambda]}$ comes before $T_j^{[\lambda]}$. For [21] of $\mathscr{S}_3$ the dictionary order is T_1, T_2. For $[21^2]$ of $\mathscr{S}_4$ we have

$$A = \begin{array}{|c|c|}\hline 1 & 2 \\ \hline 3 \\ \cline{1-1} 4 \\ \cline{1-1}\end{array} \qquad B = \begin{array}{|c|c|}\hline 1 & 4 \\ \hline 2 \\ \cline{1-1} 3 \\ \cline{1-1}\end{array} \qquad C = \begin{array}{|c|c|}\hline 1 & 3 \\ \hline 2 \\ \cline{1-1} 4 \\ \cline{1-1}\end{array}$$

Comparison of A and B gives $4-2>0$ so B comes after A. Comparison of B and C gives $3-4<0$ so B comes after C. Comparison of A and C gives $3-2>0$ so C comes after A. Thus the dictionary order of the tableaux is $T_1 = A$; $T_2 = C$; $T_3 = B$.

We now associate with each standard tableau T_i a function $\Phi_i(1, 2, \ldots, n)$. If we choose $\Phi_1 = \Phi(1, 2, \ldots, n)$ then the remaining $(n_\lambda - 1)$ functions Φ_i; $i = 2, 3, \ldots, n_\lambda$ are obtained by writing $\Phi_i = \hat{P}_i\Phi_1$ where P_i is the permutation which takes the standard tableau T_1 into the standard tableau T_i. Since we can obtain $n!$ functions from $\Phi(1, 2, \ldots, n)$ by permutations of the coordinates it is not surprising to find that the n_λ functions Φ_i do not in general form a basis for the IR $[\lambda]$. However we can construct functions of definite permutational symmetry by application of the Young operators. Let

$$\Psi_i = \hat{Y}_i\Phi_i \qquad i = 1, 2, \ldots, n_\lambda \tag{6.3.7}$$

where Φ_i is defined as above and where $\hat{Y}_i$ is the Young operator associated with the standard tableau T_i. It can be shown that the n_λ functions Ψ_i do form a basis for the IR $[\lambda]$. (See Hamermesh, 1962).

Let us consider now some examples of how all this works in practice. For

[21] of $\mathscr{S}_3$ we have from (6.3.5) the functions

$$\Phi_1 = \Phi(1, 2, 3) \qquad \Phi_2 = \Phi(1, 3, 2)$$

We now apply (6.3.6) and (6.3.7) to find

$$\Psi_1 = \hat{Y}_1\Phi_1 = [E-(13)][E+(12)]\Phi(1, 2, 3)$$
$$\Psi_2 = \hat{Y}_2\Phi_2 = [E-(12)][E+(13)]\Phi(1, 3, 2)$$

which expands to yield

$$\Psi_1 = \Phi(1, 2, 3) + \Phi(2, 1, 3) - \Phi(3, 2, 1) - \Phi(2, 3, 1) \tag{6.3.8}$$
$$\Psi_2 = \Phi(1, 3, 2) + \Phi(3, 1, 2) - \Phi(2, 3, 1) - \Phi(3, 2, 1) \tag{6.3.9}$$

The generators for $\mathscr{S}_3$ are (12) and (13). We have

$$(12)\Psi_1 = \Phi(2, 1, 3) + \Phi(1, 2, 3) - \Phi(3, 1, 2) - \Phi(1, 3, 2) = \Psi_1 - \Psi_2$$
$$(12)\Psi_2 = \Phi(2, 3, 1) + \Phi(3, 2, 1) - \Phi(1, 3, 2) - \Phi(3, 1, 2) = -\Psi_2$$

which yields the matrix representation

$$(12) = \begin{bmatrix} 1 & 0 \\ -1 & -1 \end{bmatrix} \tag{6.3.10}$$

Similarly we find

$$(13) = \begin{bmatrix} -1 & -1 \\ 0 & 1 \end{bmatrix} \tag{6.3.11}$$

The matrix for (132) is

$$(132) = (12)(13) = \begin{bmatrix} 0 & +1 \\ -1 & -1 \end{bmatrix}$$

Since conjugate elements have the same cycle structure and the same character it can readily be seen that the above matrix representation is irreducible.

As a second example let us take the 3-dimensional IR $[21^2]$ of $\mathscr{S}_4$. The standard tableaux in dictionary order are

$$T_1 = \begin{array}{|c|c|} \hline 1 & 2 \\ \hline 3 \\ \cline{1-1} 4 \\ \cline{1-1} \end{array} \qquad T_2 = \begin{array}{|c|c|} \hline 1 & 3 \\ \hline 2 \\ \cline{1-1} 4 \\ \cline{1-1} \end{array} \qquad T_3 = \begin{array}{|c|c|} \hline 1 & 4 \\ \hline 2 \\ \cline{1-1} 3 \\ \cline{1-1} \end{array}$$

and we have the basis functions

$$\Psi_1 = \hat{Y}_1\Phi(1234) \qquad \Psi_2 = \hat{Y}_2\Phi(1324) \qquad \Psi_3 = \hat{Y}_3\Phi(1423) \tag{6.3.12}$$

with

$$\hat{Y}_1 = \{E-(13)-(14)-(34)+(134)+(143)\}\{E+(12)\}$$
$$\hat{Y}_2 = \{E-(12)-(14)-(24)+(124)+(142)\}\{E+(13)\}$$
$$\hat{Y}_3 = \{E-(12)-(13)-(23)+(123)+(132)\}\{E+(14)\}$$

In the basis (Ψ_1, Ψ_2, Ψ_3) we now find

$$(12)=\begin{bmatrix} 1 & 0 & 0 \\ -1 & -1 & 0 \\ 1 & 0 & -1 \end{bmatrix} \tag{6.3.13}$$

$$(13)=\begin{bmatrix} -1 & -1 & 0 \\ 0 & 1 & 0 \\ 0 & -1 & -1 \end{bmatrix} \tag{6.3.14}$$

$$(14)=\begin{bmatrix} -1 & 0 & 1 \\ 0 & -1 & -1 \\ 0 & 0 & 1 \end{bmatrix} \tag{6.3.15}$$

for the generators of $\mathscr{S}_4$. We also have

$$(132)=(12)(13)=\begin{bmatrix} 0 & +1 & 0 \\ -1 & -1 & 0 \\ 0 & 1 & 1 \end{bmatrix}$$

$$(1432)=(12)(13)(14)=\begin{bmatrix} 0 & 0 & 1 \\ 1 & 0 & -1 \\ 0 & +1 & 1 \end{bmatrix}$$

$$(14)(23)=(14)(132)(12)=\begin{bmatrix} 0 & -1 & -1 \\ -1 & 0 & 1 \\ 0 & 0 & -1 \end{bmatrix}$$

Using the characters of these matrices and the results in Table 13 we see that the representation spanned by (Ψ_1, Ψ_2, Ψ_3) is irreducible.

The dual representation [31] has standard tableaux

$$T_1=\begin{array}{|c|c|c|}\hline 1 & 3 & 4 \\ \hline 2 \\ \cline{1-1}\end{array} \qquad T_2=\begin{array}{|c|c|c|}\hline 1 & 2 & 4 \\ \hline 3 \\ \cline{1-1}\end{array} \qquad T_3=\begin{array}{|c|c|c|}\hline 1 & 2 & 3 \\ \hline 4 \\ \cline{1-1}\end{array}$$

If we denote the basis functions for this representation by $\tilde{\Psi}_i$ then we can write

$$\tilde{\Psi}_1=\hat{\tilde{Y}}_1\Phi(1234) \qquad \tilde{\Psi}_2=\hat{\tilde{Y}}_2\Phi(1324) \qquad \tilde{\Psi}_3=\hat{\tilde{Y}}_3\Phi(1423) \tag{6.3.16}$$

with

$$\hat{\tilde{Y}}_1=[E-(12)][E+(13)+(14)+(34)+(134)+(143)]$$

and similar expressions for $\tilde{Y}_2$ and $\tilde{Y}_3$. It should be noted that the standard tableaux for the dual representation are in reverse dictionary order. We could equally well regard $[21^2]$ as the dual of [31]. Thus we could take the tableaux of [31] in dictionary order and define basis functions Ψ_i by writing

$$\Psi_1=\hat{Y}_1\Phi(1234) \qquad \Psi_2=\hat{Y}_2\Phi(1243) \qquad \Psi_3=\hat{Y}_3\Phi(1342)$$

with $\hat{Y}_1 = \tilde{\hat{Y}}_3$; $\hat{Y}_2 = \tilde{\hat{Y}}_2$; $\hat{Y}_3 = \tilde{\hat{Y}}_1$. If we use the dual basis $(\tilde{\Psi}_1, \tilde{\Psi}_2, \tilde{\Psi}_3)$ we find

$$(12) = \begin{bmatrix} -1 & -1 & -1 \\ 0 & 1 & 0 \\ 0 & 0 & 1 \end{bmatrix} \tag{6.3.17}$$

$$(13) = \begin{bmatrix} 1 & 0 & 0 \\ -1 & -1 & -1 \\ 0 & 0 & 1 \end{bmatrix} \tag{6.3.18}$$

$$(14) = \begin{bmatrix} 1 & 0 & 0 \\ 0 & 1 & 0 \\ -1 & -1 & -1 \end{bmatrix} \tag{6.3.19}$$

The construction of the representation matrices by the above method becomes very involved as n increases. Furthermore each symmetric group has to be treated individually. We note however that $\mathcal{S}_{n-1}$ is a subgroup of $\mathcal{S}_n$. This observation leads us to consider the possibility of obtaining the IR's of $\mathcal{S}_n$ by a building up process.

6.4 The standard irreducible representations of $\mathcal{S}_n$

Consider the IR of $\mathcal{S}_n$ corresponding to a partition $[\lambda]$. If we remove a cell from the shape of $[\lambda]$ in such a way as to obtain another allowed shape then we say that this is a *regular removal* of a cell. By an *allowed* shape of course we mean one that satisfies (6.2.6). It is clear that the new shape corresponds to an IR of $\mathcal{S}_{n-1}$. By using the theory of group characters it can be shown that the branching rules for the reduction $\mathcal{S}_n \to \mathcal{S}_{n-1}$ are obtained by regular removal of a cell from the shape of the IR $[\lambda]$ of $\mathcal{S}_n$ in all possible ways (see Hamermesh, 1962). As an example we have the branching rules

$$\begin{aligned}
[4] &\to [3] \\
[1^4] &\to [1^3] \\
[2^2] &\to [21] \\
[31] &\to [21] \oplus [3] \\
[21^2] &\to [21] \oplus [1^3]
\end{aligned}$$

for the reduction $\mathcal{S}_4 \to \mathcal{S}_3$.

If we have a standard tableau T for some IR of $\mathcal{S}_n$ then we can obtain standard tableaux for a string of IR's corresponding to the subgroups $\mathcal{S}_{n-1}$, $\mathcal{S}_{n-2}, \ldots, \mathcal{S}_1$ by removing from T in succession the cells which contain the symbols $n, n-1, n-2, \ldots, 1$. Thus for example we have

$$\begin{array}{|c|c|c|}\hline 1 & 2 & 4 \\ \hline 3 & 5 \\ \cline{1-2} 6 \\ \cline{1-1}\end{array} \quad \begin{array}{|c|c|c|}\hline 1 & 2 & 4 \\ \hline 3 & 5 \\ \cline{1-2}\end{array} \quad \begin{array}{|c|c|c|}\hline 1 & 2 & 4 \\ \hline 3 \\ \cline{1-1}\end{array} \quad \begin{array}{|c|c|}\hline 1 & 2 \\ \hline 3 \\ \cline{1-1}\end{array} \quad \begin{array}{|c|c|}\hline 1 & 2 \\ \hline\end{array} \quad \begin{array}{|c|}\hline 1 \\ \hline\end{array} \tag{6.4.1}$$

This reduction process can be described by assigning to T a *Yamanouchi symbol* (Y-symbol) which is denoted by $(r_n r_{n-1} r_{n-2} \dots r_1)$. The Y-symbol indicates that the number i occurs in row r_i of T. The reduction to subgroups is then given by reading the Y-symbol from left to right. Thus the Y-symbol corresponding to (6.4.1) is (321211).

We can now state without proof the following important result which derives from a consideration of the branching rules (see Hamermesh, 1962).

Theorem (*Young-Yamanouchi*)

Consider the IR of $\mathscr{S}_n$ which is labelled by the partition $[\lambda] = [\lambda_1 \lambda_2 \dots \lambda_r \dots \lambda_n]$. Let $(r) = (r_n r_{n-1} \dots r_1)$ be the Y-symbol for a given standard tableaux T of $[\lambda]$. The first part of the theorem states that, for each (r), there exists some function $\Phi_{(r)}(1, 2, \dots, n)$ of the coordinates which serves as a basis function not only for the IR $[\lambda]$ of $\mathscr{S}_n$ but also for the string of IR's of $\mathscr{S}_{n-1}, \mathscr{S}_{n-2}, \dots, \mathscr{S}_1$ which are obtained by successive deletion of cells from T according to the prescription given by the Y-symbol (r). The IR's of $\mathscr{S}_n$ which are obtained from such basis functions are called *standard* IR's. Even although the explicit functional form of $\Phi_{(r)}(1, 2, \dots, n)$ is not used nevertheless it is possible to obtain the representation matrices by referring to group character theory. Suppose we have the standard IR's of $\mathscr{S}_{n-1}$ and we require those for $\mathscr{S}_n$. It is only necessary to obtain the matrix representatives of the generators (12, (13), ..., $(1n)$. Now

$$(in) = (n-1n)(in-1)(n-1n)$$

and by supposition we already know the standard matrix representatives of (12), (13), ..., (1n−1). Thus all we require is the matrix for $(n-1n)$. Let us introduce the notation

$$|[\lambda](r_n r_{n-1} \dots r_1)\rangle \equiv \Phi^{[\lambda]}_{(r)}(1, 2, \dots, n)$$

for the standard basis functions. The second part of the theorem shows that the matrix of $(n-1n)$ is given by

$$\left.\begin{aligned}
&\text{I.} \quad (n-1n)\,|[\lambda](rrr_{n-2}\dots r_1)\rangle = +|[\lambda](rrr_{n-2}\dots r_1)\rangle \\
&\text{II.} \quad (n-1n)\,|[\lambda](rr-1r_{n-2}\dots r_1)\rangle = -|[\lambda](rr-1r_{n-2}\dots r_1)\rangle \\
&\qquad \text{if } (r-1rr_{n-2}\dots 1) \text{ does not exist} \\
&\text{III.} \quad (n-1n)\,|[\lambda](rsr_{n-2}\dots r_1)\rangle = \sigma_{rs}\,|[\lambda](rsr_{n-2}\dots r_1)\rangle \\
&\qquad\qquad + \sqrt{1-\sigma_{rs}^2}\,|[\lambda](srr_{n-2}\dots r_1)\rangle
\end{aligned}\right\} \quad (6.4.2)$$

when both $(rsr_{n-2} \dots r_1)$ and $(srr_{n-2} \dots r_1)$ exist $(r \neq s)$.

The quantity σ_{rs} is given by

$$\sigma_{rs} = \frac{1}{\lambda_r - \lambda_s + s - r}$$

for the IR $[\lambda_1\lambda_2 \ldots \lambda_r \ldots \lambda_s \ldots \lambda_n]$. Finally it should be noted that the matrices in the standard representation are real orthogonal (real and unitary).

We now endeavour to clarify this lengthy theorem by using it to build up the standard IR's for the first few symmetric groups.

We start with $\mathscr{S}_2$. The IR [2] has the standard tableau $\begin{array}{|c|c|}\hline 1 & 2 \\ \hline\end{array}$ with basis function $\|[2](11)\rangle$. From I of (6.4.2) we find

$$(12)\,\|[2](11)\rangle = \|[2](11)\rangle$$

The IR $[1^2]$ has standard tableau $\begin{array}{|c|}\hline 1 \\ \hline 2 \\ \hline\end{array}$ with basis function $\|[1^2](21)\rangle$. Since the Y-symbol (12) does not exist we have from II of (6.4.2)

$$(12)\,\|[1^2](21)\rangle = -\|[1^2](21)\rangle$$

and we can write

$$(12)[\|[2](11)\rangle\ \|[1^2](21)\rangle] = [\|[2](11)\rangle\ \|[1^2](21)\rangle]\begin{bmatrix} 1 & 0 \\ 0 & -1 \end{bmatrix} \qquad (6.4.3)$$

The first non-trivial case arises when we consider the IR [21] of $\mathscr{S}_3$. The standard tableaux are $T_1 = \begin{array}{|c|c|}\hline 1 & 2 \\ \hline 3 \\ \cline{1-1}\end{array}$ and $T_2 = \begin{array}{|c|c|}\hline 1 & 3 \\ \hline 2 \\ \cline{1-1}\end{array}$ with basis functions $\|[21](211)\rangle$ and $\|[21](121)\rangle$ respectively. On deletion of cell 3 from T_1 we see that $\|[21](211)\rangle$ is a basis function for [2] of $\mathscr{S}_2$. Similarly $\|[21](121)\rangle$ is a basis function for $[1^2]$ of $\mathscr{S}_2$. It follows from (6.4.3) that in the IR [21] of $\mathscr{S}_3$ we have

$$(12) = \begin{bmatrix} 1 & 0 \\ 0 & -1 \end{bmatrix} \qquad (6.4.4)$$

Since both (211) and (121) exist we have from III of (6.4.2)

$$(23)\,\|[21](211)\rangle = \sigma_{21}\,\|[21](211)\rangle + \sqrt{1-\sigma_{21}^2}\,\|[21](121)\rangle$$

$$(23)\,\|[21](121)\rangle = \sigma_{12}\,\|[21](121)\rangle + \sqrt{1-\sigma_{12}^2}\,\|[21](211)\rangle$$

Now in [21] we find $\lambda_1 = 2$ and $\lambda_2 = 1$ so

$$\sigma_{12} = \tfrac{1}{2} \qquad \sigma_{21} = -\tfrac{1}{2}$$

and we have

$$(23) = \begin{bmatrix} -\dfrac{1}{2} & \dfrac{\sqrt{3}}{2} \\[2mm] \dfrac{\sqrt{3}}{2} & \dfrac{1}{2} \end{bmatrix} \qquad (6.4.5)$$

The rest of the calculation is straightforward.

$$(13) = (23)(12)(23) = \begin{bmatrix} -\frac{1}{2} & -\frac{\sqrt{3}}{2} \\ -\frac{\sqrt{3}}{2} & \frac{1}{2} \end{bmatrix} \qquad (6.4.6)$$

$$(123) = (13)(12) = \begin{bmatrix} -\frac{1}{2} & \frac{\sqrt{3}}{2} \\ -\frac{\sqrt{3}}{2} & -\frac{1}{2} \end{bmatrix} \qquad (6.4.7)$$

$$(132) = (12)(13) = \begin{bmatrix} -\frac{1}{2} & -\frac{\sqrt{3}}{2} \\ \frac{\sqrt{3}}{2} & -\frac{1}{2} \end{bmatrix}$$

For the IR $[2^2]$ of $\mathscr{S}_4$ we have basis functions $|[2^2](2211)\rangle$ and $|[2^2](2121)\rangle$. Since (1221) does not exist it follows that

$$(34) = \begin{bmatrix} 1 & 0 \\ 0 & -1 \end{bmatrix}$$

The matrices for (12) and (13) are given by (6.4.4) and (6.4.6) respectively while the matrix for the other generator (14) is given by

$$(14) = (34)(13)(34)$$

As a final example we take the IR $[21^2]$ of $\mathscr{S}_4$. The standard tableaux are

$$T_1 = \begin{array}{|c|c|}\hline 1 & 2 \\ \hline 3 \\ \cline{1-1} 4 \\ \cline{1-1}\end{array} \qquad T_2 = \begin{array}{|c|c|}\hline 1 & 3 \\ \hline 2 \\ \cline{1-1} 4 \\ \cline{1-1}\end{array} \qquad T_3 = \begin{array}{|c|c|}\hline 1 & 4 \\ \hline 2 \\ \cline{1-1} 3 \\ \cline{1-1}\end{array}$$

with corresponding basis functions $|[21^2](3211)\rangle$, $|[21^2](3121)\rangle$ and $|[21^2](1321)\rangle$ respectively. Application of (6.4.2) yields

$$(34)\,|[21^2](3211)\rangle = -|[21^2](3211)\rangle$$

$$(34)\,|[21^2](3121)\rangle = \sigma_{31}\,|[21^2](3121)\rangle + \sqrt{1-\sigma_{31}^2}\,|[21^2](1321)\rangle$$

$$(34)\,|[21^2](1321)\rangle = \sigma_{13}\,|[21^2](1321)\rangle + \sqrt{1-\sigma_{13}^2}\,|[21^2](3121)\rangle$$

with

$$\sigma_{31} = -\sigma_{13} = \tfrac{1}{3}$$

6.5 Standard Young operators

The Young-Yamanouchi theorem demonstrates the existence of a set of standard basis functions for the IR's of $\mathscr{S}_n$. These functions are labelled by Y-symbols. The explicit functional form of the standard basis functions can be found by using the projection operators (5.1.7). Since the Y-symbols serve to label the rows and columns of the matrices it is conventional to write the projection operators in the form

$$\hat{\omega}_{rs}^{[\lambda]} = \sqrt{\frac{n_\lambda}{n!}} \sum_P \langle[\lambda](r)|\, \hat{P}\, |[\lambda](s)\rangle\, \hat{P} \tag{6.5.1}$$

The factor $\sqrt{n_\lambda/n!}$ is a normalizing factor and n_λ is the dimension of the IR $[\lambda]$. The projection operators $\hat{\omega}_{rs}^{[\lambda]}$ are called *standard Young operators.* A standard Young operator $\hat{\omega}$ is clearly similar to a Young operator $\hat{Y}$ as defined by (6.3.4). However, in general, if $\Phi(1, 2, \ldots, n)$ is an arbitrary function then $\hat{Y}\Phi$ is a basis function for a non-standard IR of $\mathscr{S}_n$ while $\hat{\omega}\Phi$ is a basis function for a standard IR of $\mathscr{S}_n$.

6.6 Dual representation

Let $[\tilde{\lambda}]$ be the IR of $\mathscr{S}_n$ which is dual to $[\lambda]$. A basis function for $[\tilde{\lambda}]$ may be labelled by a Y-symbol $(\tilde{r})$ which is obtained from the standard tableau corresponding to (r) by interchanging rows and columns. Thus in $\mathscr{S}_4$ we have [31] and $[21^2]$ as dual IR's. The basis functions for [31] are $|[31](1121)\rangle$, $|[31](1211)\rangle$ and $|[31](2111)\rangle$. They correspond respectively to the standard tableaux

$$T_1 = \begin{array}{|c|c|c|} \hline 1 & 3 & 4 \\ \hline 2 \\ \cline{1-1} \end{array} \qquad T_2 = \begin{array}{|c|c|c|} \hline 1 & 2 & 4 \\ \hline 3 \\ \cline{1-1} \end{array} \qquad T_3 = \begin{array}{|c|c|c|} \hline 1 & 2 & 3 \\ \hline 4 \\ \cline{1-1} \end{array}$$

The dual standard tableaux are

$$\tilde{T}_1 = \begin{array}{|c|c|} \hline 1 & 2 \\ \hline 3 \\ \cline{1-1} 4 \\ \cline{1-1} \end{array} \qquad \tilde{T}_2 = \begin{array}{|c|c|} \hline 1 & 3 \\ \hline 2 \\ \cline{1-1} 4 \\ \cline{1-1} \end{array} \qquad \tilde{T}_3 = \begin{array}{|c|c|} \hline 1 & 4 \\ \hline 2 \\ \cline{1-1} 3 \\ \cline{1-1} \end{array}$$

and the basis functions for $[21^2]$ are $|[21^2](3211)\rangle = |[\widetilde{31}](\widetilde{1121})\rangle$, $|[21^2](3121)\rangle = |[\widetilde{31}](\widetilde{1211})\rangle$ and $|[21^2](1321)\rangle = |[\widetilde{31}](\widetilde{2111})\rangle$ respectively.

With the basis functions for $[\lambda]$ and $[\tilde{\lambda}]$ ordered in this way it can be shown, as a corollary to the Young-Yamanouchi theorem, that the matrix elements of $(n-1n)$ in $[\tilde{\lambda}]$ and $[\lambda]$ are related as follows (see Hamermesh, 1962)

$$\begin{aligned} \langle[\tilde{\lambda}](\tilde{i})|\,(n-1n)\,|[\tilde{\lambda}](\tilde{j})\rangle &= \langle[\lambda](i)|\,(n-1n)\,|[\lambda](j)\rangle \qquad i \neq j \\ \langle[\tilde{\lambda}](\tilde{i})|\,(n-1n)\,|[\tilde{\lambda}](\tilde{i})\rangle &= -\langle[\lambda](i)|\,(n-1n)\,|[\lambda](i)\rangle \end{aligned} \tag{6.6.1}$$

From this result we see that in dual representations the characters of $(n-1n)$ are opposite in sign. This observation can be generalized and it is found as a further corollary to the Young-Yamanouchi theorem that, in dual representations, the characters of even permutations are the same while the characters of odd permutations are opposite in sign. This result can be formulated as

$$\chi^{[\tilde{\lambda}]}(P) = \chi^{[\lambda]}(P) \cdot \chi^{[1^n]}(P) \tag{6.6.2}$$

6.7 Direct products

In this section we consider the decomposition into IR's of the direct product $[\lambda] \otimes [\mu]$ where $[\lambda]$ and $[\mu]$ are IR's of $\mathscr{S}_n$. If we write

$$[\lambda] \otimes [\mu] = \sum_{\nu}^{\oplus} a_\nu [\nu]$$

then we are required to determine the coefficients a_ν for given $[\lambda]$ and $[\mu]$. We shall not consider the general case but instead we single out one particular decomposition which is of central importance for applications to quantum chemistry. This is concerned with the totally antisymmetric representation $[1^n]$.

From (6.6.2) we have

$$[\lambda] \otimes [1^n] = [\tilde{\lambda}] \tag{6.7.1}$$

and from the definition of direct product it is clear that

$$\begin{aligned}([\lambda] \otimes [\mu]) \otimes [\nu] &= [\lambda] \otimes ([\mu] \otimes [\nu]) \\ &= ([\lambda] \otimes [\nu]) \otimes [\mu]\end{aligned}$$

When $[\nu] = [1^n]$ we find

$$\widetilde{[\lambda] \otimes [\mu]} = [\lambda] \otimes [\tilde{\mu}] = [\tilde{\lambda}] \otimes [\mu]$$

and this can be written as

$$[\tilde{\lambda}] \otimes [\tilde{\mu}] = [\lambda] \otimes [\mu]$$

Now suppose that $[\lambda] \otimes [\mu]$ contains $[1^n]$ a number k times. According to the result established at the end of Section 5.2 it follows that $[\lambda] \otimes [1^n]$ contains $[\mu]$ a number k times. However $[\lambda] \otimes [1^n] = [\tilde{\lambda}]$ and we have the result that $[\lambda] \otimes [\mu]$ contains the totally antisymmetric representation $[1^n]$ once if and only if $[\mu] = [\tilde{\lambda}]$.

Now let $\Phi_{(r)}^{[\lambda]}$ be a set of n_λ basis functions for the standard IR $[\lambda]$ of $\mathscr{S}_n$ where (r) denotes the Y-symbol $(r_n\, r_{n-1} \ldots r_1)$. Similarly let $\Theta_{(\tilde{s})}^{[\tilde{\lambda}]}$ be a set of n_λ basis functions for the dual representation $[\tilde{\lambda}]$. We have just seen that

$[1^n]$ is contained in $[\lambda]\otimes[\tilde{\lambda}]$ and it follows that we can construct from n_λ^2 functions $\Phi_{(r)}^{[\lambda]}\,\Theta_{(\tilde{s})}^{[\tilde{\lambda}]}$ one totally antisymmetric function $\Psi\equiv\Psi_{(n\,n-1\ldots 1)}^{[1^n]}$. This function will be given by

$$\Psi=\sum_{r=1}^{n_\lambda}\sum_{s=1}^{n_\lambda} c_{rs}\Phi_{(r)}^{[\lambda]}\,.\,\Theta_{(\tilde{s})}^{[\tilde{\lambda}]} \tag{6.7.2}$$

where the coefficients c_{rs} are to be determined so that Ψ is totally antisymmetric. The coefficients c_{rs} are examples of what are called coupling coefficients. We shall consider coupling coefficients more generally in later chapters.

To see how the coefficients c_{rs} are determined we consider as an example $[21]\otimes[\widetilde{21}]$ of $\mathcal{S}_3$. We have

$$\Psi=c_{11}\Phi_{(211)}\Theta_{(211)}+c_{22}\Phi_{(121)}\Theta_{(121)}+c_{12}\Phi_{(211)}\Theta_{(121)}+c_{21}\Phi_{(121)}\Theta_{(211)}$$

Now for Ψ to be totally antisymmetric we must have

$$(12)\Psi=-\Psi \qquad \text{and} \qquad (13)\Psi=-\Psi$$

From (6.4.4) we have

$$(12)\Psi=c_{11}\Phi_{(211)}\Theta_{(211)}+c_{22}\Phi_{(121)}\Theta_{(121)}-c_{12}\Phi_{(211)}\Theta_{(121)}-c_{21}\Phi_{(121)}\Theta_{(211)}$$

and it follows that $c_{11}=c_{22}=0$.

From (6.4.6) we have

$$(13)\Psi=c_{12}\left(-\frac{1}{2}\Phi_{(211)}-\frac{\sqrt{3}}{2}\Phi_{(121)}\right)\left(-\frac{\sqrt{3}}{2}\Theta_{(211)}+\frac{1}{2}\Theta_{(121)}\right)$$
$$+c_{21}\left(-\frac{\sqrt{3}}{2}\Phi_{(211)}+\frac{1}{2}\Phi_{(121)}\right)\left(-\frac{1}{2}\Theta_{(211)}-\frac{\sqrt{3}}{2}\Theta_{(121)}\right)$$

and it follows that $c_{12}+c_{21}=0$. The normalized totally antisymmetric function is thus given by

$$\begin{aligned}\Psi&=\frac{1}{\sqrt{2}}(\Phi_{(211)}\Theta_{(121)}-\Phi_{(121)}\Theta_{(211)})\\&=\frac{1}{\sqrt{2}}(\Phi_{(211)}\Theta_{(\widetilde{211})}-\Phi_{(121)}\Theta_{(\widetilde{121})})\end{aligned} \tag{6.7.3}$$

The above procedure can be applied in general and it is found that the totally antisymmetric function is given by (Hamermesh, 1962)

$$\Psi=\frac{1}{\sqrt{n_\lambda}}\sum_{r=1}^{n_\lambda}(-1)^{R_r}\Phi_{(r)}^{[\lambda]}\Theta_{(\tilde{r})}^{[\tilde{\lambda}]} \tag{6.7.4}$$

where R_r is the number of transpositions required to bring the letters in the symbol (r) to dictionary order. Thus (6.7.3) is obtained from (6.7.4) by noting that the Y-symbol (211) is brought to dictionary order by two transpositions while (121) is brought to dictionary order by one transposition.

We can also obtain a result similar to (6.7.4) but written in terms of non standard basis functions. Consider again $[21] \otimes [\widetilde{21}]$. Let Ψ_1 and Ψ_2 be the basis functions for [21] as given by (6.3.8) and (6.3.9) respectively. The tableaux T_1 and T_2 corresponding to Ψ_1 and Ψ_2 respectively are in dictionary order. Let $\tilde{\Psi}_1$ and $\tilde{\Psi}_2$ be basis functions for $[\widetilde{21}] \equiv [21]$. The tableaux $\tilde{T}_1$ and $\tilde{T}_2$ are now in reverse dictionary order. Thus the analogues of (6.3.8) and (6.3.9) are

$$\tilde{\Psi}_1 = \Phi(123) + \Phi(321) - \Phi(213) - \Phi(312) \tag{6.7.5}$$

$$\tilde{\Psi}_2 = \Phi(132) + \Phi(231) - \Phi(312) - \Phi(213) \tag{6.7.6}$$

In the basis $(\tilde{\Psi}_1, \tilde{\Psi}_2)$ we find

$$(12) = \begin{bmatrix} -1 & -1 \\ 0 & 1 \end{bmatrix} \qquad (13) = \begin{bmatrix} 1 & 0 \\ -1 & -1 \end{bmatrix} \tag{6.7.7}$$

The totally antisymmetric function is given by

$$\Psi = c_{11}\Psi_1\tilde{\Psi}_1 + c_{22}\Psi_2\tilde{\Psi}_2 + c_{12}\Psi_1\tilde{\Psi}_2 + c_{21}\Psi_2\tilde{\Psi}_1$$

Application of (6.3.10), (6.3.11) and (6.7.7) yields

$$c_{12} = c_{21} = 0 \qquad c_{22} = -c_{11}$$

Thus we have

$$\Psi = \frac{1}{\sqrt{2}}(\Psi_1\tilde{\Psi}_1 - \Psi_2\tilde{\Psi}_2) \tag{6.7.8}$$

which should be compared with (6.7.3).

As a second example we consider $\mathcal{S}_4$ and the totally antisymmetric function contained in $[21^2] \otimes [31]$. The basis functions for $[21^2]$ are (Ψ_1, Ψ_2, Ψ_3) as given by (6.3.12). The basis functions for [31] are $(\tilde{\Psi}_1, \tilde{\Psi}_2, \tilde{\Psi}_3)$ as given by (6.3.16). The totally antisymmetric function is

$$\Psi = \sum_{i,j=1}^{3} c_{ij}\Psi_i\tilde{\Psi}_j$$

Application of (6.3.13), (6.3.14) and (6.3.17), (6.3.19) yields

$$\Psi = \frac{1}{\sqrt{3}}(\Psi_1\tilde{\Psi}_1 - \Psi_2\tilde{\Psi}_2 + \Psi_3\tilde{\Psi}_3) \tag{6.7.9}$$

These examples lead us to anticipate that in the general case we have the totally antisymmetric function given by

$$\Psi = \frac{1}{\sqrt{n_\lambda}} \sum_{p=1}^{n_\lambda} (-1)^{P_p} \Psi_p^{[\lambda]} \tilde{\Psi}_p^{[\tilde{\lambda}]} \tag{6.7.10}$$

where P_p is the number of transpositions required to bring the standard tableau T_p into the standard tableau T_1.

6.8 Outer product representations

In this section we use the notation $\mathscr{S}_n\ (\alpha_1\alpha_2\ldots\alpha_n)$ to denote the symmetric group of degree n with the permutations being amongst the n symbols $\alpha_1, \alpha_2, \ldots, \alpha_n$.

Consider two isolated atoms one with n_1 electrons labelled $1, 2, \ldots, n_1$ and the other with n_2 electrons labelled $n_1+1, n_1+2, \ldots, n_1+n_2$. The permutational symmetries of the electron coordinates appropriate to these isolated atoms are given by the IR's of the groups $\mathscr{S}_{n_1}(1, 2, \ldots, n_1)$ and $\mathscr{S}_{n_2}(n_1+1, n_1+2, \ldots, n_1+n_2)$ respectively. If the atoms are now brought together to form an (n_1+n_2) electron molecule then the permutational symmetries of the electron coordinates are given by the IR's of the group $\mathscr{S}_{n_1+n_2}(1, 2, \ldots, n_1, n_1+1, \ldots, n_1+n_2)$.

Let $\Phi_i^{[\lambda]}(1, 2, \ldots, n_1)$ be a set of n_λ basis functions for the IR $[\lambda]$ of $\mathscr{S}_{n_1}(1, 2, \ldots, n_1)$. Let $\Psi_j^{[\mu]}(n_1+1, n_1+2, \ldots, n_1+n_2)$ be a set of n_μ basis functions for the IR $[\mu]$ of $\mathscr{S}_{n_2}(n_1+1, \ldots, n_1+n_2)$. Consider now a product function $\Phi_i^{[\lambda]}\Psi_j^{[\mu]}$ and permutations amongst the coordinates of all (n_1+n_2) electrons. The n_1 coordinates which occur in $\Phi_i^{[\lambda]}$ can be chosen from the (n_1+n_2) coordinates in

$$ {}^{n_1+n_2}C_{n_1} = \frac{(n_1+n_2)!}{n_1!\ n_2!} \quad \text{ways} $$

For each of these ways the coordinates in $\Psi_j^{[\mu]}$ are then fixed. The total number of such product functions is thus

$$ N = \frac{(n_1+n_2)!}{n_1!\ n_2!}\, n_\lambda n_\mu \tag{6.8.1} $$

These N product functions form the basis for a representation of $\mathscr{S}_{n_1+n_2}(1, 2, \ldots, n_1, n_1+1, \ldots, n_1+n_2)$. This representation is called the *outer product* of $[\lambda]$ and $[\mu]$ and is denoted by $[\lambda]\odot[\mu]$. The outer product must not be confused with the direct product (inner product) which refers to products of functions spanning IR's of the same group.

The question now arises as to what IR's of $\mathscr{S}_{n_1+n_2}$ are contained in $[\lambda]\odot[\mu]$. To begin with let $\phi(1)$ be a basis function for the IR $[1]$ of $\mathscr{S}_1(1)$ and let $\psi(2)$ be a basis function for the IR $[1]$ of $\mathscr{S}_1(2)$. The product functions which form the basis for $[1]\odot[1]$ of $\mathscr{S}_2(1, 2)$ are $\phi(1)\psi(2)$ and $\psi(1)\phi(2)$. We know that the IR $[2]$ of $\mathscr{S}_2(1, 2)$ is spanned by a symmetric function while the IR $[1^2]$ is spanned by an antisymmetric function. Now the linear combinations $\phi(1)\psi(2)\pm\psi(1)\phi(2)$ are respectively symmetric

and antisymmetric. It follows that

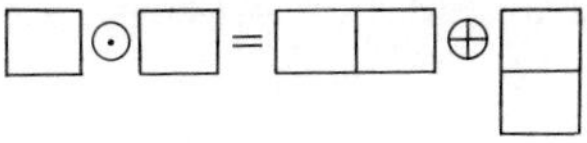

$$\square \odot \square = \square\,\square \oplus {\square \atop \square} \tag{6.8.2}$$

or

$$[1]\odot[1]=[2]\oplus[1^2]$$

Consider next $[2]\odot[1]$ for $\mathscr{S}_2(1, 2)$ and $\mathscr{S}_1(3)$. The basis function corresponding to $[1]$ is not subject to any symmetry restrictions and it follows that the decomposition of the outer product is obtained by adding the cell $\square$ to $\square\,\square$ in all possible ways which lead to allowed shapes for the IR's of $\mathscr{S}_3(1, 2, 3)$. Thus

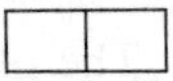

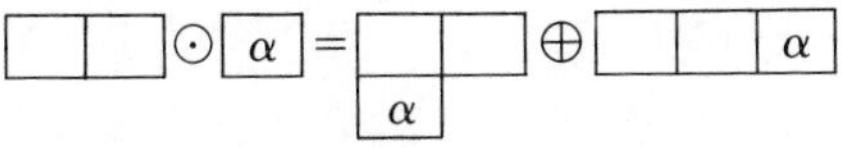

$$\square\,\square \odot \boxed{\alpha} = {\square\,\square \atop \boxed{\alpha}} \oplus \square\,\square\,\boxed{\alpha} \tag{6.8.3}$$

It is clear that $[21]\odot[1]=[1]\odot[21]$. From (6.8.3) we see that the decomposition of the outer product can also be obtained by adding the two cells in $[2]$ to the cell in $[1]$ in all possible ways subject to allowed shapes being obtained and with the further restriction that both cells are not placed in the same column. This last condition is in fact a consequence of the symmetry requirement that the basis function for $[2]$ be symmetric in 1 and 2. Thus

$$\square \odot \boxed{\alpha\mid\beta} = {\square\,\boxed{\alpha} \atop \boxed{\beta}} \oplus \square\,\boxed{\alpha\mid\beta}$$

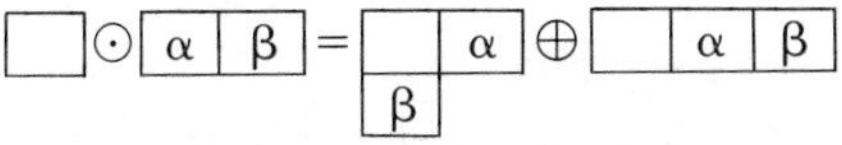

If we continue this process we arrive at the following general result (see Hamermesh, 1962).

In order to find the decomposition of the outer product $[\lambda]\odot[\mu]$ where $[\mu]$ has a shape consisting of a single row or column we draw the shape $[\lambda]$ and then add the cells of $[\mu]$ to $[\lambda]$ in all possible ways subject to the restrictions that

    I. the resulting shape must have standard form
   II. no two cells may be placed in the same column if $[\mu]$ consists of a single row of cells
  III. no two cells may be placed in the same row if $[\mu]$ consists of a single column of cells.

As an example we have

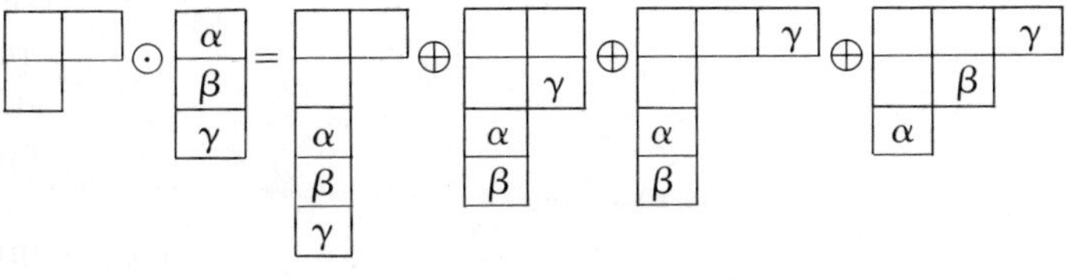

In a case such as $[21]\odot[21]$ where neither shape consists of a single row or column we can use a building up process to obtain the decomposition. Thus we have

$$[2]\odot[1]=[3]\oplus[21]$$

or

$$[21]=([2]\odot[1])\ominus[3]$$

Using this we now write

$$\begin{aligned}[21]\odot[21]&=[21]\odot([2]\odot[1])\ominus([21]\odot[3])\\&=([21]\odot[1])\odot[2]\ominus([21]\odot[3])\\&=([31]\oplus[2^2]\oplus[21^2])\oplus[2]\ominus([51]\oplus[42]\oplus[41^2]\oplus[321])\\&=([31]\odot[2])\oplus([2^2]\odot[2])\oplus([21^2]\odot[2])\\&\quad\ominus([51]\oplus[42]\oplus[41^2]\oplus[321])\\&=[3^2]\oplus[42]\oplus 2[321]\oplus[2^3]\oplus[41^2]\oplus[31^3]\oplus[2^21^2]\end{aligned}$$

6.9 The antisymmetrizer

Of singular importance in quantum chemistry is the Young operator given by (6.3.2). In normalized form it is called the *antisymmetrizer* $\hat{A}^{(n)}$ and we have

$$\hat{A}^{(n)}(1, 2, \ldots, n)=\frac{1}{n!}\sum_{P}(-1)^{p_P}\hat{P} \qquad (6.9.1)$$

where the operator $\hat{P}$ is a permutation of the n coordinates $1, 2, \ldots, n$.

If $\hat{Q}$ is any permutation in $\mathcal{S}_n$ we have

$$\hat{Q}\hat{A}^{(n)}=(-1)^{p_Q}\hat{A}^{(n)} \qquad (6.9.2)$$

and it follows that

$$\hat{A}^{(n)}\hat{A}^{(n)}=\hat{A}^{(n)} \qquad (6.9.3)$$

Now let R denote any permutation of the first n_1 coordinates $1, 2, \ldots, n_1$ and let S denote any permutation of the last n_2 coordinates $n_1+1,\ n_1+2, \ldots,\ n_1+n_2=n$. The set of all permutations RS forms the direct product subgroup $\mathcal{S}_{n_1}\otimes\mathcal{S}_{n_2}$ of order $n_1!\ n_2!$ of the group $\mathcal{S}_n$. In the case where $\mathcal{G}=\mathcal{S}_n$ and $\mathcal{H}=\mathcal{S}_{n_1}\otimes\mathcal{S}_{n_2}$ we have from (1.2.4)

$$P=Q_iRS$$

and it follows that

$$\hat{A}^{(n)}=\frac{1}{n!}\sum_{Q_iRS}(-1)^{p_{Q_iRS}}\hat{Q}_i\hat{R}\hat{S}=\frac{n_1!\ n_2!}{n!}\sum_{Q_i}(-1)^{p_{Q_i}}\hat{Q}_i\hat{A}^{(n_1)}\hat{A}^{(n_2)} \qquad (6.9.4)$$

This result can clearly be generalized and we find

$$\hat{A}^{(n)} = \frac{n_1!\ n_2! \ldots n_t!}{n!} \sum_{Q_i} (-1)^{pQ_i} \hat{Q}_i \hat{A}^{(n_1)} \hat{A}^{(n_2)} \ldots \hat{A}^{(n_t)} \tag{6.9.5}$$

where now $\hat{Q}_i$ is an element of the coset

$$Q_i(\mathcal{S}_{n_1} \otimes \mathcal{S}_{n_2} \otimes \ldots \otimes \mathcal{S}_{n_t})$$

and $n = n_1 + n_2 + \ldots + n_t$.

As an example we consider $\mathcal{H} = \mathcal{S}_2(1, 2) \otimes \mathcal{S}_2(3, 4) = \{1, (12), (34), (12)(34)\}$. The cosets are obtained by multiplying the elements of $\mathcal{H}$ by those elements of $\mathcal{G} = \mathcal{S}_4$ which do not occur in $\mathcal{H}$ until the whole group is exhausted. Thus we have

$\mathcal{H}$:	1	(12)	(34)	(12)(34)
(13)$\mathcal{H}$:	(13)	(123)	(134)	(1234)
(23)$\mathcal{H}$:	(23)	(132)	(234)	(1342)
(14)$\mathcal{H}$:	(14)	(124)	(143)	(1243)
(24)$\mathcal{H}$:	(24)	(142)	(243)	(1432)
(13)(24)$\mathcal{H}$:	(13)(24)	(1432)	(1324)	(14)(23)

The choice of the elements Q_i which occur in (6.9.4) is clearly arbitrary so long as we select one element from each coset. Thus we may write

$$\hat{A}^{(4)} = \frac{2!\ 2!}{4!}[1 - (12) - (23) - (14) - (24) + (13)(24)]\hat{A}^{(2)}\hat{A}^{(2)} \tag{6.9.6}$$

It is more convenient however to choose the permutations Q_i that maintain the dictionary order of coordinates in each part of a function $\Psi(12/34)$ which is antisymmetric in 1,2 and antisymmetric in 3,4. Such a choice can be found by inspection and we have

$$\hat{A}^{(4)} = \frac{2!\ 2!}{4!}[1 + (123) - (23) - (1243) + (243) + (13)(24)]\hat{A}^{(2)}\hat{A}^{(2)} \tag{6.9.7}$$

As a second example we take $\mathcal{H} = \mathcal{S}_1(1) \otimes \mathcal{S}_3(2, 3, 4)$ and we have

$\mathcal{H}$:	1	(23)	(24)	(34)	(234)	(243)
(12)$\mathcal{H}$:	(12)	(123)	(124)	(12)(34)	(1234)	(1243)
(13)$\mathcal{H}$:	(13)	(132)	(13)(24)	(134)	(1342)	(1324)
(14)$\mathcal{H}$:	(14)	(14)(23)	(142)	(143)	(1423)	(1432)

We can now write

$$\hat{A}^{(4)} = \frac{1!\ 3!}{4!}[1 - (12) - (13) - (14)]\hat{A}^{(3)} \tag{6.9.8}$$

or

$$\hat{A}^{(4)} = \frac{1!\ 3!}{4!}[1-(12)+(132)-(1432)]\hat{A}^{(3)} \tag{6.9.9}$$

if we wish to maintain dictionary order in $\Psi(1/234)$.

For some purposes it is desirable to express the antisymmetrizer in the form of a two-sided operator. If we evaluate

$$[1-(23)][1-(12)-(13)][1-(23)]$$

we see that

$$\hat{A}^{(3)} = \frac{2!}{3!}\hat{A}^{(2)}[1-(12)-(13)]\hat{A}^{(2)} \tag{6.9.10}$$

This result can clearly be generalized (see Hassitt, 1955) and we have

$$\hat{A}^{(n)} = \frac{(n-1)!}{n!}\hat{A}^{(n-1)}\left[1-\sum_{j=2}^{n}(1j)\right]\hat{A}^{(n-1)} \tag{6.9.11}$$

We now return to (6.9.7). If we write

$$(123)=(12)(23) \qquad (1243)=(12)(24)(34)$$
$$(243)=(24)(34)$$

then we readily find

$$\hat{A}^{(4)} = \frac{2!\ 2!}{4!}[1+(13)(24)-4\hat{A}^{(2)}\hat{A}^{(2)}(23)]\hat{A}^{(2)}\hat{A}^{(2)}$$

We now use (6.9.3) and the relations

$$(\alpha\beta)(\alpha\gamma)=(\beta\gamma)(\alpha\beta)$$
$$(\alpha\beta)(\gamma\beta)=(\gamma\beta)(\gamma\alpha)$$
$$(\alpha\beta)(\gamma\delta)=(\alpha\delta)(\gamma\beta)(\alpha\gamma)(\delta\beta)$$

Thus

$$(13)(24)\hat{A}^{(2)}\hat{A}^{(2)}=(12)(34)(23)(14)\hat{A}^{(2)}\hat{A}^{(2)}$$

and

$$(23)(14)\hat{A}^{(2)}\hat{A}^{(2)}=\hat{A}^{(2)}\hat{A}^{(2)}(14)(23)$$

Now

$$(12)(34)\hat{A}^{(2)}\hat{A}^{(2)}=\hat{A}^{(2)}\hat{A}^{(2)}$$

and we have established the result

$$\hat{A}^{(4)} = \frac{2!\ 2!}{4!}\hat{A}^{(2)}\hat{A}^{(2)}[1-4(23)+(14)(23)]\hat{A}^{(2)}\hat{A}^{(2)} \tag{6.9.12}$$

A similar analysis carried out on (6.9.9) leads to the result

$$\hat{A}^{(4)} = \frac{3!}{4!}\hat{A}^{(3)}[1-3(12)]\hat{A}^{(3)} \tag{6.9.13}$$

The above idea concerning rearrangement of permutations is capable of being completely generalized to deal with any Young operator. The formula however are cumbersome and the reader is referred to Horie (1964) for details.

References

Boerner, H. (1963) "Representations of Groups", North-Holland, Chapter 4.

Hamermesh, M. (1962) "Group Theory and its Application to Physical Problems", Addison-Wesley, Chapter 7.

Horie, H. (1964) *J. Phys. Soc. Japan*, **19,** 1783.

Hassitt, A. (1955) *Proc. Roy. Soc. (London)*, **A229,** 110.

Jahn, H. A. & Van Wieringen, H. (1951) *Proc. Roy. Soc. (London)*, **A209,** 502.

Ledermann, W. (1957) "Introduction to the theory of finite groups", Oliver and Boyd, Chapters 2 and 3.

Littlewood, D. E. (1940) "The Theory of Group Characters", Oxford.

Rutherford, D. E. (1948) "Substitutional Analysis", Edinburgh.

Weyl, H. (1931) "The Theory of Groups and Quantum Mechanics", Dover, Chapter 5.

CHAPTER 7

Spin Free Quantum Chemistry

In this chapter we shall be concerned with some of the applications of permutational symmetry to problems in quantum chemistry. The mathematical apparatus for this is the representation theory of the symmetric group which we developed in the last chapter.

7.1 Spin values in a many electron system

If $j = j_1 + j_2$ is the resultant of two quantum mechanical angular momenta j_1 and j_2 then it can readily be shown using elementary methods that the possible values of j are given by (see, for example, Newing and Cunningham, 1967)

$$j = j_1 + j_2,\ j_1 + j_2 - 1,\ j_1 + j_2 - 2, \ldots, |j_1 - j_2| \qquad (7.1.1)$$

This result in itself has group theoretical significance which we shall explore later in connection with continuous groups.

It is well known that an electron possesses an intrinsic angular momentum $j = s$ whose magnitude is $\frac{1}{2}$.‡ This is called electron spin. The three space coordinates x, y, z are insufficient to give a complete specification of the state of an electron and we introduce a fourth coordinate. This fourth coordinate is an internal coordinate called the spin coordinate and denoted by σ. The range of σ consists of only two points: $\sigma = \pm 1$. This is the Pauli spin postulate. A function $\xi(\sigma)$ is defined at only two points and it follows that the general spin function is

$$\xi(\sigma) = c_1\, \delta_{\sigma,+1} + c_2\, \delta_{\sigma,-1} \qquad (7.1.2)$$

It is conventional to write

$$\delta_{\sigma,+1} = \alpha \qquad \delta_{\sigma,-1} = \beta \qquad (7.1.3)$$

and to call the spin functions α and β the components of an *elementary spinor*. If $\xi(\sigma)$ and $\eta(\sigma)$ are arbitrary spin functions then the scalar product is clearly defined by

$$\langle \xi(\sigma) \mid \eta(\sigma) \rangle = \xi^*(-1)\eta(-1) + \xi^*(1)\eta(1) \qquad (7.1.4)$$

‡ We use dimensionless operators whose eigenvalues give angular momenta in units of $\hbar$.

It follows that α and β are orthonormal functions. If $\hat{S}_z$ is the operator corresponding to the z component of spin then the eigenvalues of $\hat{S}_z$ are $\pm\frac{1}{2}$ a.u. The operator $\hat{S}_z$ is defined by the relations

$$\hat{S}_z\alpha = \tfrac{1}{2}\alpha \qquad \hat{S}_z\beta = -\tfrac{1}{2}\beta \tag{7.1.5}$$

Consider now n electrons which are isolated and non-interacting. Each electron has spin angular momentum $\frac{1}{2}$. When we bring the electrons together to form a single n-electron system it is clear that we can define the total spin S of the system by combining the spins of the individual electrons. This can be done by using (7.1.1) repeatedly. Thus for two electrons we have $j_1 = \frac{1}{2}$ and $j_2 = \frac{1}{2}$ in (7.1.1). The possible total spin is

$$j = S = \tfrac{1}{2}+\tfrac{1}{2}, \tfrac{1}{2}-\tfrac{1}{2} = 1, 0$$

When we have a three electron system we use (7.1.1) with $j_1 = 0, 1$ and $j_2 = \frac{1}{2}$ where the j_1 values correspond to possible spin for two electrons. Thus

$$j = S = 0+\tfrac{1}{2} = \tfrac{1}{2} \qquad \text{and} \qquad j = S = 1+\tfrac{1}{2}, 1-\tfrac{1}{2} = \tfrac{3}{2}, \tfrac{1}{2}$$

and the possible total spin S for three electrons is $S = \frac{1}{2}, \frac{3}{2}$. The result of applying this building up method to systems containing up to seven electrons is shown in Table 14.

TABLE 14. Spin values

Number of electrons n	Possible total spin S
1	$\frac{1}{2}$
2	0, 1
3	$\frac{1}{2}, \frac{3}{2}$
4	0, 1, 2
5	$\frac{1}{2}, \frac{3}{2}, \frac{5}{2}$
6	0, 1, 2, 3
7	$\frac{1}{2}, \frac{3}{2}, \frac{5}{2}, \frac{7}{2}$

7.2 Permutational symmetry of spin functions

The spin function α (or β) can evidently be taken as a basis function for the IR [1] of $\mathscr{S}_1$. When we combine two 1-electron systems to form a 2-electron system the possible permutational symmetry in the resultant spin functions will be given by the IR's of $\mathscr{S}_2$ which are contained in the outer product $[1]\odot[1]$. From (6.8.2) we see that the spin functions are either symmetric or antisymmetric. From Table 14 we see that the possible spin values are $S=0$ and $S=1$. It is well known that the three spin functions corresponding to $S=1$ are the symmetric combinations $\alpha(1)\alpha(2)$, $\alpha(1)\beta(2)+\beta(1)\alpha(2)$ and $\beta(1)\beta(2)$ where for example $\alpha(i)$ means that electron labelled i has z-component of spin $+\frac{1}{2}$. It now follows that we can associate with the permutational symmetry [2] the spin value $S=1$. Also we can associate with the permutational symmetry $[1^2]$ the spin value $S=0$.

Consider next a 3-electron system. The spin values are $S=\frac{1}{2}$ and $S=\frac{3}{2}$. We can consider the system to be formed by bringing together a 2-electron system and a 1-electron system. If the 2-electron system has $S=0$ it follows that the permutational symmetry of the 3-electron spin function is determined by the IR's of $\mathscr{S}_3$ which occur in the outer product $[1^2]\odot[1]$. Now

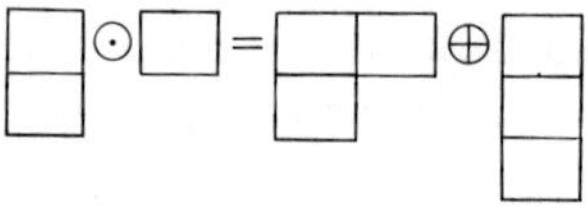

A basis function for $[1^3]$ must be totally antisymmetric. Since there are only two basic spin functions, namely α and β, it is clearly impossible to construct from them a non-zero totally antisymmetric product function. Thus the Young shapes for spin functions are limited to those with no more than two rows. It is now clear that we can associate with the permutational symmetry [21] the spin value $S=\frac{1}{2}$. If the 2-electron system has $S=1$ then the 3-electron system can have $S=\frac{1}{2}$ or $S=\frac{3}{2}$. Also

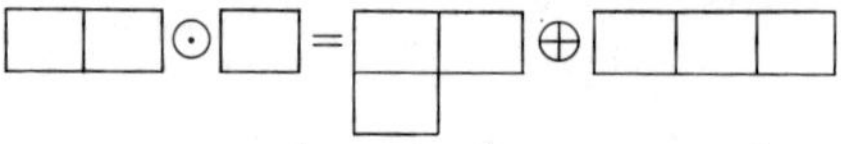

$S=1 \quad S=\frac{1}{2} \quad S=\frac{1}{2} \qquad S=\frac{3}{2}$

and we find that we can associate with the permutational symmetry [3] the spin value $S=\frac{3}{2}$.

We can continue this process and consider a 4-electron system to be formed by bringing together a 3-electron system and a 1-electron system. The two possibilities are given by

(a)

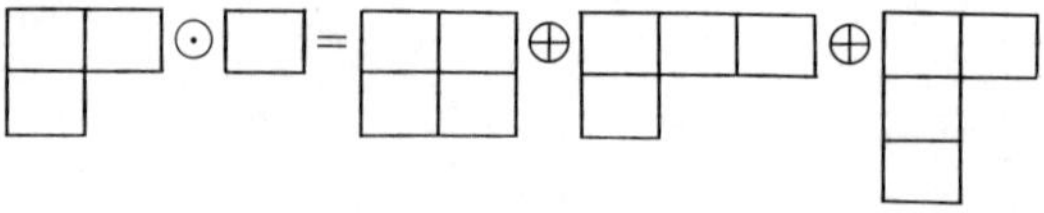

$S=\frac{1}{2} \quad S=\frac{1}{2} \qquad S=0, 1$

(b)

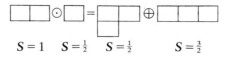

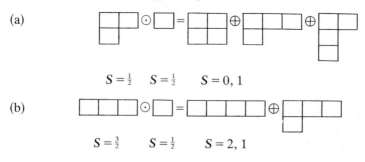

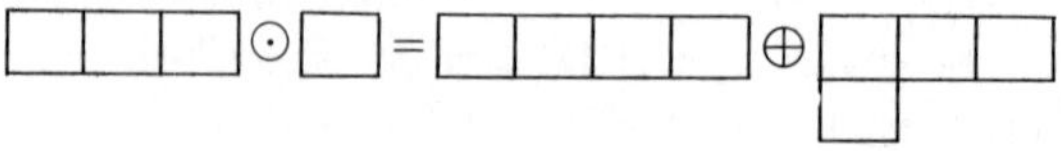

$S=\frac{3}{2} \qquad S=\frac{1}{2} \qquad S=2, 1$

The shape of $[21^2]$ is inadmissible for spin functions. Comparison of (a) and (b) now shows that we can associate with the permutational symmetries $[2^2]$, [31] and [4] the spin values $S=0$, 1 and 2 respectively. This process can be carried on for higher values of n and we arrive at the results shown in Table 15. From Table 15 we see that in general the permutational symmetry $[\lambda]=[\lambda_1\lambda_2]$ is associated with a spin value

$$S=\tfrac{1}{2}(\lambda_1-\lambda_2)$$

Since $\lambda_1+\lambda_2=n$ we can write the converse result. Thus in an n-electron system with total spin S the spin function has permutational symmetry $[\lambda]=[\lambda_1\lambda_2]$ where

$$\begin{aligned}\lambda_1&=\frac{n}{2}+S\\ \lambda_2&=\frac{n}{2}-S\end{aligned}\qquad(7.2.1)$$

TABLE 15. Permutational symmetry of spin functions

Number of electrons	Total spin	Permutational symmetry
1	$\frac{1}{2}$	[1]
2	0	$[1^2]$
	1	[2]
3	$\frac{1}{2}$	[21]
	$\frac{3}{2}$	[3]
4	0	$[2^2]$
	1	[31]
	2	[4]
5	$\frac{1}{2}$	[32]
	$\frac{3}{2}$	[41]
	$\frac{5}{2}$	[5]
6	0	$[3^2]$
	1	[42]
	2	[51]
	3	[6]
7	$\frac{1}{2}$	[43]
	$\frac{3}{2}$	[52]
	$\frac{5}{2}$	[61]
	$\frac{7}{2}$	[7]

7.3 The spin-free Hamiltonian

Consider the Hamiltonian $\hat{H}$ for the non-relativistic motion of n electrons in the field of m nuclei. If Z_a is the charge on the ath nucleus we have (in atomic units)*

$$\hat{H} = -\frac{1}{2}\sum_{i=1}^{n} \nabla_i^2 - \sum_{i=1}^{n}\sum_{a=1}^{m} \frac{Z_a}{r_{ia}} + \sum\sum_{i>j=1}^{n} \frac{1}{r_{ij}} \tag{7.3.1}$$

where r_{ia} is the distance between electron i and nucleus a and r_{ij} is the distance between electrons i and j. In the Schrödinger equation

$$\hat{H}\Psi = E\Psi \tag{7.3.2}$$

the wave function Ψ depends on both the space and spin coordinates of the electrons. The Hamiltonian $\hat{H}$ however does not depend upon the spin coordinates. It follows that we can write Ψ in the form

$$\Psi(r_1\sigma_1, r_2\sigma_2, \ldots, r_n\sigma_n) = \Phi(r_1, r_2, \ldots, r_n)\Theta(\sigma_1, \sigma_2, \ldots, \sigma_n) \tag{7.3.3}$$

Here the space function Φ depends only upon the space coordinates r_i whilst the spin function Θ depends only upon the spin coordinates σ_i.

We now take into account the Pauli principle which states that Ψ must be totally antisymmetric. In the last section we found that in an n-electron system with spin S the spin function Θ is a basis function for the IR $[\lambda] = [(n/2)+S, (n/2)-S]$ of $\mathscr{S}_n$. If Ψ is to be totally antisymmetric it follows from Section 6.7 that the space function Φ must be a basis function for the dual IR $[\tilde{\lambda}]$. Thus we can take the Pauli principle into account by using (6.7.4) or (6.7.10) to replace (7.3.3) by the antisymmetric combination

$$\Psi = \frac{1}{\sqrt{n_\lambda}}\sum_i (-1)^{P_i}\Phi_i^{[\tilde{\lambda}]}\Theta_i^{[\lambda]} \equiv \frac{1}{\sqrt{n_\lambda}}\sum_j (-1)^{P_j}\Phi_j^{[\Phi]}\Theta_j^{[\tilde{\lambda}]} \tag{7.3.4}$$

Since a definite permutational symmetry in the space function Φ is associated with a given spin value it follows that we can dispense entirely with spin functions when discussing a many electron system which is described by a spinless Hamiltonian. This is called the *spin-free formalism*. If we substitute (7.3.4) into (7.3.3) we have

$$\sum_j (-1)^{P_j}(\hat{H}\Phi_j^{[\lambda]})\Theta_j^{[\tilde{\lambda}]} = E\sum_j (-1)^{P_j}\Phi_j^{[\tilde{\lambda}]}\Theta_j^{[\tilde{\lambda}]} \tag{7.3.5}$$

The energy E now refers to states of the system in which the spin has the definite value $S = \frac{1}{2}(\lambda_1 - \lambda_2)$ when $[\tilde{\lambda}] = [\lambda_1\lambda_2]$. The functions $\Theta_j^{[\tilde{\lambda}]}$ are linearly independent and it follows that we can equate their coefficients. Thus we find

$$\hat{H}\Phi_k^{[\lambda]} = E\Phi_k^{[\lambda]} \qquad k = 1, 2, \ldots, n_\lambda \tag{7.3.6}$$

There will be one such set of spin-free Schrödinger equations for each

* In atomic units e, m, $\hbar$, $4\pi\varepsilon_0$ (ε_0 = permittivity of free space) all have unit values.

possible spin value. If the functions $\Phi_k^{[\lambda]}$ are normalized we can write

$$E=\langle\Phi_k^{[\lambda]}|\,\hat{H}\,|\Phi_k^{[\lambda]}\rangle \tag{7.3.7}$$

7.4 Examples of the use of the spin-free formalism

Let us consider the ground state of lithium. If we neglect the interaction between the electrons this state is described by an electron configuration $(1s)^2(2s)$. The electron configuration is, in perturbation theory, a zeroth order wave function $u(1)u(2)v(3)$. Here we have used u and v respectively to denote the $1s$ and $2s$ hydrogenic atomic orbitals. We have also used the abbreviation

$$a(i)\equiv a(r_i)$$

According to the exclusion principle the spin function is given by $\alpha(1)\beta(2)\alpha(3)$ (or $\alpha(1)\beta(2)\beta(3)$). It follows that the total spin value for the ground state is $S=\frac{1}{2}$. The totally antisymmetric function which describes the ground state is thus given in unnormalized form by (6.7.8) as

$$\Psi=\Phi_1\Theta_1-\Phi_2\Theta_2 \tag{7.4.1}$$

From (6.3.8), (6.3.9) we have

$$\begin{aligned}\Phi_1&=\Phi(123)+\Phi(213)-\Phi(321)-\Phi(231)\\ \Phi_2&=\Phi(132)+\Phi(312)-\Phi(231)-\Phi(321)\end{aligned} \tag{7.4.2}$$

while from (6.7.5), (6.7.6) we have

$$\begin{aligned}\Theta_1&=\Theta(123)+\Theta(321)-\Theta(213)-\Theta(312)\\ \Theta_2&=\Theta(132)+\Theta(231)-\Theta(312)-\Theta(213)\end{aligned} \tag{7.4.3}$$

In the particular case of the zeroth order wave function for the ground state of Li we also have

$$\begin{aligned}\Phi(123)&=u(1)u(2)v(3)\\ \Theta(123)&=\alpha(1)\beta(2)\alpha(3)\end{aligned} \tag{7.4.4}$$

If we substitute (7.4.4) into (7.4.2) and (7.4.3) and then use (7.4.1) we find the well known Slater determinant

$$\Psi=\begin{vmatrix}u(1)\alpha(1) & u(1)\beta(1) & v(1)\alpha(1)\\ u(2)\alpha(2) & u(2)\beta(2) & v(2)\alpha(2)\\ u(3)\alpha(3) & u(3)\beta(3) & v(3)\alpha(3)\end{vmatrix} \tag{7.4.5}$$

One of the big advantages of using (7.4.1) instead of (7.4.5) is that (7.4.1) yields a totally antisymmetric function in "product form" even when the Φ's are not given as a product of orbitals. This is very valuable when we consider the construction of correlated wave functions in which the orbital approximation no longer holds.

Consider next the calculation of the first order energy for the ground state of lithium. In perturbation theory the first order energy is given by

$$E_1 = \langle \Psi_0 | \hat{H}_1 | \Psi_0 \rangle \tag{7.4.6}$$

where Ψ_0 is the zeroth order approximation to Ψ and $\hat{H}_1$ is the perturbation potential energy. In our case

$$\hat{H}_1 = \frac{1}{r_{12}} + \frac{1}{r_{13}} + \frac{1}{r_{23}}$$

If we do not use the spin free formalism the simplification of (7.4.6) is relatively cumbersome. In the spin free formalism however we have from (7.3.7)

$$E_1 = \langle \Phi_1^{[21]} | \frac{1}{r_{12}} + \frac{1}{r_{13}} + \frac{1}{r_{23}} | \Phi_1^{[21]} \rangle \tag{7.4.7}$$

Here the normalized function $\Phi_1^{[21]}$ is given by (7.4.2) and (7.4.4), as

$$\Phi_1^{[21]} = \frac{1}{\sqrt{2}} u(2)\{u(1)v(3) - v(1)u(3)\} \tag{7.4.8}$$

When (7.4.8) is substituted in (7.4.7) we find that the resulting expression simplifies almost immediately to

$$E_1 = \langle u(1)u(2) | \frac{1}{r_{12}} | u(1)u(2) \rangle + 2\langle u(1)v(2) | \frac{1}{r_{12}} | u(1)v(2) \rangle - \langle u(1)v(2) | \frac{1}{r_{12}} | v(1)u(2) \rangle$$

As a final example we consider the molecule LiH. This four electron molecule has a ground state which is described, in zeroth order, by the electron configuration a^2b^2 where a and b are respectively 1σ and 2σ molecular orbitals. The exclusion principle shows that this ground state has a resultant spin $S = 0$. From (7.3.7) and Table 15 it follows that the energy is given by

$$E = \langle \Phi_k^{[2^2]} | \hat{H} | \Phi_k^{[2^2]} \rangle \qquad k = 1, 2$$

since $[2^2]$ is self dual. A basis function $\Phi_1^{[2^2]}$ can be obtained from the standard tableau

1	2
3	4

Thus

$$\Phi_1^{[2^2]} = [E - (13)][E - (24)][E + (12)][E + (34)]a(1)a(2)b(3)b(4)$$

The normalized form of this function is

$$\Phi_1^{[2^2]} = \tfrac{1}{2}[a(1)b(3) - b(1)a(3)][a(2)b(4) - b(2)a(4)]$$

It is clear that this function gives a more immediate result for the energy expression than does a 4×4 determinant.

Further applications of the spin free formalism can be found in Matsen (1960).

References

Matsen, F. A. (1960) *Adv. in Quantum Chem.*, **1**, 60.
Newing, R. A. and Cunningham, J. (1967) "Quantum Mechanics", Oliver and Boyd, Chapter 5.

CHAPTER 8

Continuous Groups
The Method of Irreducible Tensors

In Section 2.14 the group manifold is defined in connection with a finite group. A group is said to be continuous if some definition of continuity is imposed on the points of the group manifold. In particular if the group manifold is specified by a finite set of parameters then the group is continuous if these parameters vary continuously in a given range.

There are two methods available for a study of continuous groups and their matrix representations. They are the method of irreducible tensors and the method of Lie algebras. These two methods are quite different in their approach and the results of one method complement the results of the other. The method of Lie algebras is based on the possibility of using the calculus whilst the method of irreducible tensors is based on the possibility of using algebraic techniques. The Lie algebra method is the more powerful for it can be applied to a very large class of continuous groups. The irreducible tensor method is limited to those continuous groups which consist of linear transformations (linear groups). Nevertheless it is these linear groups which are of particular value for applications to quantum chemistry.

8.1 The full linear group GL(M)

Let $\mathscr{L}$ be an M-dimensional linear space of covariant vectors. Let $e_1, e_2, \ldots, e_M$ be a basis in $\mathscr{L}$ and consider the change of basis given by

$$e'_{j_1} = \sum_{i_1=1}^{M} e_{i_1} a_{i_1 j_1} \qquad j_1 = 1, 2, \ldots, M \tag{8.1.1}$$

This basis change is a linear transformation in $\mathscr{L}$. The set of all such linear transformations clearly forms a group. The numbers $a_{i_1 j_1}$ are in general complex and it follows that the matrix $\mathbf{A} = [a_{i_1 j_1}]$ is a function of $2M^2$ real parameters each of which can vary continuously over an infinite range. Thus the set of all linear transformations constitutes a continuous group called the full linear group in M dimensions. This group is denoted by $GL(M)$.

It is evident that the vectors $e_1, e_2, \ldots, e_M$ provide a basis for an M-dimensional representation of $GL(M)$ which is given by the matrices $[a_{i_1 j_1}]$

themselves. This representation is clearly irreducible since the linear transformations are not in any way restricted and no invariant subspace of $\mathscr{L}$ can be found. For the moment we shall denote this IR by Γ.

If ψ is a vector in $\mathscr{L}$ with components $\psi_1, \psi_2, \ldots, \psi_M$ then according to (3.1.4) and (3.1.6) we have

$$\psi'_{i_1} = \sum_{j_1=1}^{M} a_{i_1j_1}\psi_{j_1} \qquad i_1 = 1, 2, \ldots, M$$

and the components $\psi_1, \psi_2, \ldots, \psi_M$ carry the M-dimensional IR Γ.

We can construct a new representation of $GL(M)$ by forming the direct product of Γ with itself. Thus $\Gamma \otimes \Gamma$ is an M^2-dimensional representation of $GL(M)$. If we write

$$e'_{j_2} = \sum_{i_2=1}^{M} e_{i_2}a_{i_2j_2} \qquad j_2 = 1, 2, \ldots, M \tag{8.1.2}$$

then the direct product is given by

$$e'_{j_1}e'_{j_2} = \sum_{i_1=1}^{M}\sum_{i_2=1}^{M} e_{i_1}e_{i_2}a_{i_1j_1}a_{i_2j_2} \qquad j_1, j_2 = 1, 2, \ldots, M$$

We now apply the result (5.2.6) which, although derived in connection with finite groups, is clearly quite general. The M^2 dimensional representation $\Gamma \otimes \Gamma$ is reducible to a $\frac{1}{2}M(M+1)$ dimensional representation $(\Gamma \otimes \Gamma)^+$ and $\frac{1}{2}M(M-1)$ dimensional representation $(\Gamma \otimes \Gamma)^-$. The representation $(\Gamma \otimes \Gamma)^+$ has symmetric basis vectors $e_{j_1}e_{j_2} + e_{j_2}e_{j_1}$ while the representation $(\Gamma \otimes \Gamma)^-$ has antisymmetric basis vectors $e_{j1}e_{j2} - e_{j2}e_{j1}$. The representations $(\Gamma \otimes \Gamma)^{\pm}$ are irreducible for the same reason that Γ is irreducible. In terms of components we have

$$\psi'_{i_1}\psi'_{i_2} = \sum_{j_1=1}^{M}\sum_{j_2=1}^{M} a_{i_1j_1}a_{i_2j_2}\psi_{j_1}\psi_{j_2} \tag{8.1.3}$$

and the set of components $\{\psi_{i_1}\psi_{i_2}\}$ carries the representation $\Gamma \otimes \Gamma$ of $GL(M)$.

We can arrive at the above result by a method which is more suited to generalization. We regard each vector e_{j_1} as a basis vector for the IR [1] of the symmetric group $\mathscr{S}_1(1)$. Similarly we regard each vector e_{j_2} as a basis vector for the IR [1] of $\mathscr{S}_1(2)$. We can then label the IR Γ of $GL(M)$ by the symbol [1] and we can denote the base vectors by $e_{\boxed{1}}, e_{\boxed{2}}, \ldots, e_{\boxed{M}}$. According to Section 6.8 the vectors e_{j_1} e_{j_2} are basis vectors for the outer product $[1] \odot [1]$ of the group $\mathscr{S}_2(1, 2)$. From (6.8.2) this outer product is reducible to $[1^2]$ and $[2]$. The reduction of the outer product $[1] \odot [1]$ for the symmetric group $\mathscr{S}_2$ clearly gives the reduction of the direct product $[1] \otimes [1]$ of $GL(M)$ since $[2]$ is spanned by a symmetric function and $[1^2]$ is spanned by an antisymmetric function.

At this point it is convenient to define any set of M^2 quantities $T_{i_1 i_2}$ which transform like (8.1.3) as

$$T'_{i_1 i_2} = \sum_{j_1=1}^{M} \sum_{j_2=1}^{M} a_{i_1 j_1} a_{i_2 j_2} T_{j_1 j_2} \tag{8.1.4}$$

to be the components of a *covariant tensor* $\mathbf{T}^{(2)}$ of *rank* 2 with respect to the group $GL(M)$ (see 2.11). It is clear that the tensor components $T_{i_1 i_2}$ carry the representation $[1] \otimes [1]$ of $GL(M)$. The components which carry the IR [2] are symmetric and we denote them by $T_{\boxed{i_1}\boxed{i_2}}$ with

$$T_{\boxed{i}\boxed{i}} = T_{ii} \qquad i = 1, 2, \ldots, M$$
$$T_{\boxed{i_1}\boxed{i_2}} = T_{i_1} T_{i_2} + T_{i_2} T_{i_1} \qquad i_1 \neq i_2 = 1, 2, \ldots, M$$

The components which carry $[1^2]$ are antisymmetric and we denote them by $T_{\begin{array}{|c|}\hline i_1\\\hline i_2\\\hline\end{array}}$ with

$$T_{\begin{array}{|c|}\hline i_1\\\hline i_2\\\hline\end{array}} = T_{i_1 i_2} - T_{i_2 i_1} \qquad i_1 \neq i_2 = 1, 2, \ldots, M$$
$$T_{\begin{array}{|c|}\hline i\\\hline i\\\hline\end{array}} = 0$$

Since $T_{\boxed{i_2}\boxed{i_1}} = T_{\boxed{i_1}\boxed{i_2}}$ it follows that a set of independent components which carry [2] is given by taking $i_1 \leqslant i_2$. Since $T_{\begin{array}{|c|}\hline i_2\\\hline i_1\\\hline\end{array}} = -T_{\begin{array}{|c|}\hline i_1\\\hline i_2\\\hline\end{array}}$ with $T_{\begin{array}{|c|}\hline i\\\hline i\\\hline\end{array}} = 0$ it follows that a set of independent components which carry $[1^2]$ is given by taking the tensor components $T_{\begin{array}{|c|}\hline i_1\\\hline i_2\\\hline\end{array}}$ with $i_1 < i_2$. Because the representation [2] is irreducible we say that the tensor $\mathbf{T}^{(2)}_{[2]}$ which has components $T_{\boxed{i_1}\boxed{i_2}}$ is an *irreducible* covariant tensor of rank 2 with respect to $GL(M)$. We also say that $\mathbf{T}^{(2)}_{[2]}$ is of symmetry type [2]. In the same way the tensor $\mathbf{T}^{(2)}_{[1^2]}$ is an irreducible covariant second rank tensor of symmetry type $[1^2]$.

It is now clear that the dimension of the IR [2] of $GL(M)$ is given by the number of standard tableaux $\boxed{i_1}\boxed{i_2}$ which can be constructed with $i_1 \leqslant i_2$. Similarly the dimension of the IR $[1^2]$ of $GL(M)$ is given by the number of standard tableaux $\begin{array}{|c|}\hline i_1\\\hline i_2\\\hline\end{array}$ which can be constructed with $i_1 < i_2$. As an example we have in $GL(3)$ the standard tableaux

$$\boxed{1}\boxed{1} \qquad \boxed{2}\boxed{2} \qquad \boxed{3}\boxed{3}$$

$$\boxed{1}\boxed{2} \qquad \boxed{1}\boxed{3} \qquad \boxed{2}\boxed{3}$$

$$\begin{array}{|c|}\hline 1\\\hline 2\\\hline\end{array} \qquad \begin{array}{|c|}\hline 1\\\hline 3\\\hline\end{array} \qquad \begin{array}{|c|}\hline 2\\\hline 3\\\hline\end{array}$$

Thus in $GL(3)$ the IR [2] is of dimension 6 whilst the IR $[1^2]$ is of dimension 3. In order to find further IR's of $GL(M)$ we proceed to consider the M^3-dimensional representation $[1]\otimes[1]\otimes[1]$. We have

$$\psi'_{i_1}\psi'_{i_2}\psi'_{i_3}=\sum_{j_1=1}^{M}\sum_{j_2=1}^{M}\sum_{j_3=1}^{M} a_{i_1j_1}a_{i_1j_2}a_{i_3j_3}\psi_{j_1}\psi_{j_2}\psi_{j_3} \qquad i_1, i_2, i_3=1, 2, \ldots, M$$

and we define the components $T_{i_1i_2i_3}$ of a third rank covariant tensor with respect to $GL(M)$ by the transformation law

$$T'_{i_1i_2i_3}=\sum_{j_1=1}^{M}\sum_{j_2=1}^{M}\sum_{j_3=1}^{M} a_{i_1j_1}a_{i_2j_2}a_{i_3j_3}T_{j_1j_2j_3} \qquad i_1, i_2, i_3=1, 2, \ldots, M$$

The reduction of $[1]\otimes[1]\otimes[1]$ to IR's of $GL(M)$ is given by the corresponding reduction of the outer product $[1]\odot[1]\odot[1]$ in the symmetric group $\mathscr{S}_3$. According to the results in Section 6.8 we have

$$[1]\odot[1]\odot[1]=[3]\oplus[1^3]\oplus 2[21]$$

and it follows that we have found IR's of $GL(M)$ which can be labelled [3], $[1^3]$ and [21]. To find the dimensions of these IR's we must determine the number of independent tensor components of each symmetry type. Let us consider [21]. According to (6.3.8) and (6.3.9) components which carry [21] considered as an IR of $\mathscr{S}_3$ are given in terms of tensor components by

$$T_{\begin{smallmatrix} i_1 & i_2 \\ i_3 & \end{smallmatrix}}=T_{i_1i_2i_3}+T_{i_2i_1i_3}-T_{i_3i_2i_1}-T_{i_2i_3i_1}$$

$$T_{\begin{smallmatrix} i_1 & i_3 \\ i_2 & \end{smallmatrix}}=T_{i_1i_3i_2}+T_{i_3i_1i_2}-T_{i_2i_3i_1}-T_{i_3i_2i_1}$$

Now

$$T_{\begin{smallmatrix} i_3 & i_2 \\ i_1 & \end{smallmatrix}}=T_{i_3i_2i_1}+T_{i_2i_3i_1}-T_{i_1i_2i_3}-T_{i_2i_1i_3}=-T_{\begin{smallmatrix} i_1 & i_2 \\ i_3 & \end{smallmatrix}}$$

and it follows that for non zero independent tensor components we must impose on $T_{\begin{smallmatrix} i_1 & i_2 \\ i_3 & \end{smallmatrix}}$ the condition $i_1<i_3$. We also have

$$\begin{aligned} T_{\begin{smallmatrix} i_2 & i_1 \\ i_3 & \end{smallmatrix}} &= T_{i_2i_1i_3}+T_{i_1i_2i_3}-T_{i_3i_1i_2}-T_{i_1i_3i_2} \\ &= T_{\begin{smallmatrix} i_1 & i_2 \\ i_3 & \end{smallmatrix}}-T_{\begin{smallmatrix} i_1 & i_3 \\ i_2 & \end{smallmatrix}} \end{aligned}$$

and $T_{\begin{smallmatrix} i_2 & i_1 \\ i_3 & \end{smallmatrix}}$ is not independent of $T_{\begin{smallmatrix} i_1 & i_2 \\ i_3 & \end{smallmatrix}}$. Thus for independent tensor components we must impose on $T_{\begin{smallmatrix} i_1 & i_2 \\ i_3 & \end{smallmatrix}}$ the condition $i_1\leqslant i_2$. This clearly exhausts the restrictive conditions.

As an example we have for [21] of $GL(3)$ the standard tableaux

$$\begin{array}{|c|c|}\hline 1 & 1 \\ \hline 2 \\ \cline{1-1}\end{array} \quad \begin{array}{|c|c|}\hline 1 & 1 \\ \hline 3 \\ \cline{1-1}\end{array} \quad \begin{array}{|c|c|}\hline 2 & 2 \\ \hline 3 \\ \cline{1-1}\end{array} \quad \begin{array}{|c|c|}\hline 1 & 2 \\ \hline 2 \\ \cline{1-1}\end{array}$$

$$\begin{array}{|c|c|}\hline 1 & 3 \\ \hline 3 \\ \cline{1-1}\end{array} \quad \begin{array}{|c|c|}\hline 1 & 2 \\ \hline 3 \\ \cline{1-1}\end{array} \quad \begin{array}{|c|c|}\hline 2 & 3 \\ \hline 3 \\ \cline{1-1}\end{array} \quad \begin{array}{|c|c|}\hline 1 & 3 \\ \hline 2 \\ \cline{1-1}\end{array}$$

It follows that the dimension of the IR [21] is eight for $GL(3)$.

We are now in a position to state without proof the following generalizations of the foregoing concepts. A set of M^k quantities $T_{i_1 i_2 \ldots i_k}$ which transform according to

$$T'_{i_1 i_2 \ldots i_k} = \sum_{j_1=1}^{M} \sum_{j_2=1}^{M} \ldots \sum_{j_k=1}^{M} a_{i_1 j_1} a_{i_2 j_2} \ldots a_{i_k j_k} T_{j_1 j_2 \ldots j_k}$$

$$i_1, i_2, \ldots, i_k = 1, 2, \ldots, M \quad (8.1.5)$$

are said to form the components of a covariant tensor $\mathbf{T}^{(k)}$ of rank k with respect to $GL(M)$. These tensor components carry the direct product representation

$$\underbrace{[1] \otimes [1] \otimes \ldots \otimes [1]}_{k \text{ factors}}$$

of $GL(M)$. This direct product can be decomposed into IR's of $GL(M)$ by considering the decomposition of the outer product

$$\underbrace{[1] \odot [1] \odot [1] \odot \ldots \odot [1]}_{k \text{ factors}}$$

into IR's $[\lambda]$ of the symmetric group $\mathscr{S}_k$. In this way the representation space of kth rank covariant tensors is decomposed into invariant subspaces. Each invariant subspace is spanned by irreducible covariant tensors $\mathbf{T}^{(k)}_{[\lambda]}$ of rank k and symmetry type $[\lambda]$. A set of IR's of $GL(M)$ is obtained by letting k take the values

$$k = 0, 1, 2, \ldots$$

Among the possible IR's of $GL(M)$ which are obtained in this way there will occur those for which the Young shape contains more than M rows. These can be discarded since the corresponding Young tableaux must have at least one symbol repeated in the first column and the corresponding tensor component vanishes. Thus the IR's of $GL(M)$ can be labelled by partitions $[\lambda_1 \lambda_2 \ldots \lambda_M]$ with no more than M rows and

$$\lambda_1 + \lambda_2 + \ldots + \lambda_M = k$$

To find the dimension of an IR $[\lambda]$ of $GL(M)$ we draw the shape of $[\lambda]$ and then insert into the cells the symbols $1, 2, \ldots, M$ in such a way that the numbers increase on going down a column and do not decrease on going along a row from left to right. The number of standard tableaux obtained in this way gives the dimension of the IR (see, for example, Hamermesh, 1962).

Thus far we have been concerned exclusively with covariant tensors defined with respect to $GL(M)$. The contravariant vectors in the dual space

$\tilde{\mathscr{L}}$ undergo the contragredient transformations. Thus in terms of components we have

$$(\psi^{i_1})' = \sum_{j_1=1}^{M} a^{\#}_{i_1 j_1} \psi^{j_1} \qquad i_1 = 1, 2, \ldots, M \tag{8.1.6}$$

where $\psi^1, \psi^2, \ldots, \psi^M$ are the components of a contravariant vector in the space $\tilde{\mathscr{L}}$ and the matrix $[a_{i_1 j_1}]^{\#}$ is the matrix contragredient to $[a_{i_1 j_1}]$. It is clear that $\tilde{\mathscr{L}}$ is thus the representation space for a matrix representation of $GL(M)$ in which the matrices are contragredient to the corresponding matrices in the representation Γ induced by the space $\mathscr{L}$. Just as Γ is irreducible so the M-dimensional representation induced by $\tilde{\mathscr{L}}$ is irreducible. We call this representation the *contragredient* representation to Γ and we denote it by $\Gamma^{\#}$. It is important to note that for general linear transformations contragredient representations are not equivalent (see Weyl, 1946).

In a completely analogous fashion to the case of covariant vectors we can construct new representations of $GL(M)$ by forming the direct product of $\Gamma^{\#}$ with itself a given number of times. A set of M^k quantities $T^{i_1 i_2 \ldots i_k}$ which transform according to

$$(T^{i_1 i_2 \ldots i_k})' = \sum_{j_1=1}^{M} \sum_{j_2=1}^{M} \ldots \sum_{j_k=1}^{M} a^{\#}_{i_1 j_1} a^{\#}_{i_2 j_2} \ldots a^{\#}_{i_k j_k} T^{j_1 j_2 \ldots j_k} \tag{8.1.7}$$

are said to form the components of a contravariant tensor of rank k with respect to $GL(M)$ (see 2.11). The contravariant tensor components carry the direct product representation

$$\underbrace{\Gamma^{\#} \otimes \Gamma^{\#} \otimes \ldots \otimes \Gamma^{\#}}_{\longleftarrow k \text{ factors} \longrightarrow} \tag{8.1.8}$$

The reduction of this direct product by means of permutational symmetry is completely analogous to the reduction of the direct product

$$\underbrace{\Gamma \otimes \Gamma \otimes \ldots \otimes \Gamma}_{\longleftarrow k \text{ factors} \longrightarrow} \tag{8.1.9}$$

which we have already discussed. In this way the representation space of kth rank contravariant tensors is decomposed into invariant subspaces. Each invariant subspace is spanned by irreducible contravariant tensors $\bar{\mathbf{T}}^{(k)}_{[\lambda]}$ of rank k and of symmetry type $[\lambda]$. We now have a further set of IR's of $GL(M)$ (the contragredient representations). These are obtained by letting k take on the values

$$k = 0, 1, 2, 3, \ldots$$

The labelling of IR's of $GL(M)$ by partitions is now no longer unique since $[\lambda_1\lambda_2 \ldots \lambda_M]$ is being used as a label not only for a given IR but also for its contragredient representation. Thus we cannot tell whether $[\lambda_1\lambda_2 \ldots \lambda_M]$ means an IR in terms of covariant tensors or in terms of contravariant tensors. There exists however a method by which this difficulty can be overcome. We use the partition $[\lambda_1\lambda_2 \ldots \lambda_M]$ with

$$\lambda_1 \geqslant \lambda_2 \geqslant \ldots \geqslant \lambda_M \geqslant 0$$

and

$$\sum_i \lambda_i = k$$

to label the IR spanned by the kth rank covariant irreducible tensor of symmetry type $[\lambda_1\lambda_2 \ldots \lambda_M]$. For the corresponding kth rank contravariant irreducible tensor of symmetry type $[\lambda_1\lambda_2 \ldots \lambda_M]$ we define a set of non-positive integers

$$0 \geqslant \lambda_1^{\#} \geqslant \lambda_2^{\#} \geqslant \ldots \geqslant \lambda_M^{\#}$$

with

$$\sum_i \lambda_i^{\#} = -k$$

Corresponding to each symbol $[\lambda_1\lambda_2 \ldots \lambda_M]$ there is now a unique symbol $[\lambda_1^{\#}\lambda_2^{\#} \ldots \lambda_M^{\#}]$ which can be used to label the contragredient representations. In fact it can be shown (see Weyl, 1946) that the non-positive integers $\lambda_i^{\#}$ are related to the non-negative integers λ_i by the equation

$$\lambda_i^{\#} = -\lambda_{M+1-i} \qquad (8.1.10)$$

We can also construct representations of $GL(M)$ by considering mixed direct products such as

$$\underbrace{\Gamma \otimes \Gamma \otimes \ldots \otimes \Gamma}_{l \text{ factors}} \otimes \underbrace{\Gamma^{\#} \otimes \Gamma^{\#} \otimes \ldots \otimes \Gamma^{\#}}_{(k-l) \text{ factors}} \qquad (8.1.11)$$

A set of M^k quantities $T_{i_1 i_2 \ldots i_l}^{i_{l+1} i_{l+2} \ldots i_k}$ which transform according to

$$(T_{i_1 i_2 \ldots i_l}^{i_{l+1} i_{l+2} \ldots i_k})' = \sum_{j_1, j_2, \ldots, j_k} a_{i_1 j_1} \ldots a_{i_l j_l} a^{\#}_{i_{l+1} j_{l+1}} \ldots a^{\#}_{i_k j_k} T_{j_1 \ldots j_l}^{j_{l+1} \ldots j_k} \qquad (8.1.12)$$

are said to form the components of a mixed tensor of rank k with respect to $GL(M)$ (see 2.11). The direct product (8.1.11) can be reduced by separately considering permutational symmetry in the covariant indices and permutational symmetry in the contravariant indices. In general however this process does not yield IR's of $GL(M)$. The reason for this is that mixed tensors can also be reduced by a method which is quite distinct from that of permutational symmetry. In order to illustrate this let us consider the simplest case namely $\Gamma \otimes \Gamma^{\#}$. In terms of partition symbols we can write

$$\Gamma \otimes \Gamma^{\#} = [1\ 0 \ldots 0] \otimes [0 \ldots 0 -1] \qquad (8.1.13)$$

Such a direct product is carried by the M^2 tensor components $T^{i_2}_{i_1}$ and we have from (8.1.12)

$$(T^{i_2}_{i_1})' = \sum_{j_1,j_2} a_{i_1j_1} a^{\#}_{i_2j_2} T^{j_2}_{j_1} \tag{8.1.14}$$

A well known process in tensor algebra is contraction (see 2.12). In this process we equate two tensor indices and then sum over all values of the common index. This leads to a tensor of rank $(k-2)$ which is often called a *pair trace* of the original tensor. For $T^{i_2}_{i_1}$ contraction with respect to i_1 and i_2 yields a scalar and from (8.1.14) we have

$$\begin{aligned}\sum_{i_1} (T^{i_1}_{i_1})' &= \sum_{i_1,j_1,j_2} a_{i_1j_1} a^{\#}_{i_1j_2} T^{j_2}_{j_1} \\ &= \sum_{j_1,j_2}\Big(\sum_{i_1} \tilde{a}_{j_1i_1} \tilde{a}^{-1}_{i_1j_2}\Big) T^{j_2}_{j_1} = \sum_{j_1,j_2} \delta^{j_2}_{j_1} T^{j_2}_{j_1} = \sum_{j_1} T^{j_1}_{j_1}\end{aligned} \tag{8.1.15}$$

It follows that under general linear transformations the pair trace $\sum_i T^i_i$ is a scalar invariant and consequently it carries the identity representation $[0\ldots0]$ of $GL(M)$. It should be clear that for general linear transformations such a reduction process does not exist either for covariant tensors $T_{i_1i_2}$ or for contravariant tensors $T^{i_1i_2}$. The M^2-dimensional space of second rank mixed tensors is reducible under the process of contraction with respect to the upper and lower index. This leaves an (M^2-1)-dimensional space which is spanned by tensor components $S^{i_2}_{i_1}$ obtained from $T^{i_2}_{i_1}$ by removing the pair trace. It is most convenient to separate the pair trace by writing

$$S^{i_2}_{i_1} = T^{i_2}_{i_1} - \frac{1}{M}\sum_i T^i_i\, \delta^{i_2}_{i_1} \tag{8.1.16}$$

for then we find

$$\sum_{i_1} S^{i_1}_{i_1} = 0 \tag{8.1.17}$$

and the tensor with components $S^{i_2}_{i_1}$ is traceless. Since there is only one covariant index and only one contravariant index in $S^{i_2}_{i_1}$ it is not difficult to see that this traceless tensor has components which carry an (M^2-1)-dimensional IR of $GL(M)$. According to our labelling convention with regard to covariant and contravariant indices this IR can be denoted by $[1\ 0\ldots0-1]$ and we have

$$[1\ 0\ldots0]\otimes[0\ldots0-1]=[0\ldots0]\oplus[1\ 0\ldots0-1] \tag{8.1.18}$$

The IR $[1\ 0\ldots0-1]$ is often referred to as the *adjoint* representation.

We have seen how covariant or contravariant tensors are reduced by permutational symmetry and how a second rank mixed tensor reduces by separation of the pair trace. For a general mixed tensor careful use of both types of operation leads to a complete reduction into IR's of $GL(M)$. Thus it can be shown that a mixed tensor is irreducible with respect to $GL(M)$ if and only if (see Edmonds, 1962; Behrends *et al.*, 1962)

(a) the covariant indices have a definite permutational symmetry
(b) the contravariant indices have a definite permutational symmetry
(c) contraction with respect to one covariant and one contravariant index gives zero (tracelessness condition).

As an example of how all this works in practice we consider the group $GL(3)$. The first few IR's are determined by considering reduction of tensor spaces of ranks $k = 0, 1, 2, 3, \ldots$. For $k = 0$ we have the identity representation [000]. For $k = 1$ the covariant tensor components T_a give the 3-dimensional IR [100] while the contravariant tensor components T^a give the contragredient 3-dimensional IR $[00-1]$. For $k = 2$ the covariant tensor components T_{ab} give a 6-dimensional IR [200] and a 3-dimensional IR [110]. The contravariant tensor components T^{ab} give the contragredient IR's $[00-2]$ and $[0-1-1]$ respectively. The mixed tensor components T^b_a give the identity representation and the 8-dimensional adjoint IR $[10-1]$. For $k = 3$ the covariant tensor components T_{abc} give a 1-dimensional IR [111], a 3-dimensional IR [300] and an 8-dimensional IR [210] (twice). Likewise the contravariant tensor components T^{abc} give the contragredient IR's $[-1-1-1]$, $[00-3]$ and $[0-1-2]$ respectively. We now come to the tensor spaces spanned by the mixed tensor components T^a_{bc} and T^{bc}_a respectively. According to the general method T^a_{bc} is reduced as follows:

(A) Definite permutational symmetry in the covariant indices. This yields subspaces $T^{a}_{\boxed{b}\boxed{c}}$ and $T^{a}_{\begin{smallmatrix}\boxed{b}\\\boxed{c}\end{smallmatrix}}$.

(B) Definite permutational symmetry in the contravariant indices. This gives nothing new here but we write $T^{\boxed{a}}_{\boxed{b}\boxed{c}}$ and $T^{\boxed{a}}_{\begin{smallmatrix}\boxed{b}\\\boxed{c}\end{smallmatrix}}$ to denote that the process has been considered.

(C) Separation of the pair traces. For $T^{\boxed{a}}_{\boxed{b}\boxed{c}}$ the pair traces are

$$P_c = \sum_a T^{\boxed{a}}_{\boxed{a}\boxed{c}} \qquad \text{and} \qquad P_b = \sum_a T^{\boxed{a}}_{\boxed{b}\boxed{a}} = \sum_a T^{\boxed{a}}_{\boxed{a}\boxed{b}}$$

Such a pair trace is a covariant tensor of rank one and thus gives the IR [100]. The remaining traceless tensor with components

$$S^{\boxed{a}}_{\boxed{b}\boxed{c}} = T^{\boxed{a}}_{\boxed{b}\boxed{c}} - \tfrac{1}{4}(P_c\,\delta^a_b + P_b\,\delta^a_c)$$

is now irreducible. This tensor clearly has

$$3 \times 6 - 3 = 15$$

independent components and thus we have found a 15-dimensional IR of $GL(3)$ which can be denoted by $[20-1]$. For $T_{\substack{\boxed{a}\\\boxed{b}\\\boxed{c}}}$ the pair traces are

$$Q_c = \sum_a T_{\substack{\boxed{a}\\\boxed{a}\\\boxed{c}}} \qquad \text{and} \qquad Q_b = \sum_a T_{\substack{\boxed{a}\\\boxed{b}\\\boxed{a}}} = -\sum_a T_{\substack{\boxed{a}\\\boxed{a}\\\boxed{b}}}$$

Such a pair trace is again a covariant rank one tensor and gives the IR $[100]$. The remaining traceless tensor with components

$$S_{\substack{\boxed{a}\\\boxed{b}\\\boxed{c}}} = T_{\substack{\boxed{a}\\\boxed{b}\\\boxed{c}}} - \tfrac{1}{2}(Q_c\,\delta^a_b - Q_b\,\delta^a_c)$$

is now irreducible. This tensor has $3\times3-3=6$ independent components and we have found a 6-dimensional IR of $GL(3)$ which is denoted by $[11-1]$.

The reduction of T^{bc}_a is evidently similar to that of T^a_{bc} and we have thus generated all the IR's of $GL(3)$ which can be constructed from tensor spaces up to and including rank three. The results are summarized in Table 16.

TABLE 16. Low rank IR's of $GL(3)$

Rank k	Dimension of tensor space 3^k	IR	Dimension of IR
0	1	[000]	1
1	3	[100]	3
		[00−1]	3
2	9	[200]	6
		[110]	3
		[00−2]	6
		[0−1−1]	3
		[000]	1
		[10−1]	8
3	27	[300]	10
		2[210]	8
		[111]	1
		[00−3]	10
		2[0−1−2]	8
		[−1−1−1]	1
		2[100]	3
		[20−1]	15
		[11−1]	6
		2[00−1]	3
		[10−2]	15
		[1−1−1]	6

8.2 The special unitary group $SU(M)$

In quantum chemistry we are usually concerned with linear transformations which take one set of orthonormal basis functions into another set of orthonormal basis functions. We have seen that the orthonormality of basis functions is preserved by unitary transformations. A unitary transformation is one which has a unitary matrix. The set of all unitary transformations in an M-dimensional space clearly constitutes a subgroup of $GL(M)$. This subgroup is called the unitary group and is denoted by $U(M)$. The elements of $U(M)$ are given by the transformations

$$e_{j_1} = \sum_{i_1=1}^{M} e_{i_1} u_{i_1 j_1} \qquad j_1 = 1, 2, \ldots, M$$

or

$$\psi'_{i_1} = \sum_{j_1=1}^{M} u_{i_1 j_1} \psi_{j_1} \qquad i_1 = 1, 2, \ldots, M \tag{8.2.1}$$

with $\mathbf{U}^{\dagger}\mathbf{U} = 1$ where $\mathbf{U} = [u_{i_1 j_1}]$.

Now

$$\det \mathbf{U}^{\dagger}\mathbf{U} = \det \mathbf{U}^{\dagger} \det \mathbf{U} = \det \tilde{\mathbf{U}}^{*} \det \mathbf{U}$$
$$= \det \mathbf{U}^{*} \det \mathbf{U} = (\det \mathbf{U})^{*}(\det \mathbf{U}) = |\det \mathbf{U}|^{2} = \det \mathbf{I} = 1$$

and it follows that

$$\det \mathbf{U} = \pm 1$$

The group $U(M)$ is still too general for applications in quantum chemistry. If the e_{j_1} are wave functions then we know that they are only determined up to an arbitrary phase factor. When a common phase factor is removed from each of the basis vectors e_{j_1} the transformations with $\det \mathbf{U} = -1$ are equivalent to transformations with $\det \mathbf{U} = +1$. It follows that we can restrict ourselves to those unitary transformations with

$$\det \mathbf{U} = +1 \tag{8.2.2}$$

Unitary transformations with the property (8.2.2) are called *unimodular* or special. It is evident that the set of all unimodular unitary transformations forms a subgroup of $GL(M)$. This subgroup is called the special unitary group and we denote it by $SU(M)$.

It can be shown that IR's of $GL(M)$ remain irreducible when considered as representations either of the subgroup $U(M)$ or of the subgroup $SU(M)$ (see Hamermesh, 1962). In the case of $SU(M)$ however the unimodularity condition (8.2.2) imposes certain equivalences amongst the IR's. In order to illustrate how this comes about let us take the group $SU(3)$. We begin by considering the third rank covariant Levi-Civita tensor. This tensor has components ε_{ijk} which are defined as follows:

(1) $\varepsilon_{ijk} = +1$ if ijk is an even permutation of 123
(2) $\varepsilon_{ijk} = -1$ if ijk is an odd permutation of 123
(3) $\varepsilon_{ijk} = 0$ otherwise.

Under $GL(3)$ we have the transformation property (8.1.5)

$$\varepsilon'_{ijk} = \sum_{l,m,n} a_{il}a_{jm}a_{kn}\varepsilon_{lmn}$$

Using the definition of ε_{ijk} it is a simple matter to show that this reduces to

$$\varepsilon'_{ijk} = (\det \mathbf{A})\varepsilon_{ijk}$$

Now if we have unimodular transformations

$$\det \mathbf{A} = +1$$

and ε_{ijk} is an invariant. Using ε_{ijk} we can construct a rank four mixed tensor with components

$$\varepsilon_{ijk}T^a$$

where T^a are the components of a contravariant vector. By contraction of this tensor with respect to the indices i and a we produce a rank two covariant tensor with components

$$T_{jk} = \sum_i \varepsilon_{ijk}T^i$$

Using the properties of ε_{ijk} we readily find that

$$T_{jk} = -T_{kj}$$

and we can write the tensor components as $T_{\boxed{j}\atop\boxed{k}}$. Now for unimodular transformations ε_{ijk} is an invariant. It is thus clear that under unimodular transformations the contravariant vector components T^i transform in the same way as the antisymmetric covariant rank two tensor components $T_{\boxed{j}\atop\boxed{k}}$.

In other words the two 3-dimensional IR's $[00-1]$ and $[110]$ are equivalent in $SU(3)$. We note that if 1 is added to all the symbols in $[00-1]$ we obtain the partition $[110]$.

In general it can be shown that two IR's $[\lambda]=[\lambda_1\lambda_2 \ldots \lambda_M]$ and $[\lambda']=[\lambda'_1\lambda'_2 \ldots \lambda'_M]$ are equivalent in $SU(M)$ if

$$\lambda'_i = \lambda_i + s \qquad i = 1, 2, \ldots, M$$

where s is any integer (see Weyl, 1946). On choosing $s = -\lambda_M$ we have

$$[\lambda_1\lambda_2 \ldots \lambda_M] \equiv [\lambda_1 - \lambda_M \ \ \lambda_2 - \lambda_M \ldots \lambda_{M-1} - \lambda_M \ \ 0] \qquad (8.2.3)$$

By using generalized Levi-Civita tensors it can be shown that the IR's of $SU(M)$ may be associated with purely covariant tensors. Thus the IR's of $SU(M)$ are labelled by partitions whose shapes have fewer than M rows (see Gourdin, 1967).

As an illustration we write down in Table 17 the first few IR's of $SU(3)$. They are of course derived from Table 16 as indicated. As we shall see later it is convenient to introduce a new labelling scheme. Thus the IR $[\lambda_1\lambda_2]$ is labelled by the symbol (nm) where

$$n = \lambda_1 - \lambda_2$$

$$m = \lambda_2$$

It should be clear from the mode of construction that when $n \neq m$ the two IR's (nm) and (mn) are contragredient. According to (2.7.15) the contragredient representations are simply complex conjugate representations when unitary transformations are considered.

According to (8.1.10) the IR's $[\lambda_1\lambda_2 \ldots \lambda_M]$ and $[-\lambda_M - \lambda_{M-1} \ldots -\lambda_1]$ are contragredient. For the group $SU(M)$ we have an equivalence like (8.2.3) which is obtained by taking $s = \lambda_1$. Thus

$$[-\lambda_M - \lambda_{M-1} \ldots -\lambda_1] \equiv [\lambda_1 - \lambda_M \lambda_1 - \lambda_{M-1} \ldots \lambda_1 - \lambda_2 0]$$

TABLE 17. Low rank IR's of $SU(3)$

IR of $GL(3)$ $[\lambda_1'\lambda_2'\lambda_3']$	IR of $SU(3)$ $[\lambda_1\lambda_2] = [\lambda_1' - \lambda_3'\lambda_2' - \lambda_3']$	IR of $SU(3)$ (nm)	Dimension
[000] [111] [−1−1−1]	[00]	(00)	1
[100] [0−1−1]	[10]	(10)	3
[00−1] [110]	[11]	(01)	3
[200] [1−1−1]	[20]	(20)	6
[00−2] [11−1]	[22]	(02)	6
[10−1] [210] [0−1−2]	[21]	(11)	8
[300]	[30]	(30)	10
[00−3]	[33]	(03)	10
[20−1]	[31]	(21)	15
[10−2]	[32]	(12)	15

and it follows that in $SU(M)$ the two IR's $[\lambda_1\lambda_2\ldots\lambda_M]$ and $[\lambda_1-\lambda_M\ \lambda_1-\lambda_{M-1}\ldots\lambda_1-\lambda_2\ 0]$ are contragredient. This result may be written as

$$[\lambda_1\lambda_2\ldots\lambda_M]^{\#}=[\lambda_1-\lambda_M\ \lambda_1-\lambda_{M-1}\ldots\lambda_1-\lambda_2\ 0] \tag{8.2.4}$$

8.3 The rotation group *R*(*M*)

A linear transformation with matrix $\mathbf{A}$ such that

$$\mathbf{A}^{-1}=\tilde{\mathbf{A}}$$

is called an *orthogonal* transformation. The set of all orthogonal transformations in M-dimensional space constitutes a subgroup of $GL(M)$ which is called the orthogonal group. This subgroup is denoted by $O(M)$. If the orthogonal transformations are also unimodular then we have the special orthogonal group $SO(M)$. A real unimodular orthogonal transformation is usually called a *rotation* in M-dimensional space. When the transformations in $SO(M)$ are restricted to be real then we have the rotation group in M-dimensions. This group is denoted by $R(M)$.

For an orthogonal transformation we have

$$\mathbf{A}^{\#}=\tilde{\mathbf{A}}^{-1}=\mathbf{A} \tag{8.3.1}$$

and the matrix $\mathbf{A}$ is self contragredient. It follows (see (2.9.15)) that there is no distinction between contravariant and covariant tensor spaces when orthogonal transformations are considered. Thus in order to obtain the IR's of $O(M)$ it is sufficient to consider the reduction of covariant tensor spaces. Unfortunately the IR's of $GL(M)$ are not in general irreducible when considered as representations of the subgroup $O(M)$. This can readily be seen as follows. The condition (8.3.1) can be written as

$$\sum_i a_{ij}a_{ik}=\sum_i a_{ji}a_{ki}=\delta_{jk} \tag{8.3.2}$$

This condition implies the existence of a reduction process for kth rank tensors by means of contraction with respect to pairs of indices. Thus if $T_{i_1i_2}$ are the components of a second rank tensor defined with respect to $O(M)$ by the transformation law

$$T'_{i_1i_2}=\sum_{j_1,j_2} a_{i_1j_1}a_{i_2j_2}T_{j_1j_2}$$

then it follows from (8.3.2) that

$$\sum_i T'_{ii}=\sum_j T_{jj}$$

The pair trace is thus invariant under orthogonal transformations. Following (8.1.16) we separate the invariant pair trace by writing

$$S_{i_1 i_2} = T_{i_1 i_2} - \frac{1}{M}\left(\sum_i T_{ii}\right)\delta_{i_1 i_2}$$

and the tensor with components $S_{i_1 i_2}$ is traceless. This reduction process by means of contraction is analogous to that which we found for mixed tensors under the full linear group $GL(M)$. It follows that a tensor is irreducible with respect to the subgroup $O(M)$ if and only if:

(A) the tensor indices have a definite permutational symmetry
(B) contraction with respect to any pair of tensor indices gives zero (tracelessness condition).

As a first example we consider the group $O(3)$. The reduction process (A) gives IR's of $GL(3)$. Thus we have (c.f. Table 16)

Rank k	$[\lambda_1\lambda_2\lambda_3]$ (IR of $GL(3)$)	Dimension
0	[000]	1
1	[100]	3
2	[200]	6
	[110]	3
3	[300]	10
	[210]	8
	[111]	1
4	[400]	15
	[310]	15
	[220]	6
	[211]	3

In order to obtain IR's of $O(3)$ we must now apply the reduction process (B) so that tensors of symmetry type $[\lambda_1\lambda_2\lambda_3]$ are traceless. For [200] we have

$$T_{\boxed{1}\boxed{1}} + T_{\boxed{2}\boxed{2}} + T_{\boxed{3}\boxed{3}} = 0$$

and it follows that there are only 5 independent tensor components when the tensor is required to be traceless. Thus [200] is the label for a 5-dimensional IR of $O(3)$. An important feature arises when we consider [211]. For a tensor of this symmetry type to be traceless we require

$$\sum_{a=1}^{3} T_{\begin{smallmatrix} \boxed{a} & \boxed{a} \\ \boxed{b} & \\ \boxed{c} & \end{smallmatrix}} = 0 \qquad b, c = 1, 2, 3$$

It is readily seen from this that

$$T_{\begin{smallmatrix}1&1\\2&\\3&\end{smallmatrix}}=T_{\begin{smallmatrix}1&2\\2&\\3&\end{smallmatrix}}=T_{\begin{smallmatrix}1&3\\2&\\3&\end{smallmatrix}}=0$$

and the tensor has no non-zero components. Thus [211] does not serve to label an IR of $O(3)$. The same result is found for [220]. The condition (B) thus eliminates a large class of partitions as labels for IR's of $O(M)$. Those partitions which do give IR's of $O(M)$ are called *permissible* partitions. If we consider the Young shape Y corresponding to the partition $[\lambda]=[\lambda_1\lambda_2\ldots\lambda_M]$ then λ_i is the length of the ith row in Y. We now introduce the length l_i of the ith column in Y. When this is done we can describe the shape Y by means of a symbol $\{l_1l_2\ldots l_k\}$. For the group $O(3)$ the first few IR's are shown in Table 18. From Table 18 we see that in $O(3)$ partitions for which

$$l_1+l_2>3$$

are not permissible. This result can be generalized and it is found that the tracelessness condition eliminates all those partitions for which (see Weyl, 1946)

$$l_1+l_2>M \tag{8.3.3}$$

Thus the irreducible representations of $O(M)$ are labelled by partitions $[\lambda_1\lambda_2\ldots\lambda_M]$ for which

$$l_1+l_2\leq M \tag{8.3.4}$$

Since $l_2\leq l_1$ it follows from (8.3.3) that

$$l_2\leq M/2$$

The case $l_2=M/2$ can occur only for even M. It follows that there is a

TABLE 18. IR's of $O(3)$

Rank k	IR; $[\lambda_1\lambda_2\lambda_3]$	$\{l_1l_2\ldots l_k\}$	Dimension
0	[000]		1
1	[100]	{1}	3
2	[200]	{11}	5
	[110]	{20}	3
3	[300]	{111}	7
	[210]	{210}	5
	[111]	{300}	1
4	[400]	{1111}	9
	[310]	{2110}	7
	[220]	{2200}	0
	[211]	{3100}	0

distinction between orthogonal groups of even and odd dimensionality. Since $l_1 \leqslant M$ this distinction between even and odd dimensionality can be described by the following technique. We have

$$l_1 = 0, 1, 2, \ldots, M-2, M-1, M$$

When M is odd we can set up a correspondence between permissible partitions. With the partition $[\lambda]$ for which $l_1 = x < (M/2)$ we associate the partition $[\lambda]'$ with $l_1' = M - x = M - l_1$. The partitions $[\lambda]$ and $[\lambda]'$ are called *associated* partitions. Apart from the length of the first column associated partitions have equal column lengths. This correspondence between partitions is schematically described by

$$l_1 = 0, 1, 2, 3, \ldots, M-3, M-2, M-1, M$$

When M is even we can set up the same correspondence between partitions. In this case however since l_1 can be exactly $M/2$ there exists, in addition to associated partitions, a set of *self associated* partitions. For a self associated partition $l_1 = M/2$ and $l_1' = M - M/2 = M/2 = l_1$. In Tables 19 and 20 we illustrate the concept of associate partitions by considering some of the IR's of $O(3)$ and $O(4)$ respectively.

TABLE 19. IR's of $O(3)$

IR	Associate IR	Dimension
[000]	[111]	1
[100]	[110]	3
[200]	[210]	5
[300]	[310]	7

TABLE 20. IR's of $O(4)$

IR	Associate IR	Dimension
[0000]	[1111]	1
[1000]	[1110]	4
[2000]	[2110]	9
[3000]	[3110]	16
[1100]	[1100]	6
[2100]	[2100] Self	16
[3100]	[3100] Associate	30
[2200]	[2200]	10

We consider next the IR's of the unimodular groups $SO(M)$ and $R(M)$. It can be shown that an IR of $SO(M)$ remains irreducible when considered as a representation of $R(M)$ (see Hamermesh, 1962). In general IR's of

$O(M)$ do not remain irreducible when considered as representations of $SO(M)$ or $R(M)$. However it can be shown that those IR's of $O(M)$ which are not self-associate do remain irreducible when considered as representations of $R(M)$ (see Weyl, 1946). As we found in the last section so we find here that the unimodularity condition imposes an equivalence among certain IR's. Thus we find that IR's corresponding to associate partitions in $O(M)$ become equivalent in $SO(M)$ or $R(M)$. The self associate IR's of $O(M)$ (which can only occur when M is even) become reducible as representations of $SO(M)$. It is possible to show that a self associate IR of $O(M)$ gives rise to two non-equivalent IR's of $SO(M)$. Furthermore those two IR's have the same dimension (see Weyl, 1946).

In the case of odd M we can now label the IR's of $R(M)$ by partitions $[\lambda]$ for which $l_1 < M/2$. Such partitions are of the form $[\lambda_1\lambda_2 \ldots \lambda_n 0 \ldots 0]$ where

$$n = \left(\frac{M-1}{2}\right)$$

Since the zeros in the partition are redundant we can label the IR's of $R(M)$ by a symbol $(\lambda_1\lambda_2 \ldots \lambda_n)$. In the case of even M we can label the representations of $R(M)$ by partitions $[\lambda]$ for which $l_1 \leq M/2$. Such partitions are of the form $[\lambda_1\lambda_2 \ldots \lambda_n 0 \ldots 0]$ where

$$n = \frac{M}{2}$$

Again the zeros are redundant and we label the representations by a symbol $(\lambda_1\lambda_2 \ldots \lambda_n)$. When this partition symbol arises from IR's of $O(M)$ which are not self associate then we can use $(\lambda_1\lambda_2 \ldots \lambda_n)$ to denote IR's of $R(M)$. The labelling of IR's of $R(M)$ which arise from self associate IR's of $O(M)$ is not so simple. Let us consider $R(4)$. When $(\lambda_1\lambda_2)$ arises from a self-associate IR of $O(4)$ it splits into two non-equivalent IR's of $R(4)$ of equal dimension. In order to label these IR's we introduce a new symbol $(j_1 j_2)$ where

$$j_1 = \tfrac{1}{2}(\lambda_1 + \lambda_2) \qquad j_2 = \tfrac{1}{2}(\lambda_1 - \lambda_2)$$

The IR's of $R(4)$ are now characterized by $(j_1 j_2)$ and $(j_2 j_1)$. The first few IR's of $R(3)$ and $R(4)$ are shown in Tables 21 and 22 respectively.

TABLE 21. IR's of $R(3)$

IR of $O(3)$ $[\lambda_1\lambda_2\lambda_3]$	IR of $R(3)$ (l)	Spectroscopic notation	Dimension
[000]; [111]	(0)	S	1
[100]; [110]	(1)	P	3
[200]; [210]	(2)	D	5
[300]; [310]	(3)	F	7

TABLE 22. IR's of $R(4)$

IR of $O(4)$ $[\lambda_1\lambda_2\lambda_3\lambda_4]$	IR of $R(4)$ (j_1j_2)	Dimension
[0000]; [1111]	(00)	1
[1000]; [1110]	$(\frac{1}{2}\frac{1}{2})$	4
[2000]; [2110]	(11)	9
[3000]; [3110]	$(\frac{3}{2}\frac{3}{2})$	16
[1100]	(10)	3
	(01)	3
[2100]	$(\frac{3}{2}\frac{1}{2})$	8
	$(\frac{1}{2}\frac{3}{2})$	8
[3100]	(21)	15
	(12)	15
[2200]	(20)	5
	(02)	5

8.4 Scalar invariants

Thus far in our discussion of the representations of linear groups we have assumed only that the original M-dimensional representation space $\mathscr{L}$ is a linear space. We now introduce a metric so that the M-dimensional representation space becomes a vector space.

If $e_1, e_2, \ldots, e_M$ is an arbitrary basis in $\mathscr{L}$ then the scalar product of two vectors ϕ and ψ in $\mathscr{L}$ is given in matrix form by

$$\langle \phi \mid \psi \rangle = \tilde{\boldsymbol{\phi}}^* \mathbf{G} \boldsymbol{\psi} \tag{8.4.1}$$

where $\mathbf{G}$ is the metric matrix with elements

$$g_{ij} = \langle e_i \mid e_j \rangle$$

Let $\mathscr{G}$ be a group of non-singular linear transformations in $\mathscr{L}$. Let $\mathbf{A}$ be the matrix of such a transformation. We have image vectors ϕ' and ψ' given in matrix form by

$$\boldsymbol{\phi}' = \mathbf{A}\boldsymbol{\phi} \qquad \boldsymbol{\psi}' = \mathbf{A}\boldsymbol{\psi}$$

The scalar product of the image vectors is

$$\langle \phi' \mid \psi' \rangle = \tilde{\boldsymbol{\phi}}'^* \mathbf{G} \boldsymbol{\psi}' = \tilde{\boldsymbol{\phi}}^* (\mathbf{A}^* \mathbf{G} \mathbf{A}) \boldsymbol{\psi}$$

It follows that the scalar product is invariant under those transformation groups for which

$$\tilde{\mathbf{A}}^* \mathbf{G} \mathbf{A} = \mathbf{G} \tag{8.4.2}$$

In particular if we use the basis (2.8.5) in $\mathscr{L}$ then $\mathbf{G} = \mathbf{I}$ and the scalar product is invariant for those groups in which

$$\tilde{\mathbf{A}}^*\mathbf{A} = \mathbf{I}$$

It is clear that the unitary groups $U(M)$ and $SU(M)$ satisfy this requirement as do the real orthogonal groups $O(M)$ and $R(M)$. However the complex orthogonal group does not qualify.

If we do not use the basis (2.8.5) in $\mathscr{L}$ then $\mathbf{G} \neq \mathbf{I}$. A matrix $\mathbf{A}$ which satisfies (8.4.2) is called a unitary matrix with respect to the metric $\mathbf{G}$. The corresponding group which leaves the scalar product (8.4.1) invariant is called a pseudo-unitary group (see Gourdin, 1967). When $\mathbf{A}$ is real we speak of the pseudo-orthogonal group. We have

$$\tilde{\mathbf{G}} = \mathbf{G}^*$$

and it follows that for the real pseudo-orthogonal group the metric matrix is symmetric. The matrix $\mathbf{G}$ is also required to be non-singular and we say that the scalar product (8.4.1) is a non-degenerate symmetric bilinear form.

We now introduce a new non-degenerate symmetric bilinear form $[\phi \mid \psi]$ which is defined by

$$[\phi \mid \psi] = \tilde{\boldsymbol{\phi}}\mathbf{G}\boldsymbol{\psi} \tag{8.4.3}$$

In the basis (2.8.5) this form becomes

$$[\phi \mid \psi] = \tilde{\boldsymbol{\phi}}\boldsymbol{\psi} \tag{8.4.4}$$

and it is almost immediately evident that $[\phi \mid \psi]$ is invariant under orthogonal transformations (either real or complex).

From the above discussion we see that it is possible to define various subgroups of the general linear group $GL(M)$ by requiring that their transformations leave invariant certain symmetric non-degenerate bilinear forms (such as $\langle\phi \mid \psi\rangle$ or $[\phi \mid \psi]$). We can also define subgroups of $GL(M)$ by requiring that their transformations leave invariant a form different from a non-degenerate symmetric bilinear one. In particular we can define a subgroup by requiring the invariance of a nondegenerate antisymmetric bilinear form. For this purpose it is convenient to define the *skew product* $\{\phi \mid \psi\}$ of two vectors ϕ and ψ by the equation

$$\{\phi \mid \psi\} = \tilde{\boldsymbol{\phi}}\mathbf{H}\boldsymbol{\psi} \tag{8.4.5}$$

Here $\mathbf{H}$ is any non-singular antisymmetric matrix whose elements are given by

$$h_{ij} = \{e_i \mid e_j\} \tag{8.4.6}$$

in the basis $e_1, e_2, \ldots, e_M$. Since $\mathbf{H}$ is assumed to be antisymmetric we have

$$\tilde{\mathbf{H}} = -\mathbf{H} \tag{8.4.7}$$

A linear transformation with matrix **A** which leaves the skew product (8.4.5) invariant is called a *symplectic* transformation.

8.5 The sympletic group $Sp(M)$

The set of all non-singular linear transformations which leave invariant the skew product $\{\phi \mid \psi\}$ constitutes a subgroup of $GL(M)$. This subgroup is called the symplectic group and it is denoted by $Sp(M)$. If **A** is the matrix of a symplectic transformation then by definition we have

$$\tilde{\mathbf{A}}\mathbf{H}\mathbf{A} = \mathbf{H} \tag{8.5.1}$$

Now

$$\det \tilde{\mathbf{H}} = \det(-\mathbf{H}) = (-1)^M \det \mathbf{H} = \det \mathbf{H}$$

When M is odd $\det \mathbf{H} = 0$ and **H** is singular. It follows that we can only define $Sp(M)$ when M is even.

Let $\mathscr{L}$ be a linear space in an even number $M = 2\mu$ dimensions. The M basis vectors (2.8.5) $\varepsilon_1, \varepsilon_2, \ldots, \varepsilon_\mu, \varepsilon_{\mu+1}, \ldots, \varepsilon_{2\mu}$ can be arranged in the form $\varepsilon_1, \varepsilon_2, \ldots, \varepsilon_\mu, \varepsilon_{1'}, \varepsilon_{2'}, \ldots \varepsilon_{\mu'}$, where

$$\begin{aligned} &\varepsilon_1 = (1, 0, \ldots, 0) && \varepsilon_{1'} = (0, 0, \ldots 0, 1) \\ &\varepsilon_2 = (0, 1, 0, \ldots, 0) && \varepsilon_{2'} = (0, 0, \ldots 0, 1, 0) \\ &\cdots\cdots\cdots\cdots && \cdots\cdots\cdots\cdots \\ &\varepsilon_\mu = (0, 0, \ldots 0, 1, 0, \ldots, 0) && \varepsilon_{\mu'} = (0, 0, \ldots 0, 0, 1, 0, \ldots 0) \end{aligned} \tag{8.5.2}$$

In terms of the basis vectors (2.8.5) the matrix **H** can now be defined by setting

$$\begin{aligned} h_{ij} &= \{e_i \mid e_j\} = 0 \qquad j \neq i' \\ h_{ii'} &= \{e_i \mid e_{i'}\} = 1 = -h_{i'i} \end{aligned} \tag{8.5.3}$$

and we have

$$\mathbf{H} = \begin{bmatrix} \mathbf{O} & \mathbf{I} \\ -\mathbf{I} & \mathbf{O} \end{bmatrix} \tag{8.5.4}$$

and

$$\{\phi \mid \psi\} = \sum_{i=1}^{\mu} (\phi_i \psi_{i'} - \phi_{i'} \psi_i) \tag{8.5.5}$$

where

$$\phi = (\phi_1, \phi_2, \ldots, \phi_\mu, \phi_{1'}, \phi_{2'}, \ldots \phi_{\mu'})$$
$$\psi = (\psi_1, \psi_2, \ldots, \psi_\mu, \psi_{1'}, \psi_{2'}, \ldots \psi_{\mu'})$$

The expression (8.5.5) is clearly analogous to the expressions

$$\langle \phi \mid \psi \rangle = \sum_{i=1}^{M} \phi_i^* \psi_i$$

and

$$[\phi \mid \psi] = \sum_{i=1}^{M} \phi_i \psi_i$$

when the basis (2.8.5) is used.

Consider now the case $M = 2$. We have

$$\{\phi \mid \psi\} = \phi_1 \psi_{1'} - \phi_{1'} \psi_1 = \begin{vmatrix} \phi_1 & \psi_1 \\ \phi_{1'} & \psi_{1'} \end{vmatrix}$$

If $\mathbf{A}$ is the matrix of a transformation in $Sp(2)$ then for the image vectors ϕ' and ψ' we have the skew product

$$\{\phi' \mid \psi'\} = \phi'_1 \psi'_{1'} - \phi'_{1'} \psi'_1$$

From this it is trivial to show that

$$\{\phi' \mid \psi'\} = (\det \mathbf{A})\{\phi \mid \psi\}$$

By definition $\{\phi \mid \psi\}$ is invariant under symplectic transformations and it follows that

$$\det \mathbf{A} = +1$$

This result can be generalized (see Hamermesh, 1962) and we find that the symplectic group is of necessity unimodular.

We turn out attention next to the representations of $Sp(M)$. Since

$$\tilde{\mathbf{A}}\mathbf{H}\mathbf{A} = \mathbf{H}$$

for symplectic $\mathbf{A}$ and antisymmetric $\mathbf{H}$ it follows that

$$\mathbf{A}^{\#} = \mathbf{H}\mathbf{A}\mathbf{H}^{-1}$$

Thus a given symplectic transformation is equivalent to its contragredient and we can limit ourselves to covariant tensor spaces as we did for the group $O(M)$.

In general IR's of $GL(M)$ do not remain irreducible when considered as representations of the symplectic subgroup $Sp(M)$. This is due to the invariance of generalized pair traces under symplectic transformations. From (8.5.1) we find the analogue (8.3.2) to be

$$\sum_{k,l} a_{ki} h_{kl} a_{lj} = h_{ij} \tag{8.5.6}$$

If $T_{i_1 i_2}$ are the components of a second rank tensor defined with respect to $Sp(M)$ by the transformation law

$$T'_{i_1 i_2} = \sum_{j_1, j_2} a_{i_1 j_1} a_{i_2 j_2} T_{j_1 j_2}$$

then we readily find

$$\sum_{i_1,i_2} h_{i_1 i_2} T'_{i_1 i_2} = \sum_{j_1,j_2} h_{j_1 j_2} T_{j_1 j_2}$$

The generalized pair trace

$$\sum_{i,j} h_{ij} T_{ij}$$

is thus invariant under symplectic transformations. Following the method used for $O(M)$ we can now assert that a tensor is irreducible with respect to $Sp(M)$ if and only if

(A). the tensor indices have a definite permutational symmetry

(B). generalized contraction with respect to any pair of tensor indices gives zero (tracelessness condition).

As an example we consider the group $Sp(4)$. The reduction process (A) gives IR's of $GL(4)$. These are labelled by partitions $[\lambda_1\lambda_2\lambda_3\lambda_4]$. In order to illustrate the reduction process (B) we take the rank two tensors described by partitions [2000] and [1100] respectively. Since $T_{\boxed{a}\boxed{b}}$ is symmetric and h_{ab} is antisymmetric it follows that

$$\sum_{a,b} h_{ab} T_{\boxed{a}\boxed{b}} = 0$$

Thus $T_{\boxed{a}\boxed{b}}$ is automatically traceless with respect to h_{ab} and the reduction process (B) gives nothing new. The partition [2000] is thus the label for a 10-dimensional IR of $Sp(4)$. For [1100] the pair trace condition

$$\sum_{a,b} h_{ab} T_{\substack{\boxed{a}\\ \boxed{b}}} = 0$$

gives one relation amongst the six components and it follows that [1100] is the label for a 5-dimensional IR of $Sp(4)$.

TABLE 23. IR's of $Sp(4)$

Rank k	IR of $GL(4)$ $[\lambda_1\lambda_2\lambda_3\lambda_4]$	Dimension	IR of $Sp(4)$ $(\lambda_1\lambda_2)$	Dimension
0	[0000]	1	(00)	1
1	[1000]	4	(10)	4
2	[2000]	10	(20)	10
	[1100]	6	(11)	5
3	[3000]	20	(30)	20
	[2100]	20	(21)	16
4	[4000]	35	(40)	35
	[3100]	45	(31)	35
	[2200]	20	(22)	14

Following a similar line of reasoning to that used for $O(M)$ it can be shown that partitions $[\lambda_1\lambda_2 \dots \lambda_M]$ whose shapes have more than $\mu = M/2$ rows do not give rise to IR's of $Sp(M)$. This is due to the tracelessness condition with respect to the antisymmetric matrix with elements h_{ab}. Thus the IR's of $Sp(4)$ can be labelled by a partition symbol $(\sigma_1\sigma_2 \dots \sigma_\mu)$. As an example we list the first few IR's of $Sp(4)$ in Table 23.

8.6 Direct products

We have seen how the IR's $[\lambda]$ of the group $SU(M)$ can be obtained by a consideration of the outer product of representations of the symmetric group. Let $[\lambda]$ and $[\mu]$ be any two IR's of $SU(M)$. We can construct the direct product $[\lambda]\otimes[\mu]$ in the usual way. From what has gone before it is not surprising to find that the method for decomposing $[\lambda]\otimes[\mu]$ into IR's of $SU(M)$ is identical to the method for decomposing the outer product $[\lambda]\odot[\mu]$ into IR's of the symmetric group (see Littlewood, 1940). This method has been discussed in Section 6.8.

As an example we consider the group $SU(2)$ and the direct product $[\lambda_1]\otimes[\lambda_2]$. Without loss of generality we may suppose $\lambda_1 \geq \lambda_2$. If we follow the method which leads to (6.8.4) we have

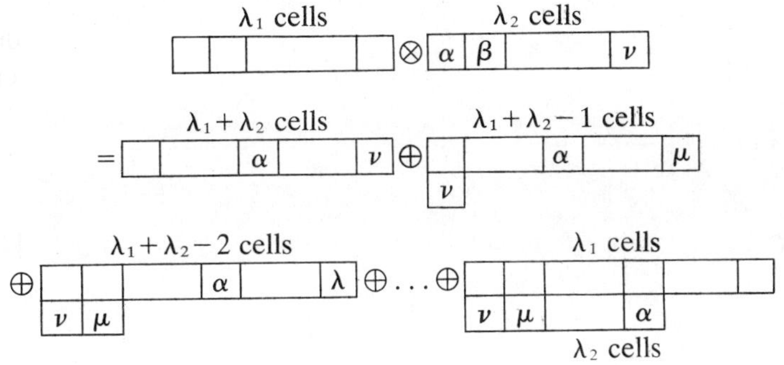

The required decomposition is thus

$$[\lambda_1]\otimes[\lambda_2] = [\lambda_1+\lambda_2]\oplus[\lambda_1+\lambda_2-1, 1]$$
$$\oplus[\lambda_1+\lambda_2-2, 2]\dots\oplus[\lambda_1, \lambda_2]$$

If we use (8.2.3) this becomes

$$[\lambda_1]\otimes[\lambda_2] = [\lambda_1+\lambda_2]\oplus[\lambda_1+\lambda_2-2]\otimes[\lambda_1+\lambda_2-4]\oplus\dots\oplus[\lambda_1-\lambda_2]$$

We can rewrite this result as

$$(j_1)\otimes(j_2) = (j_1+j_2)\oplus(j_1+j_2-1)\oplus(j_1+j_2-2)\oplus\dots\oplus(j_1-j_2) \quad (8.6.1)$$

where

$$j_1 = \frac{\lambda_1}{2} \qquad j_2 = \frac{\lambda_2}{2} \qquad \text{and} \qquad (j) \equiv [2j]$$

As we shall see later the equation (8.6.1) is the group theoretic version of the well known result given by (7.1.1).

References

Behrends, R. E., Dreitlein, J., Fronsdal, C. and Lee, W. (1962) *Rev. Mod. Phys.*, **34,** 22.
Edmonds, A. R. (1962) *Proc. Roy. Soc. (Lond.)*, **A268,** 567.
Gourdin, M. (1967) "Unitary Symmetries," North-Holland, Chapter 17.
Hamermesh, M. (1962) "Group Theory and its application to Physical Problems," Addison-Wesley, Chapter 10.
Littlewood, D. E. (1940) "The Theory of Group Characters," Oxford.
Weyl, H. (1946) "The Classical Groups," Princeton, Chapters 4, 5 and 6.

CHAPTER 9

Continuous Groups
The Method of Lie Algebras

In this chapter we consider specific continuous groups which are useful in the study of certain quantum mechanical systems. The method we employ here differs from that of the previous chapter in that we make explicit use of the continuity of the group manifold. The groups we consider are examples of what are called *Lie groups* of transformations. Such groups can be investigated by consideration of infinitesimal transformations which are transformations in a neighbourhood of the identity. The study of infinitesimal transformations leads to the concept of what is called the *Lie algebra* of the group. The theory of a large class of Lie algebras and their representations can be made very elegant and completely general. Because of the rather indigestible nature of this general treatment we shall be content in this chapter to simply quote results from it. However for easy reference to the interested reader we summarize the general treatment in Appendix 1.

9.1 A one-parameter Lie group of transformations

In order to illustrate the general method we shall consider a very simple example. The rotation of a plane vector $r = (x, y)$ about the z-axis by angle ω gives the rotated vector $r' = (x', y')$. From Fig. 12 we see that

$$\begin{aligned} x' &= r \cos(\theta + \omega) = x \cos \omega - y \sin \omega \\ y' &= r \sin(\theta + \omega) = x \sin \omega + y \cos \omega \end{aligned} \tag{9.1.1}$$

If we denote the rotation by $R(\omega)$ we have

$$\hat{R}(\omega) r = r'$$

which in matrix notation becomes

$$\mathbf{R}(\omega)\begin{bmatrix} x \\ y \end{bmatrix} = \begin{bmatrix} x' \\ y' \end{bmatrix}$$

with

$$\mathbf{R}(\omega) = \begin{bmatrix} \cos \omega & -\sin \omega \\ \sin \omega & \cos \omega \end{bmatrix} \tag{9.1.2}$$

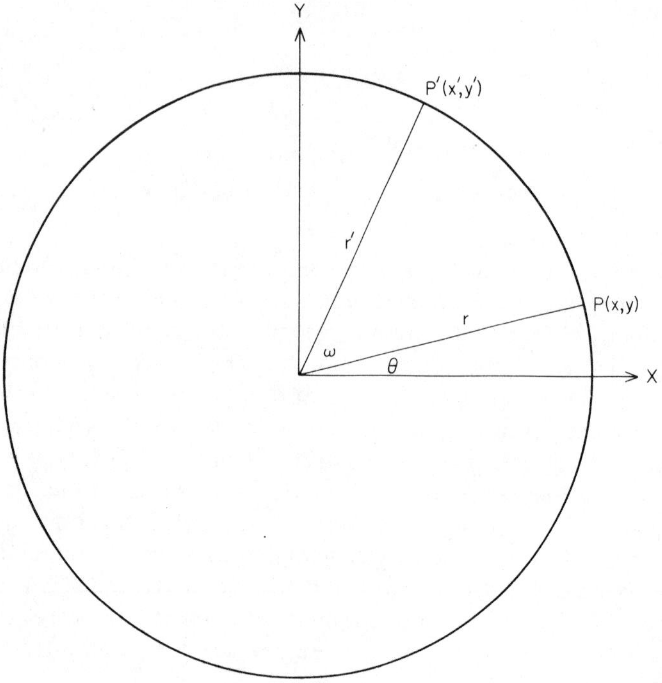

FIG. 12. Plane rotation.

It is evident that

$$\tilde{\mathbf{R}}\mathbf{R} = \mathbf{I} \qquad \det \mathbf{R} = +1$$

which is in agreement with the general definition of a rotation as a unimodular real orthogonal transformation (see Section 8.3). The set of rotations $R(\omega)$ about the Z-axis by all angles ω in the range $0 \leq \omega \leq 2\pi$ clearly constitutes a group. This is called the 2-dimensional rotation group or the axial group and is denoted by $R(2)$. The group $R(2)$ can also be considered as arising from the group $\mathscr{C}_n$ in the limit $n \to \infty$ and for this reason it is sometimes denoted by $\mathscr{C}_\infty$.

If $R(\omega_1)$ and $R(\omega_2)$ are any two group elements we have

$$R(\omega_1)R(\omega_2) = R(\omega_1 + \omega_2) = R(\omega_2)R(\omega_1) \tag{9.1.3}$$

The group $R(2)$ is thus abelian and it follows from Schur's lemma that all the IR's are 1-dimensional. The spherical harmonic $Y_{lm}(\theta, \phi)$ is clearly a basis function for a 1-dimensional representation since by (2.15.15)

$$\hat{R}(\omega)Y_{lm}(\theta, \phi) = Y_{lm}(\theta, \phi - \omega) = e^{-im\omega}Y_{lm}(\theta, \phi) \qquad (9.1.4)$$

$$m = 0, \pm 1, \pm 2, \ldots \qquad (9.1.5)$$

We denote the IR $e^{-im\omega}$ by $\Gamma^{(m)}$.

It can be shown that there are no other IR's of $R(2)$ (see Hamermesh, 1962). We note that the IR $\Gamma^{(m)}$ is unitary since $\Gamma^{\dagger(m)}\Gamma^{(m)} = e^{im\omega}e^{-im\omega} = 1$. The IR's $\Gamma^{(m)}$ and $\Gamma^{(-m)}$ are complex conjugate and it is convenient to combine them into a single 2-dimensional representation E_m (see Section 3.3) given by

$$E_m = \Gamma^{(m)} \oplus \Gamma^{(-m)}$$

The character of $R(\omega)$ in the representation E_m is

$$\chi^{(m)}(\omega) = e^{-im\omega} + e^{im\omega} = 2\cos m\omega \qquad (9.1.6)$$

Now the elements of the group manifold for $R(2)$ are labelled by a single parameter ω which varies continuously in the range $0 \leqslant \omega \leqslant 2\pi$. For this reason we say that $R(2)$ is an example of a 1-parameter Lie group of transformations.

At this point we introduce the concept of a general 1-parameter Lie group of transformations in a single variable x. In general a 1-parameter Lie group of transformations is given by the set of transformations

$$x' = f(x; a) \qquad (9.1.7)$$

in which f is an analytic function of a parameter a which varies continuously over some range of values. The set of transformations must satisfy the group postulates. Thus if we perform in succession two transformations

$$x' = f(x; a)$$

$$x'' = f(x'; b)$$

we require that the resulting transformation also belongs to the set. Thus there must exist a parameter value c for which

$$x'' = f(x; c)$$

and the group property is expressed by the condition

$$c = \phi(a; b) \qquad (9.1.8)$$

with ϕ some analytic function of a and b. Also the inverse transformation must exist which means there is a parameter value a' such that

$$x'' = f(x'; a') = f(f(x, a); a') = x$$

Finally we must have a parameter value a^0 which gives the identity transformation

$$x = f(x; a^0) = x$$

For the group $R(2)$ we have the successive rotations

$$r' = \hat{R}(\omega_1)r$$

$$r'' = \hat{R}(\omega_2)r'$$

and it is easy to see that

$$r'' = \hat{R}(\omega_1 + \omega_2)r = \hat{R}(\omega_3)r$$

so the group property (9.1.8) is given by

$$\omega_3 = \phi(\omega_1 + \omega_2) = \omega_1 + \omega_2$$

In this case the inverse is $a' = -\omega$ and the identity is $a^0 = 0$.

Returning to the general 1-parameter group the transformation $x' = f(x; a)$ with parameter value a takes x to x'. Since a is a continuous parameter the transformation with neighbouring value $a + da$ takes x to $x' + dx'$. However there also exists a transformation with infinitesimal parameter value δa close to the identity (for which we choose $a = 0$) which takes x' to $x' + dx'$. This is shown schematically in Fig. 13.
Thus we have

$$x' + dx' = f(x; a + da)$$

or

$$x' + dx' = f(x'; \delta a) \tag{9.1.9}$$

The group property (9.1.8) is expressed by the functional relation

$$a + da = \phi(a; \delta a) \tag{9.1.10}$$

For the group $R(2)$ equation (9.1.9) can be written as the infinitesimal rotation

$$\begin{aligned} x + dx &= x \cos \delta\omega - y \sin \delta\omega \\ y + dy &= x \sin \delta\omega + y \cos \delta\omega \end{aligned} \tag{9.1.11}$$

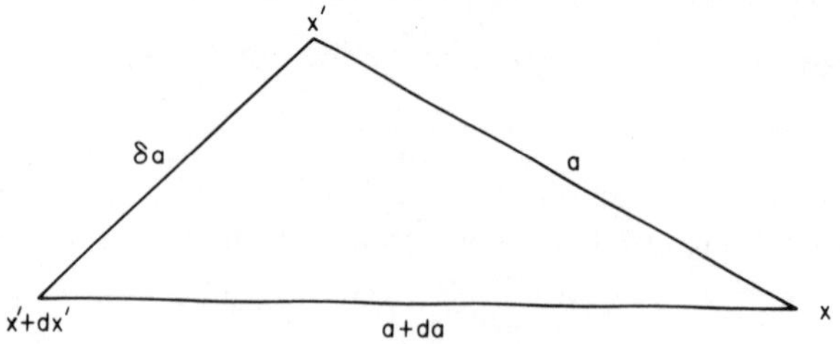

FIG. 13. Infinitesimal transformation.

Since $\delta\omega$ is infinitesimal we have in the neighbourhood of the identity

$$\begin{aligned} dx &= -y\,\delta\omega \\ dy &= x\,\delta\omega \end{aligned} \tag{9.1.12}$$

Consider next an arbitrary function $F(x, y)$. We have the differential

$$dF = \frac{\partial F}{\partial x}\,dx + \frac{\partial F}{\partial y}\,dy \tag{9.1.13}$$

Substitution of (9.1.12) in (9.1.13) leads to

$$dF = \delta\omega\left(x\frac{\partial}{\partial y} - y\frac{\partial}{\partial x}\right)F \tag{9.1.14}$$

The operator

$$\hat{X} = \left(x\frac{\partial}{\partial y} - y\frac{\partial}{\partial x}\right) \tag{9.1.15}$$

is called an *infinitesimal* operator. For the elementary functions $F = x$ and $F = y$ we have

$$\hat{X}x = -y \qquad \hat{X}y = x$$

and equations (9.1.12) become

$$\begin{aligned} dx &= \delta\omega\hat{X}x \\ dy &= \delta\omega\hat{X}y \end{aligned} \tag{9.1.16}$$

From (9.1.10) the group property for $R(2)$ is simply

$$\omega + d\omega = \phi(\omega;\, \delta\omega) = \omega + \delta\omega$$

and for this group we have the trivial result

$$d\omega = \delta\omega \tag{9.1.17}$$

From (9.1.16) we now have

$$\begin{aligned} \frac{dx}{d\omega} &= \hat{X}x = -y \\ \frac{dy}{d\omega} &= \hat{X}y = x \end{aligned} \tag{9.1.18}$$

and it follows that we can write

$$\hat{X} = \frac{d}{d\omega}$$

We now have

$$\hat{X}\hat{R}(\omega) = \frac{d\hat{R}(\omega)}{d\omega} \tag{9.1.19}$$

and since $\hat{R}(0) = \hat{1}$ it follows that

$$\hat{X} = \left[\frac{d\hat{R}(\omega)}{d\omega}\right]_{\omega=0} \tag{9.1.20}$$

and $\hat{X}$ is essentially independent of ω. We can now formally integrate (9.1.19) to obtain

$$\hat{R}(\omega) = \exp(\omega\hat{X}) \tag{9.1.21}$$

where the exponential operator is defined by

$$\exp(\omega\hat{X}) = 1 + \omega\hat{X} + \frac{\omega^2\hat{X}^2}{2!} + \frac{\omega^3\hat{X}^3}{3!} + \ldots$$

This important result establishes the link between the infinitesimal operator $\hat{X}$ and the finite rotation operator $\hat{R}$. If we can find a matrix representation $\mathbf{X}$ of $\hat{X}$ then from (9.1.21) we immediately have a matrix representation $\mathbf{R}$ of $\hat{R}$ which is given by

$$\mathbf{R} = \exp(\omega\mathbf{X}) \tag{9.1.22}$$

For a square matrix $\mathbf{A}$ the exponential matrix is defined by

$$\exp(\mathbf{A}) = \mathbf{I} + \mathbf{A} + \frac{\mathbf{A}^2}{2!} + \frac{\mathbf{A}^3}{3!} + \ldots$$

Now the 1-dimensional IR's λ of $\hat{X}$ are given by

$$\hat{X}\psi_\lambda = \lambda\psi_\lambda \tag{9.1.23}$$

where ψ_λ is a basis function for the λth IR. If we require the group representations to be unitary then

$$\mathbf{R}^\dagger\mathbf{R} = \exp(\omega\lambda^\dagger)\exp(\omega\lambda) = \exp[\omega(\lambda + \lambda^\dagger)] = 1$$

It follows that

$$\lambda + \lambda^\dagger = \lambda + \lambda^* = 0 \tag{9.1.24}$$

and λ must be pure imaginary. If we write $\lambda = -im$ then (9.1.23) becomes

$$\frac{d\psi_m}{d\omega} = -im\psi_m$$

which can be solved to give

$$\psi_m = Ae^{-im\omega}$$

where A is a constant. Since, by definition, representations are single valued it is clear that we must require

$$\psi_m(\omega) = \psi_m(\omega + 2\pi)$$

and the values of m are restricted to

$$m = 0, \pm 1, \pm 2, \ldots$$

The unitary IR's of $R(2)$ are thus given by (9.1.22) as

$$\mathbf{R}^{(m)}(\omega) = e^{-im\omega} \tag{9.1.25}$$

This is in accord with our earlier findings obtained by a method not involving the concept of infinitesimal rotation.

If we write $z = x + iy$ and $z' = x' + iy'$ then (9.1.1) becomes

$$z' = e^{i\omega} z \tag{9.1.26}$$

This is a unimodular unitary transformation in 1-dimension and it follows that the groups $R(2)$ and $SU(1)$ are isomorphic. The IR's of $SU(1)$ are given by the irreducible tensor method as $[k]$ with $k = 0, 1, 2, 3, \ldots$. For given k only one standard tableau can be constructed namely

1	1		1

k cells

and again we see that the IR's of $SU(1)$ or $R(2)$ are all 1-dimensional.

9.2 The molecular symmetry groups $\mathscr{C}_{\infty v}$ and $\mathscr{D}_{\infty h}$

Having discussed the group $R(2)$ it is appropriate now to consider the molecular symmetry groups $\mathscr{C}_{\infty v}$ and $\mathscr{D}_{\infty h}$. We have a linear molecule lying on the Z-axis. The symmetry transformations of such a molecule are of two types (a) rotations $R(\omega)$ about the Z-axis by any angle ω and (b) mirror reflections σ_v in any plane containing the Z-axis. These symmetry transformations form a group which is denoted by $\mathscr{C}_{\infty v}$. It is clear that all rotations $R(\omega)$ (apart from the identity) belong to a single class. Similarly all reflections σ_v belong to a single class.

The IR's of $\mathscr{C}_{\infty v}$ can be obtained by considering the effect of typical class elements $R(\omega)$ and $\sigma_v(xz)$ on the spherical harmonic $Y_{lm}(\theta, \phi)$. We have

$$\hat{R}(\omega) Y_{lm}(\theta, \phi) = e^{-im\omega} Y_{lm}(\theta, \phi)$$

$$\hat{\sigma}_v(xz) Y_{lm}(\theta, \phi) = Y_{lm}(\theta, -\phi) = Y_{l-m}(\theta, \phi)$$

It follows that when $m \neq 0$ the spherical harmonics (Y_{lm}, Y_{l-m}) form the basis for a 2-dimensional representation of $\mathscr{C}_{\infty v}$. This representation can easily be seen to be irreducible. We now have

$$\hat{R}(\omega)[Y_{lm} Y_{l-m}] = [Y_{lm} Y_{l-m}] \begin{bmatrix} e^{-im\omega} & 0 \\ 0 & e^{im\omega} \end{bmatrix} \tag{9.2.1}$$

$$\hat{\sigma}_v(xz)[Y_{lm} Y_{l-m}] = [Y_{lm} Y_{l-m}] \begin{bmatrix} 0 & 1 \\ 1 & 0 \end{bmatrix} \tag{9.2.2}$$

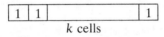

and the characters are

$$\chi^{(m)}(R(\omega)) = e^{-im\omega} + e^{im\omega} = 2\cos m\omega$$

$$\chi^{(m)}(\sigma_v) = 0$$

When $m = 0$ it is clear that Y_{10} forms a basis for the identity representation. There is however a further 1-dimensional representation. If ψ is a basis for a 1-dimensional representation then

$$\hat{\sigma}_v \psi = a\psi$$

Now $\hat{\sigma}_v^2 \psi = \hat{E}\psi = \psi = a^2\psi$ and it follows that there exists two types of basis functions. We denote them by ψ_+ and ψ_- respectively where

$$\hat{\sigma}_v \psi_+ = \psi_+ \qquad \hat{\sigma}_v \psi_- = -\psi_-$$

If $\hat{R}(\omega)$ is a rotation then $\hat{R}(\omega)\hat{\sigma}_v$ is one of the other reflections say $\hat{\sigma}_v'$. It follows that

$$\hat{R}(\omega)\psi_\pm = \psi_\pm$$

The function ψ_+ is clearly a basis function for the identity representation whilst the function ψ_- is a basis function for a new 1-dimensional representation.

It is customary to denote the IR's as follows

Value of $\lvert m \rvert$	0	1	2	3...
Notation for IR	Σ	Π	Δ	Φ

The two 1-dimensional IR's are denoted by Σ^+ and Σ^- respectively where Σ^+ is the identity. The character table is shown in Table 24. If the linear molecule also possesses an inversion centre then the appropriate symmetry group is

$$\mathscr{D}_{\infty h} = \mathscr{C}_{\infty v} \otimes \mathscr{C}_i$$

and the character table is readily obtained from Table 24. It is shown in Table 25.

TABLE 24

$\mathscr{C}_{\infty v}$	E	$R(\omega)$	σ_v
Σ^+; z, z^2	1	1	1
Σ^-	1	1	−1
Π; $(x, y)(xz, yz)$	2	$2\cos\omega$	0
Δ; (x^2-y^2, xy)	2	$2\cos 2\omega$	0

TABLE 25

$\mathcal{D}_{\infty h}$	E	$R(\omega)$	σ_v	I	$IR(\omega)$	$I\sigma_v$
Σ_g^+	1	1	1	1	1	1
Σ_u^+	1	1	1	−1	−1	−1
Σ_g^-	1	1	−1	1	1	−1
Σ_u^-	1	1	−1	−1	−1	1
Π_g	2	$2\cos\omega$	0	2	$2\cos\omega$	0
Π_u	2	$2\cos\omega$	0	−2	$-2\cos\omega$	0
Δ_g	2	$2\cos 2\omega$	0	2	$2\cos 2\omega$	0
Δ_u	2	$2\cos 2\omega$	0	−2	$-2\cos 2\omega$	0

9.3 Invariant integration

Many of the important results in the representation theory of finite groups depend upon the possibility of summation over the group manifold. Examples are (3.4.2), (3.4.3) and (5.1.7). If $\{R(1), R(2), \ldots, R(g)\}$ are the elements of a finite group $\mathcal{G}$ of order g then the group manifold consists of the set of points 1, 2, . . . , g. Let $f(R(i))$ be a function defined on the group manifold and consider the sum

$$S = \sum_{i=1}^{g} f(R(i))$$

Since $R(j)R(i) = R(k)$ is another element of the group $\mathcal{G}$ it is clear that

$$S = \sum_{i=1}^{g} f(R(i)) = \sum_{i=1}^{g} f(R(j)R(i)) \tag{9.3.1}$$

for any element $R(j)$ of $\mathcal{G}$. We say that S is *invariant* with respect to *left translation* by $R(j)$.

In order to establish, for continous groups, results analogous to (3.4.2); (3.4.3) and (5.1.7) we must evidently investigate the possibility of integration over the continuous group manifold. Such an integration process is required to be invariant in the sense of (9.3.1) and we speak of *invariant* integration (see Hamermesh, 1962). In this book we shall consider only the very simplest case of invariant integration. This occurs when we are dealing with axial groups such as $R(2)$, $\mathcal{C}_{\infty v}$ and $\mathcal{D}_{\infty h}$. The group manifold consists of the set of all points ω for which $0 \leqslant \omega \leqslant 2\pi$ and it can be shown that the required volume element in the group manifold is simply $d\omega$ (see Hamermesh, 1962).

For $R(2)$ the analogue of (3.4.3) is

$$\int_0^{2\pi} \chi^{(\mu)}(\omega)\chi^{(\nu)*}(\omega)\, d\omega = \int_0^{2\pi} e^{i(\mu-\nu)\omega}\, d\omega = 2\pi\, \delta_{\mu\nu} \tag{9.3.2}$$

and from this we see that the order g of a finite group has been replaced by the "*volume*" V of a continous group.

$$V = \int_0^{2\pi} d\omega = 2\pi$$

The analogue of (3.4.7) and (3.4.8) are clearly given by

$$a_\mu = \frac{1}{2\pi}\int_0^{2\pi} \chi^{*(\mu)}(\omega)\chi(\omega)\, d\omega \tag{9.3.3}$$

and

$$\int_0^{2\pi} |\chi(\omega)|^2\, d\omega = 2\pi \tag{9.3.4}$$

respectively for the group $R(2)$.

For the group $\mathscr{C}_{\infty v}$ we have in addition to the rotations $R(\omega)$ the mirror reflections $\sigma_v(\omega)$. Here $\sigma_v(\omega)$ is the mirror reflection in a plane passing through the Z-axis and set at an angle ω from the xz plane. Clearly every point ω in the group manifold gives rise to two distinct group elements $R(\omega)$ and $\sigma_v(\omega)$. Thus we take the volume of the group $\mathscr{C}_{\infty v}$ to be 4π. Similarly the volume of the group $\mathscr{D}_{\infty h}$ is taken to be 8π.

If we consider the group $\mathscr{D}_{\infty h}$ then from (9.2.1) we find the following matrix representatives in the IR's Π_g and Π_u.

$$\begin{aligned}
\mathbf{R}^{(\Pi_u)} = -\mathbf{S}^{(\Pi_u)} = \mathbf{R}^{(\Pi_g)} = \mathbf{S}^{(\Pi_g)} &= \begin{bmatrix} e^{-i\omega} & 0 \\ 0 & e^{i\omega} \end{bmatrix} \\
\boldsymbol{\sigma}_v^{(\Pi_u)} = -\boldsymbol{\tau}^{(\Pi_u)} = \boldsymbol{\sigma}_v^{(\Pi_g)} = \boldsymbol{\tau}^{(\Pi_g)} &= \begin{bmatrix} 0 & e^{i\omega} \\ e^{-i\omega} & 0 \end{bmatrix}
\end{aligned} \tag{9.3.5}$$

In (9.3.5) we have used the shorthand notation

$$\hat{S} = \hat{I}\hat{R} \qquad \hat{\tau} = \hat{I}\hat{\sigma}_v$$

Symmetry species for the first few IR's of $\mathscr{D}_{\infty h}$ can be obtained by using Table 25 together with (9.3.5) and projection operators analogous to (5.1.7). Thus if ϕ is a function of arbitrary symmetry we may construct the following symmetry species $\chi_\gamma^{(\Gamma)}$.

$$\chi^{(\Sigma_g^+)} = \frac{1}{8\pi}\int_0^{2\pi} (\hat{R} + \hat{\sigma}_v + \hat{S} + \hat{\tau})\phi\, d\omega \tag{9.3.6}$$

$$\chi^{(\Sigma_g^-)} = \frac{1}{8\pi}\int_0^{2\pi} (\hat{R} - \hat{\sigma}_v + \hat{S} - \hat{\tau})\phi\, d\omega \tag{9.3.7}$$

$$\chi^{(\Sigma_u^+)} = \frac{1}{8\pi}\int_0^{2\pi} (\hat{R} + \hat{\sigma}_v - \hat{S} - \hat{\tau})\phi\, d\omega \tag{9.3.8}$$

$$\chi^{(\Sigma_u^-)} = \frac{1}{8\pi}\int_0^{2\pi} (\hat{R} - \hat{\sigma}_v - \hat{S} + \hat{\tau})\phi\, d\omega \tag{9.3.9}$$

$$\chi_x^{(\Pi_u)} = \frac{1}{4\pi}\int_0^{2\pi} e^{i\omega}(\hat{R} - \hat{S})\phi \, d\omega \tag{9.3.10}$$

$$\chi_x^{(\Pi_g)} = \frac{1}{4\pi}\int_0^{2\pi} e^{i\omega}(\hat{R} + \hat{S})\phi \, d\omega \tag{9.3.11}$$

As an example we consider the linear triatomic molecule AB_2

$$\underset{2}{B}—\underset{1}{A}—\underset{3}{B}$$

Let h_i denote an s-type atomic orbital centered on atom i. In the basis (h_1, h_2, h_3) we find a representation Γ of $\mathscr{D}_{\infty h}$ given by

$$\mathbf{R} = \begin{bmatrix} 1 & 0 & 0 \\ 0 & 1 & 0 \\ 0 & 0 & 1 \end{bmatrix} \qquad \boldsymbol{\sigma} = \begin{bmatrix} 1 & 0 & 0 \\ 0 & 1 & 0 \\ 0 & 0 & 1 \end{bmatrix}$$

$$\mathbf{S} = \begin{bmatrix} 1 & 0 & 0 \\ 0 & 0 & 1 \\ 0 & 1 & 0 \end{bmatrix} \qquad \boldsymbol{\tau} = \begin{bmatrix} 1 & 0 & 0 \\ 0 & 0 & 1 \\ 0 & 1 & 0 \end{bmatrix}$$

and we see that

$$\Gamma = 2\Sigma_g^+ \oplus \Sigma_u^+$$

One of the σ_g symmetry orbitals is simply h_1 whilst the other is given by (9.3.6) as

$$\chi^{(\sigma_g^+)} = \frac{1}{8\pi}\int_0^{2\pi} (\hat{R} + \hat{\sigma}_v + \hat{S} + \hat{\tau})h_2 \, d\omega = \tfrac{1}{2}(h_2 + h_3)$$

The σ_u^+ symmetry orbital is given by (9.3.8) as

$$\chi^{(\sigma_u^+)} = \tfrac{1}{2}(h_2 - h_3)$$

Next let X_i, Y_i and Z_i denote respectively p_x, p_y and p_z type atomic orbitals centered on atom i.

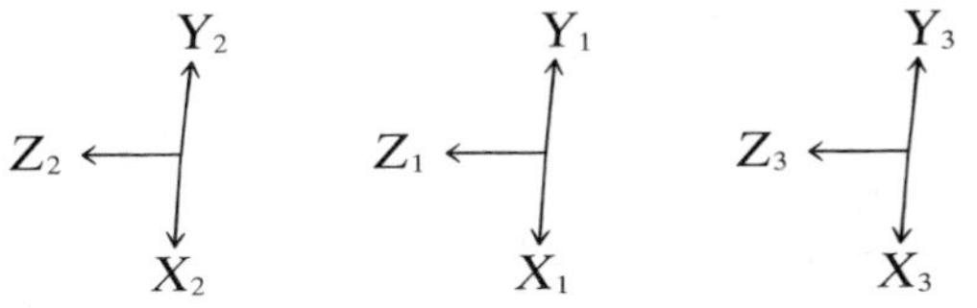

In the basis (Z_1, Z_2, Z_3) we find a representation Γ which can be decomposed into $2\Sigma_u^+ \oplus \Sigma_g^+$. One of the σ_u^+ symmetry orbitals is Z_1 whilst the other is $\frac{1}{2}(Z_2 + Z_3)$. The σ_g^+ symmetry orbital is $\frac{1}{2}(Z_2 - Z_3)$. The functions (X_1, Y_1) form a basis for Π_u. Finally the basis (X_2, X_3, Y_2, Y_3) gives a representation of $\mathscr{D}_{\infty h}$ which decomposes into $\Pi_u \oplus \Pi_g$. The x component of the Π_u symmetry orbital is given by (9.3.10) as

$$\chi_x^{(\Pi_u)} = \frac{1}{4\pi}\int_0^{2\pi} e^{i\omega}(\hat{R}-\hat{S})X_2\, d\omega$$

Now

$$\hat{R}X_2 = X_2 \cos\omega - Y_2 \sin\omega \qquad \hat{I}X_2 = -X_3$$

$$\cos\omega = \text{Re}\, e^{-i\omega} \qquad -\sin\omega = \text{Im}\, e^{-i\omega}$$

and we thus find

$$\chi_x^{(\Pi_u)} = \tfrac{1}{2}(X_2 + X_3)$$

In a similar way we find

$$\chi_x^{(\Pi_g)} = \tfrac{1}{2}(X_2 - X_3)$$

9.4 The rotation group *R*(3)

If we ignore improper rotations then the symmetry group of an isolated atom is the 3-dimensional rotation group $R(3)$. This group consists of all rotations $R(\omega, n)$ by an angle ω about all axes n where n is a unit vector passing through the origin (nucleus). The vector n specifies the direction of the axis. The rotations can of course also be defined as unimodular real orthogonal transformations given by

$$\begin{aligned} x' &= a_{11}x + a_{12}y + a_{13}z \\ y' &= a_{21}x + a_{22}y + a_{23}z \\ z' &= a_{31}x + a_{32}y + a_{33}z \end{aligned} \tag{9.4.1}$$

with $\tilde{\mathbf{A}}\mathbf{A} = \mathbf{I}$; $\det \mathbf{A} = +1$ where $\mathbf{A} = [a_{ij}]$. Now a rotation must leave $x^2 + y^2 + z^2$ invariant. This invariance condition imposes six conditions on the nine real parameters a_{ij}. Thus $R(3)$ is a three-parameter Lie group.

It is not difficult to show that rotations about all axes n through the same angle ω belong to a single class (see Judd, 1963). In Section 8.3 we saw how the IR's of $R(3)$ can be labelled by a symbol (l) with

$$l = 0, 1, 2, \ldots$$

Using the irreducible tensor method of the last chapter it can easily be shown that the dimension of the IR (l) is $(2l+1)$. Now we know that for fixed l the spherical harmonics $Y_{lm}(\theta, \phi)$ span a $(2l+1)$-dimensional space. It is thus not surprising to find that the spherical harmonics Y_{lm} form a basis for the $(2l+1)$-dimensional IR (l) of $R(3)$. Indeed it can be shown that all the IR's of $R(3)$ are given by taking the various Y_{lm} (of given l) as representation spaces (see Hamermesh, 1962).

If we consider the particular class element $R(\omega)$ where $R(\omega)$ is a rotation about the Z-axis then

$$\hat{R}(\omega)Y_{lm}(\theta, \phi) = e^{-im\omega}Y_{lm}(\theta, \phi) \qquad m = l, l-1, \ldots, -l$$

It follows that the character of $\hat{R}(\omega, n)$ in the IR (l) is given by

$$\chi^{(l)}(\omega) = \sum_{m=l}^{-l} e^{-im\omega} = \frac{\sin(l+\frac{1}{2})\omega}{\sin\frac{1}{2}\omega} \tag{9.4.2}$$

The representation matrices can be given in terms of the Euler angles of a rotation but we shall not consider this here (see Hamermesh, 1962).

We now consider $R(3)$ from the standpoint of Lie groups. The infinitesimal rotations in the neighbourhood of the identity are given from (9.4.1) by

$$\begin{aligned} x + dx &= (1+\alpha_{11})x + \alpha_{12}y + \alpha_{13}z \\ y + dy &= \alpha_{21}x + (1+\alpha_{22})y + \alpha_{23}z \\ z + dz &= \alpha_{31}x + \alpha_{32}y + (1+\alpha_{33})z \end{aligned} \tag{9.4.3}$$

where the matrix elements α_{ij} are infinitesimal (compare with (9.1.11)). If we write $\mathbf{A} = \mathbf{I} + \mathbf{B}$ with $\mathbf{B} = [\alpha_{ij}]$ then

$$\tilde{\mathbf{A}}\mathbf{A} = \mathbf{I} + \mathbf{B} + \tilde{\mathbf{B}}$$

to first order. Since $\tilde{\mathbf{A}}\mathbf{A} = \mathbf{I}$ we have

$$\mathbf{B} + \tilde{\mathbf{B}} = 0$$

and thus $\alpha_{ij} = -\alpha_{ji}$ with $\alpha_{ii} = 0$. If we write

$$\zeta = \alpha_{12} \qquad \eta = -\alpha_{13} \qquad \xi = \alpha_{23}$$

then from (9.4.3) we find

$$\begin{aligned} dx &= \zeta y - \eta z \\ dy &= -\zeta x + \xi z \\ dz &= \eta x - \xi y \end{aligned} \tag{9.4.4}$$

For a function $F(x, y, z)$ we have

$$dF = \frac{\partial F}{\partial x}dx + \frac{\partial F}{\partial y}dy + \frac{\partial F}{\partial z}dz$$

Substitution of (9.4.4) yields

$$dF = \left\{\xi\left(z\frac{\partial}{\partial y} - y\frac{\partial}{\partial z}\right) + \eta\left(x\frac{\partial}{\partial z} - z\frac{\partial}{\partial x}\right) + \zeta\left(y\frac{\partial}{\partial x} - x\frac{\partial}{\partial y}\right)\right\}F \tag{9.4.5}$$

The three linearly independent operators

$$\hat{X}_1 = \left(z\frac{\partial}{\partial y} - y\frac{\partial}{\partial z}\right) \qquad \hat{X}_2 = \left(x\frac{\partial}{\partial z} - z\frac{\partial}{\partial x}\right) \qquad \hat{X}_3 = \left(y\frac{\partial}{\partial x} - x\frac{\partial}{\partial y}\right) \tag{9.4.6}$$

are called the infinitesimal operators for the Lie group $R(3)$. In this book we shall refer to these infinitesimal operators (IO's) as the *natural* IO's since they arise directly from the infinitesimal transformations. Later we shall take specific linear combinations of the natural IO's and these linear combinations are then called *standard* IO's. An important connection with quantum mechanics is obtained when we notice that

$$\hat{X}_1 = -i\hat{l}_x \qquad \hat{X}_2 = -i\hat{l}_y \qquad \hat{X}_3 = -i\hat{l}_z \tag{9.4.7}$$

where $\hat{l}_x$, $\hat{l}_y$ and $\hat{l}_z$ are the operators corresponding to the components of orbital angular momentum.

The infinitesimal operators (9.4.6) satisfy the commutation relations

$$[\hat{X}_1, \hat{X}_2] = \hat{X}_3 \qquad [\hat{X}_2, \hat{X}_3] = \hat{X}_1 \qquad [\hat{X}_3, \hat{X}_1] = \hat{X}_2 \tag{9.4.8}$$

where as usual if $\hat{A}$ and $\hat{B}$ are two operators then the commutator is defined by $[\hat{A}, \hat{B}] \equiv \hat{A}\hat{B} - \hat{B}\hat{A}$.

We see that there is associated with this 3-parameter Lie group of rotations a set of three linearly independent infinitesimal operators (IO) $\hat{X}_1$, $\hat{X}_2$, $\hat{X}_3$. These three IO's can be regarded as forming a basis for a real three-dimensional linear space $\mathscr{L}$. We shall refer to this basis as the *natural* basis for $\mathscr{L}$. Now equations (9.4.8) can be written concisely as

$$[\hat{X}_\rho, \hat{X}_\sigma] = \sum_{\tau=1}^{3} c_{\rho\sigma}^{\tau} X_\tau \tag{9.4.9}$$

for all choices ρ, σ, τ in the set 1, 2, 3. The coefficients $c_{\rho\sigma}^{\tau}$ in (9.4.9) are called *structure constants*. The only non-vanishing structure constants for $R(3)$ are clearly

$$c_{12}^{3} = c_{23}^{1} = c_{31}^{2} = 1 \tag{9.4.10}$$

The binary relation of commutation defines a product in the linear space $\mathscr{L}$. Furthermore we see from (9.4.8) that $\mathscr{L}$ is closed with respect to this binary operation. According to Section 2.14 the linear space $\mathscr{L}$ is now a 3-dimensional algebra. This algebra is called the *Lie algebra* of the group $R(3)$ and it is denoted by B_1 (see Rowlatt, 1966).

We introduce next the shift operators

$$\hat{l}_\pm = \hat{l}_x \pm i\hat{l}_y \tag{9.4.11}$$

which also play an important role in the theory of angular momentum. They satisfy the commutation relations (CR)

$$\begin{aligned} [\hat{l}_z, \hat{l}_\pm] &= \pm\hat{l}_\pm \\ [\hat{l}_+, \hat{l}_-] &= 2\hat{l}_z \end{aligned} \tag{9.4.12}$$

These CR's can be made more symmetrical by writing

$$\hat{H} = \frac{1}{\sqrt{2}}\,\hat{l}_z \qquad \hat{E}_\pm = \tfrac{1}{2}\hat{l}_\pm \tag{9.4.13}$$

We now have

$$\begin{aligned}
[\hat{H}, \hat{E}_\pm] &= \pm\frac{1}{\sqrt{2}}\,\hat{E}_\pm \\
[\hat{E}_+, \hat{E}_-] &= \frac{1}{\sqrt{2}}\,\hat{H}
\end{aligned} \tag{9.4.14}$$

In the language of the general theory of Lie algebras (see (A1.4)) we say that $\hat{E}_+$, $\hat{E}_-$, $\hat{H}$ constitute a *standard* basis for B_1. The CR's (9.4.14) are then called *canonical* CR's.

Another operator which plays an important role in the theory of angular momentum is $\hat{l}^2$. This is given by

$$\hat{l}^2 = \hat{l}_x^2 + \hat{l}_y^2 + \hat{l}_z^2 = 2\hat{C} \tag{9.4.15}$$

with

$$\hat{C} = \hat{H}^2 + \hat{E}_+\hat{E}_- + \hat{E}_-\hat{E}_+ \tag{9.4.16}$$

From the general theory of Lie algebras (see (A1.10.8)) we call $\hat{C}$ the *Casimir operator* for B_1. $\hat{C}$ commutes with all the IO's of B_1. Since the operators $\hat{l}_z$ and $\hat{l}^2$ commute it follows that there exists states $|l, m\rangle$ which are simultaneously eigenstates of both operators. It is well known from the theory of angular momentum that the eigenvalues of these operators are given by

$$\hat{l}^2\,|l, m\rangle = l(l+1)\,|l, m\rangle \tag{9.4.17}$$

$$\hat{l}_z\,|l, m\rangle = m\,|l, m\rangle \tag{9.4.18}$$

with

$$l = 0, 1, 2, \ldots \tag{9.4.19}$$

$$m = l, l-1, \ldots, -l+1, -l \tag{9.4.20}$$

The eigenfunctions which represent the eigenstates $|l, m\rangle$ are just the spherical harmonics $Y_{lm}(\theta, \phi)$.

Another familiar result from the theory of angular momentum is expressed by the equation

$$\hat{l}_\pm\,|l, m\rangle = \sqrt{l(l+1) - m(m\pm 1)}\,|l, m\pm 1\rangle \tag{9.4.21}$$

The significance of these results, at least for our present purpose, is that the states $|l, m\rangle$ for given l form the basis for a $(2l+1)$-dimensional matrix representation of the IO's $\hat{H}$, $\hat{E}_\pm$. This representation which we denote by

(l) is given by

$$\sqrt{2}\hat{H}[|l, l\rangle\, |l, l-1\rangle \ldots |l, -l\rangle]$$

$$= [|l, l\rangle\, |l, l-1\rangle \ldots |l, -l\rangle] \begin{bmatrix} l & & & & & \\ & l-1 & & & & \\ & & \cdot & & & \\ & & & \cdot & & \\ & & & & -l+1 & \\ & & & & & -l \end{bmatrix} \tag{9.4.22}$$

$$\langle l, m \pm 1|\, \hat{E}_{\pm}\, |l, m\rangle = \tfrac{1}{2}\sqrt{l(l+1) - m(m \pm 1)} \tag{9.4.23}$$

We note that the matrix **H** which represents $\hat{H}$ is diagonal

$$\mathbf{H} = \operatorname{diag}\left[\frac{l}{\sqrt{2}}, \frac{l-1}{\sqrt{2}}, \ldots, \frac{-l+1}{\sqrt{2}}, \frac{-l}{\sqrt{2}}\right] \tag{9.4.24}$$

Using Schur's Lemma (see (2.5.1)) it can be shown that the representation (l) is irreducible.

Consider next rotations $R(\omega)$ about the Z-axis. These are plane rotations and are given by (see (9.1.1))

$$\begin{aligned} x' &= x \cos \omega - y \sin \omega \\ y' &= x \sin \omega + y \cos \omega \\ z' &= z \end{aligned} \tag{9.4.25}$$

According to (9.1.21) we have

$$\hat{R}(\omega) = \exp(\omega\hat{X}) = \exp(-\omega\hat{X}_3) = \exp(i\omega\hat{l}_z) = \exp(i\omega\sqrt{2}\,\hat{H}) \tag{9.4.26}$$

The generalization of this result to the case of a rotation through angle ω about an axis in the direction of a unit vector n is

$$\hat{R}(\omega, n) = \exp(-\omega\;\, n\,.\,\hat{X}) \tag{9.4.27}$$

where

$$\hat{X} = (\hat{X}_1, \hat{X}_2, \hat{X}_3)$$

The equation (9.4.27) gives the important connection between a finite rotation ω about axis n and the IO's. If

$$n = (n_x, n_y, n_z)$$

then we can write (9.4.27) in the forms

$$\begin{aligned} \hat{R}(\omega, n) &= \exp\{i\omega(n_x\hat{l}_x + n_y\hat{l}_y + n_z\hat{l}_z)\} \\ &= \exp\{i\omega[\hat{E}_+(n_x - in_y) + \hat{E}_-(n_x + in_y) + \hat{H}(\sqrt{2}\, n_z)]\} \end{aligned} \tag{9.4.28}$$

If $\mathbf{E}_{\pm}$, $\mathbf{H}$ constitutes a matrix representation of the Lie algebra B_1 then (9.4.28) gives a matrix representation of the Lie group $R(3)$. (Possibly only in a neighbourhood of the identity but we shall not go into topological details here.)

Before proceeding we consider the properties of the exponential matrix. If $\mathbf{A}$ is an $n\times n$ matrix and we define exp $\mathbf{A}$ by

$$\exp \mathbf{A}=\mathbf{I}+\mathbf{A}+\frac{\mathbf{A}^2}{2!}+\frac{\mathbf{A}^3}{3!}+\ldots \tag{9.4.29}$$

then exp $\mathbf{A}$ has the following properties (see Boerner, 1963)

(a) the series (9.4.29) is convergent
(b) $\exp \mathbf{A} . \exp \mathbf{B}=\exp(\mathbf{A}+\mathbf{B})$ only if $[\mathbf{A}, \mathbf{B}]=0$
(c) $\exp(\mathbf{B}\mathbf{A}\mathbf{B}^{-1})=\mathbf{B}(\exp \mathbf{A})\mathbf{B}^{-1}$
(d) if $\lambda_1, \lambda_2, \ldots, \lambda_n$ are the eigenvalues of $\mathbf{A}$ then $e^{\lambda_1}, e^{\lambda_2}, \ldots e^{\lambda_n}$ are the eigenvalues of exp $\mathbf{A}$.
(e) $\det(\exp \mathbf{A})=e^{\operatorname{tr}\mathbf{A}}$

These properties can be used in conjunction with Schur's Lemma to show that if $\mathbf{A}_i$ is an irreducible system of matrices then exp $\mathbf{A}_i$ is also an irreducible system of matrices. It is now clear that (9.4.22), (9.4.23) and (9.4.28) give the IR's of $R(3)$. Using the method of Lie algebras we have established that the IR's of $R(3)$ can be labelled by a symbol (l) with $l=0, 1, 2, \ldots$ and that the dimension of the IR (l) is $(2l+1)$.

The character $\chi^{(l)}(\omega)$ of $\hat{R}(\omega, n)$ in the IR (l) is obtained by substitution of (9.4.24) in (9.4.26). Thus

$$\begin{aligned} R^{(l)}(\omega) &= \exp(i\omega\sqrt{2}\,H) \\ &= \exp(i\omega \operatorname{diag}[l, l-1, \ldots, -l+1, -l]) \end{aligned}$$

and

$$\chi^{(l)}(\omega)=e^{i\omega l}+e^{i\omega(l-1)}+\ldots+e^{-i\omega l}=\frac{\sin(l+\frac{1}{2})\omega}{\sin\frac{1}{2}\omega} \tag{9.4.30}$$

This result is in agreement with (9.4.2).

It is customary to introduce the notation of Table 26 for the IR's of $R(3)$. We can use (9.4.2) to obtain the decomposition of the direct product

TABLE 26

IR	(0)	(1)	(2)	(3)	(4)	(5)	(6)	(7)	(8)	...
Notation	S	P	D	F	G	H	I	K	L	...

$(l_1) \otimes (l_2)$ where (l_1) and (l_2) are IR's of $R(3)$. Thus

$$\begin{aligned}\chi^{(l_1)}(\omega)\chi^{(l_2)}(\omega) &= (e^{il_1\omega}+e^{i(l_1-1)\omega}+\ldots+e^{-il_1\omega})\\ &\quad\times(e^{il_2\omega}+e^{i(l_2-1)\omega}+\ldots+e^{-il_2\omega})\\ &= (e^{i(l_1+l_2)\omega}+e^{i(l_1+l_2-1)\omega}+\ldots+e^{-i(l_1+l_2)\omega})\\ &\quad+(e^{i(l_1+l_2-1)\omega}+e^{i(l_1+l_2-2)\omega}+\ldots+e^{-i(l_1+l_2-1)\omega})\\ &\quad+\ldots+(e^{i|l_1-l_2|\omega}+e^{i(|l_1-l_2-1|)\omega}+\ldots+e^{-i|l_1-l_2|\omega})\\ &= \chi^{(l_1+l_2)}(\omega)+\chi^{(l_1+l_2-1)}(\omega)+\ldots+\chi^{(|l_1-l_2|)}(\omega)\end{aligned}$$

and it follows that

$$(l_1)\otimes(l_2) = (l_1+l_2)\oplus(l_1+l_2-1)\oplus\ldots\oplus(|l_1-l_2|) \qquad (9.4.31)$$

This result is known as the Clebsch-Gordan theorem. For orbital angular momentum it is the group theoretical equivalent of (7.1.1). We also note the similarity of (9.4.31) and (8.6.1).

When improper rotations are admitted the symmetry group for a free atom (which has a centre of symmetry) becomes the rotation-reflection group $R^+(3) = R(3) \otimes \mathscr{C}_i$. It follows that $R^+(3)$ has twice as many IR's as $R(3)$. The IR's of $R^+(3)$ are denoted by S_g, S_u, P_g, P_u, D_g, D_u, etc.

We have seen that the spherical harmonics Y_{lm} of given l form a basis for the $(2l+1)$-dimensional IR (l) of $R(3)$. Now within the central field approximation the electrons in an atom are described by atomic orbitals of the form

$$\psi(r, \theta, \phi) = R(r)Y_{lm}(\theta, \phi)$$

where $R(r)$ is some radial function. It follows that the atomic orbitals form bases for the IR's of $R(3)$. Thus s orbitals span the IR (0), p orbitals the IR (1), d orbitals the IR (2) etc. When electron spin is admitted we have the two spin functions α and β spanning 2-dimensional spin space. Since there is no 2-dimensional IR of $R(3)$ the spin functions do not form a basis for any IR of $R(3)$. To get from one orthogonal basis in 2-dimensional spin space to another requires a unitary transformation. We are therefore led to consider the unitary unimodular group $SU(2)$.

9.5 The group $SU(2)$

This group consists of all unitary unimodular transformations

$$\begin{aligned}u' &= au+bv\\ v' &= -b^*u+a^*v\end{aligned} \qquad (9.5.1)$$

$$|a|^2+|b|^2 = 1 \qquad (9.5.2)$$

The condition (9.5.2) imposes one condition on the four real parameters

Re a, Im a, Re b, Im b and therefore $SU(2)$ is a three parameter Lie group. The infinitesimal transformations are

$$u + du = (1+\alpha)u + \beta v$$
$$v + dv = -\beta^* u + (1+\alpha^*)v$$

where α and β are infinitesimal. Let

$$\alpha = p + iq \qquad \beta = r + is$$

The condition (9.5.2) requires that $p = 0$ since p, q, r, s are infinitesimal. We now have

$$du = iqu + (r+is)v$$
$$dv = (-r+is)u - iqv$$

If $F(u, v)$ is some function of u, v we find

$$dF = \left\{iq\left(u\frac{\partial}{\partial u} - v\frac{\partial}{\partial v}\right) + r\left(v\frac{\partial}{\partial u} - u\frac{\partial}{\partial v}\right) + is\left(v\frac{\partial}{\partial u} + u\frac{\partial}{\partial v}\right)\right\}F$$

and thus the natural IO's for the Lie algebra of $SU(2)$ are

$$i\left(u\frac{\partial}{\partial u} - v\frac{\partial}{\partial v}\right) = \hat{I}_3$$
$$\left(v\frac{\partial}{\partial u} - u\frac{\partial}{\partial v}\right) = \hat{I}_1 \qquad (9.5.3)$$
$$i\left(v\frac{\partial}{\partial u} + u\frac{\partial}{\partial v}\right) = \hat{I}_2$$

This Lie algebra is denoted by A_1 (see Rowlatt, 1966). If we write

$$\hat{X}_1 = \tfrac{1}{2}\hat{I}_1 \qquad \hat{X}_2 = \tfrac{1}{2}\hat{I}_2 \qquad \hat{X}_3 = -\tfrac{1}{2}\hat{I}_3 \qquad (9.5.4)$$

the CR's are found to be

$$[\hat{X}_1, \hat{X}_2] = \hat{X}_3 \qquad [\hat{X}_2, \hat{X}_3] = \hat{X}_1 \qquad [\hat{X}_3, \hat{X}_1] = \hat{X}_2$$

These CR's are identical to (9.4.8) and it follows that the Lie algebra A_1 of $SU(2)$ is isomorphic with the Lie algebra B_1 of $R(3)$.

Although the Lie algebras A_1 and B_1 are isomorphic the Lie groups $SU(2)$ and $R(3)$ are not. Consider the subgroup $\mathcal{H}$ of $SU(2)$ for which in (9.5.1) we set $a = e^{i\omega/2}$, $b = 0$. We have

$$u' = e^{i\omega/2}u \qquad (9.5.5)$$
$$v' = e^{-i\omega/2}v$$

or

$$\begin{bmatrix} u' \\ v' \end{bmatrix} = \mathbf{U}(\omega)\begin{bmatrix} u \\ v \end{bmatrix} \qquad \text{with} \qquad \mathbf{U}(\omega) = \begin{bmatrix} e^{i\omega/2} & 0 \\ 0 & e^{-i\omega/2} \end{bmatrix}$$

Consider next a mapping of the (u, v) plane onto the (x, y) plane which is defined by

$$x = \tfrac{1}{2}(u^2 - v^2) \qquad y = \frac{1}{2i}(u^2 + v^2) \tag{9.5.6}$$

The unitary transformation (9.5.5) becomes

$$\begin{aligned} x' &= x \cos \omega - y \sin \omega \\ y' &= x \sin \omega + y \cos \omega \end{aligned} \tag{9.5.7}$$

or

$$\begin{bmatrix} x' \\ y' \end{bmatrix} = \mathbf{R}(\omega) \begin{bmatrix} x \\ y \end{bmatrix} \qquad \text{with} \qquad \mathbf{R}(\omega) = \begin{bmatrix} \cos \omega & -\sin \omega \\ \sin \omega & \cos \omega \end{bmatrix}$$

This is a rotation about the Z-axis by angle ω (see (9.1.1)). We have thus produced a mapping given by (9.5.6) of a subgroup $\mathscr{H}$ of $SU(2)$ onto the subgroup $R(2)$ of $R(3)$. Now it is clear that the two distinct unitary matrices

$$\mathbf{U}(0) = \begin{bmatrix} 1 & 0 \\ 0 & 1 \end{bmatrix} \qquad \text{and} \qquad \mathbf{U}(2\pi) = \begin{bmatrix} -1 & 0 \\ 0 & -1 \end{bmatrix} = -\mathbf{U}(0)$$

are mapped onto the unit element $\mathbf{R}(0)$ in $R(2)$. The groups $\mathscr{H}$ and $R(2)$ are thus not isomorphic but are homomorphic. Each element $\mathbf{R}(\omega)$ in $R(2)$ arises from the two elements $\pm\mathbf{U}(\omega)$ in $\mathscr{H}$.

By using the Euler angles of a 3-dimensional rotation the above argument can be generalized and it is found that the groups $SU(2)$ and $R(3)$ are homomorphic (see Hamermesh, 1962; Griffith, 1964). In this homomorphism each element $R(\omega, n)$ in $R(3)$ arises from two distinct elements $\pm A$ in $SU(2)$. In particular the unit matrix in $R(3)$ arises from the two matrices

$$\mathbf{I} = \begin{bmatrix} 1 & 0 \\ 0 & 1 \end{bmatrix} \qquad \mathbf{Q} = \begin{bmatrix} -1 & 0 \\ 0 & -1 \end{bmatrix}$$

in $SU(2)$. The matrices $\mathbf{R}(\omega, X)$, $\mathbf{R}(\omega, Y)$, $\mathbf{R}(\omega, Z)$ for rotation by angle ω about the X, Y, Z axes respectively can be shown to arise from pairs of matrices in $SU(2)$. These are given in (9.5.8) (see Griffith, 1964; Hamermesh, 1962).

$$\begin{aligned} &\pm\begin{bmatrix} \cos \omega/2 & i \sin \omega/2 \\ i \sin \omega/2 & \cos \omega/2 \end{bmatrix} \to \mathbf{R}(\omega, X) \\ &\pm\begin{bmatrix} \cos \omega/2 & \sin \omega/2 \\ -\sin \omega/2 & \cos \omega/2 \end{bmatrix} \to \mathbf{R}(\omega, Y) \\ &\pm\begin{bmatrix} e^{i\omega/2} & 0 \\ 0 & e^{-i\omega/2} \end{bmatrix} \to \mathbf{R}(\omega, Z) \end{aligned} \tag{9.5.8}$$

For applications in quantum chemistry it is convenient to construct from $R(2)$ a group which is isomorphic with the subgroup $\mathscr{H}$ of $SU(2)$. This group is called the axial *double* group and is denoted by $R^*(2)$. The group $R^*(2)$ has twice as many elements as $R(2)$ but it is not a rotation group in the usual sense. To obtain $R^*(2)$ from $R(2)$ we make the formal requirement that a plane vector is brought back to its original position only after a rotation about the Z-axis by 4π. The rotation by 2π is taken as a new group element Q with matrix

$$\mathbf{Q} = \begin{bmatrix} -1 & 0 \\ 0 & -1 \end{bmatrix}$$

The elements of $R^*(2)$ are then given by $R(\omega)$ and $QR(\omega)$ where $R(\omega)$ is an element of $R(2)$. The isomorphism of $R^*(2)$ and $\mathscr{H}$ is exhibited in the form

$$\begin{aligned} \mathbf{U}(\omega) &\rightarrow \mathbf{R}(\omega) \\ -\mathbf{U}(\omega) &\rightarrow \mathbf{QR}(\omega) \end{aligned} \tag{9.5.9}$$

In an entirely analogous fashion we can construct the rotation double group $R^*(3)$ which is isomorphic with $SU(2)$. Each element $R(\omega, n)$ in $R(3)$ gives rise to two elements $R(\omega, n)$ and $QR(\omega, n)$ in $R^*(3)$ where Q is a rotation about the n axis by 2π.

We now consider the determination of the IR's for the group $SU(2)$. It is clear that the functions $\phi(10) = u$ and $\phi(01) = v$ form the basis for a 2-dimensional unitary IR of $SU(2)$. Thus we have the self-representation given by

$$[\phi'(10)\phi'(01)] = [\phi(10)\phi(01)]\begin{bmatrix} a & -b^* \\ b & a^* \end{bmatrix}$$

Consider next the three functions

$$\phi(20) = u^2 \qquad \phi(11) = uv \qquad \phi(02) = v^2$$

Under the transformation (9.5.1) we readily find

$$[\phi'(20)\phi'(11)\phi'(02)] = [\phi(20)\phi(11)\phi(02)]\begin{bmatrix} a^2 & -ab^* & b^{*2} \\ 2ab & aa^* - bb^* & -2a^*b^* \\ b^2 & a^*b & a^{*2} \end{bmatrix}$$

and we have produced a 3-dimensional representation of $SU(2)$. This representation is not however unitary. It can be made so by taking the basis functions as

$$\psi(20) = \frac{u^2}{\sqrt{2}} \qquad \psi(11) = uv \qquad \psi(02) = \frac{v^2}{\sqrt{2}} \tag{9.5.10}$$

We can proceed in the general case by taking the $(n+1)$ homogeneous polynomials of the nth degree in u and v

$$\phi(pq) = u^p v^q$$
$$p+q = n = 0, 1, 2, \ldots$$
$$q = 0, 1, 2, \ldots, n$$

Following for example Hamermesh (1962) it is convenient to replace p,q by j,m where

$$p = j+m \qquad q = j-m$$

Then $p+q = 2j = n$ and $j = 0, \frac{1}{2}, 1, \frac{3}{2}, \ldots$. We now take the $(2j+1)$ homogeneous polynomials

$$\psi(jm) = \frac{u^{j+m} v^{j-m}}{\sqrt{(j+m)!\,(j-m)!}} \tag{9.5.11}$$

Since $j-m = q = 0, 1, 2, \ldots, 2j$ we have

$$m = j, j-1, \ldots, -j+1, -j \tag{9.5.12}$$

The factor $\sqrt{(j+m)!\ (j-m)!}$ is included as in (9.5.10) to make the representations unitary. It is clear that for a given j the $(2j+1)$ functions $\psi(jm)$ are transformed among themselves by the transformation (9.5.1). We thus have a $(2j+1)$-dimensional representation of $SU(2)$ spanned by the functions $\psi(jm)$. The matrix elements in this representation which we denote by $\Gamma^{(j)}_{m'm}$ are given by Hamermesh (1962) as

$$\Gamma^{(j)}_{m'm}(a, b) = \sum_{\mu=0} \frac{[(j+m)!\,(j-m)!\,(j+m')!\,(j-m')!]^{\frac{1}{2}}}{(j+m-\mu)!\,\mu!\,(j-m'-\mu)!\,(m'-m+\mu)!} \times a^{j+m-\mu} a^{*j-m'-\mu} b^{\mu} (-b^*)^{m'-m+\mu} \tag{9.5.13}$$

These matrix elements are quite complicated. However it can be shown by elementary methods that every unitary unimodular matrix can be diagonalized by a unitary unimodular transformation. Furthermore the eigenvalues are a pair of complex conjugate numbers which we take to be $e^{i\omega/2}$ and $e^{-i\omega/2}$. Thus a transformation equivalent to (9.5.1) is (9.5.5). For this case only the term in (9.5.13) with $\mu = 0$ survives and we have

$$\Gamma^{(j)}_{m'm}(e^{i\omega/2}, 0) = e^{im\omega}\,\delta_{m'm} \tag{9.5.14}$$

It follows from (9.5.12) and (9.5.14) that the character in the jth representation is

$$\chi^{(j)}(\omega) = \sum_{m=-j}^{j} e^{im\omega} = \frac{\sin (j+\frac{1}{2})\omega}{\sin \frac{1}{2}\omega} \tag{9.5.15}$$

It can be shown (see Hamermesh, 1962) that the representation matrices $\boldsymbol{\Gamma}^{(j)}$ are irreducible and further that no other IR's of $SU(2)$ can be found. It follows that the IR's of the group $SU(2)$ can be labelled by a symbol (j) with

$$j = 0, \tfrac{1}{2}, 1, \tfrac{3}{2}, \ldots$$

The IR (j) is of dimension $(2j+1)$ and the character in the jth IR is given by (9.5.15). These conclusions can also be reached directly using the general theory of Lie algebras (see (A1.7)).

For the homomorphic rotation group $R(3)$ the parameter ω is an angle and we thus require

$$\chi^{(j)}(\omega + 2\pi) = \chi^{(j)}(\omega)$$

When j is a positive integer l this requirement is satisfied and the IR's (l) of $SU(2)$ or $R^*(3)$ are also IR's of $R(3)$ (see (9.4.30)). When j is half an odd integer we have

$$\chi^{(j)}(\omega + 2\pi) = -\chi^{(j)}(\omega)$$

and it follows that the IR's $(\frac{1}{2})$, $(\frac{3}{2})$, $(\frac{5}{2}), \ldots$ of $SU(2)$ or $R^*(3)$ are not representations of $R(3)$. These representations are sometimes referred to as double-valued representations of $R(3)$ but in reality they are not representations at all since by definition a representation is single-valued.

For the IR's (j) of $SU(2)$ or $R^*(3)$ we have as in (9.4.31) the result

$$(j_1) \otimes (j_2) = (j_1 + j_2) \oplus (j_1 + j_2 - 1) \oplus \ldots \oplus (|j_1 - j_2|) \tag{9.5.16}$$

which has already been derived using the irreducible tensor method.

Finally in this section we consider the Casimir operator for the Lie algebra A_1 of $SU(2)$. We begin by writing in (9.5.4)

$$\hat{j}_x = i\hat{X}_1 \qquad \hat{j}_y = i\hat{X}_2 \qquad \hat{j}_z = i\hat{X}_3$$

It then follows that $\hat{j}_x, \hat{j}_y, \hat{j}_z$ satisfy the CR's for a general angular momentum (i.e. not necessarily orbital angular momentum). Now it is a well known result (see, for example, Eyring, Walter & Kimball, 1944) that the eigenvalues of $\hat{j}^2 = \hat{j}_x^2 + \hat{j}_y^2 + \hat{j}_z^2$ are given by

$$\hat{j}^2 |j, m\rangle = j(j+1) |j, m\rangle$$
$$j = 0, \tfrac{1}{2}, 1, \tfrac{3}{2}, \ldots$$

By analogy with (9.4.15) and the fact that A_1 and B_1 are isomorphic the Casimir operator for A_1 is given by $\hat{C} = \frac{1}{2}\hat{j}^2$ and we have the result

$$\hat{C} |j, m\rangle = \tfrac{1}{2}j(j+1) |j, m\rangle \tag{9.5.17}$$

9.6 The rotation group in four dimensions $R(4)$

As we shall see in the next chapter this group plays an important role in the theory of the hydrogen atom. The group $R(4)$ consists of all real orthogonal

4-dimensional linear transformations with unit determinant

$$x_i' = \sum_{j=1}^{4} a_{ij}x_j \qquad i = 1, 2, 3, 4 \tag{9.6.1}$$

The infinitesimal transformations are

$$x_i + dx_i = \sum_{j=1}^{4} (\delta_{ij} + \alpha_{ij})x_j \qquad i = 1, 2, 3, 4$$

where the infinitesimal parameters α_{ij} satisfy the property $\alpha_{ji} = -\alpha_{ij}$ due to the orthogonality condition. It can be seen then that $R(4)$ is a six parameter Lie group. If we write

$$x_1 = x \qquad x_2 = y \qquad x_3 = z \qquad x_4 = t$$

then the natural IO's are found to be

$$\begin{aligned} \hat{A}_1 &= z\frac{\partial}{\partial y} - y\frac{\partial}{\partial z} \qquad \hat{A}_2 = x\frac{\partial}{\partial z} - z\frac{\partial}{\partial x} \qquad \hat{A}_3 = y\frac{\partial}{\partial x} - x\frac{\partial}{\partial y} \\ \hat{B}_1 &= x\frac{\partial}{\partial t} - t\frac{\partial}{\partial x} \qquad \hat{B}_2 = y\frac{\partial}{\partial t} - t\frac{\partial}{\partial y} \qquad \hat{B}_3 = z\frac{\partial}{\partial t} - t\frac{\partial}{\partial z} \end{aligned} \tag{9.6.2}$$

and the CR's are

$$\begin{aligned} &[\hat{A}_1, \hat{B}_1] = [\hat{A}_2, \hat{B}_2] = [\hat{A}_3, \hat{B}_3] = 0 \\ &[\hat{A}_1, \hat{A}_2] = \hat{A}_3 \qquad [\hat{A}_2, \hat{A}_3] = \hat{A}_1 \qquad [\hat{A}_3, \hat{A}_1] = \hat{A}_2 \\ &[\hat{B}_1, \hat{B}_2] = \hat{A}_3 \qquad [\hat{B}_2, \hat{B}_3] = \hat{A}_1 \qquad [\hat{B}_3, \hat{B}_1] = \hat{A}_2 \\ &[\hat{A}_1, \hat{B}_2] = \hat{B}_3 \qquad [\hat{A}_1, \hat{B}_3] = -\hat{B}_2 \\ &[\hat{A}_2, \hat{B}_1] = \hat{B}_3 \qquad [\hat{A}_2, \hat{B}_3] = \hat{B}_1 \\ &[\hat{A}_3, \hat{B}_1] = \hat{B}_2 \qquad [\hat{A}_3, \hat{B}_2] = -\hat{B}_1 \end{aligned} \tag{9.6.3}$$

We now write

$$\hat{J}_i = \tfrac{1}{2}(\hat{A}_i + \hat{B}_i) \qquad \hat{K}_i = \tfrac{1}{2}(\hat{A}_i - \hat{B}_i) \qquad i = 1, 2, 3, \tag{9.6.4}$$

and then we find that the Lie algebra of $R(4)$, which is denoted by D_2 (see Rowlatt, 1966), is given by the CR's

$$\begin{aligned} &[\hat{J}_1, \hat{J}_2] = \hat{J}_3 \qquad [\hat{J}_2, \hat{J}_3] = \hat{J}_1 \qquad [\hat{J}_3, \hat{J}_1] = \hat{J}_2 \\ &[\hat{K}_1, \hat{K}_2] = \hat{K}_3 \qquad [\hat{K}_2, \hat{K}_3] = \hat{K}_1 \qquad [\hat{K}_3, \hat{K}_1] = \hat{K}_2 \\ &[\hat{J}_i, \hat{K}_j] = 0 \qquad i, j = 1, 2, 3 \end{aligned} \tag{9.6.5}$$

These CR's clearly demonstrate that the Lie algebra D_2 is of a very special nature. The operators $\hat{J}_i$ and $\hat{K}_i$ span two disjoint (i.e. $[\hat{J}_i, \hat{K}_j] = 0$) subalgebras of D_2. From (9.4.8) we see that each of these subalgebras is just the Lie algebra B_1 of $R(3)$. In general terms (see (A1.1) and Section 2.14) this situation is described by saying that D_2 is the direct sum of B_1 with itself and we write

$$D_2 = B_1 \oplus B_1 \tag{9.6.6}$$

In order to obtain the canonical CR's (see (A1.3.22)–(A1.3.25)) we introduce the standard basis given by

$$\begin{aligned} \hat{H}_1 &= \frac{i}{\sqrt{2}}\hat{J}_3 \qquad & \hat{H}_2 &= \frac{i}{\sqrt{2}}\hat{K}_3 \\ \hat{E}_{\pm\alpha} &= \frac{i}{2}(\hat{J}_1 \pm i\hat{J}_2) \qquad & \hat{E}_{\pm\beta} &= \frac{i}{2}(\hat{K}_1 \pm i\hat{K}_2) \end{aligned} \tag{9.6.7}$$

and we find in this basis

$$\begin{aligned} [\hat{H}_1, \hat{H}_2] &= 0 \qquad & [\hat{E}_\alpha, \hat{E}_\beta] &= 0 \\ [\hat{H}_1, \hat{E}_{\pm\alpha}] &= \pm\frac{1}{\sqrt{2}}\hat{E}_{\pm\alpha} \qquad & [\hat{H}_2, \hat{E}_{\pm\alpha}] &= 0\,\hat{E}_{\pm\alpha} \\ [\hat{H}_1, \hat{E}_{\pm\beta}] &= 0\hat{E}_{\pm\beta} \qquad & [\hat{H}_2, \hat{E}_{\pm\beta}] &= \pm\frac{1}{\sqrt{2}}\hat{E}_{\pm\beta} \\ [\hat{E}_\alpha, \hat{E}_{-\alpha}] &= \frac{1}{\sqrt{2}}\hat{H}_1 \qquad & [\hat{E}_\beta, \hat{E}_{-\beta}] &= \frac{1}{\sqrt{2}}\hat{H}_2 \end{aligned} \tag{9.6.8}$$

The operators $\hat{E}_{\pm\alpha}$, $\hat{E}_{\pm\beta}$ are clearly analogous to the shift operators (9.4.13) which we introduced for the group $R(3)$.

We consider now the IR's of $R(4)$. In the case of $R(3)$ we obtained the IR's of the group from those of the algebra B_1 by using the properties of orbital angular momentum operators. This approach is not strictly accurate because as we have seen the Lie algebras B_1 and A_1 are isomorphic. Thus if we had used the properties of general angular momentum operators we would have obtained from B_1 not the IR's of $R(3)$ but instead those of the double group $R^*(3)$. We have seen that associated with a Lie group $\mathcal{G}$ there is a uniquely determined Lie algebra $\mathcal{L}$. It can be shown (see Hausner and Schwartz, 1968) that associated with each Lie algebra $\mathcal{L}$ there is a uniquely determined Lie group. In general however this is not the group $\mathcal{G}$ from which we started. This is due to certain topological properties of the group manifold which we do not consider in this book. The group which we do determine from $\mathcal{L}$ is called the *universal covering group* of $\mathcal{G}$. For the rotation group in n dimensions $R(n)$ the universal covering group is the double group $R^*(n)$. Thus the Lie algebra D_2 gives rise to the rotation double group $R^*(4)$. We shall not consider the elements of $R^*(4)$ here because from (9.6.6) $D_2 = A_1 \oplus A_1$ (since A_1 and B_1 are isomorphic) and this result can be used to show that $R^*(4)$ is isomorphic with the direct product group $SU(2)\otimes SU(2)$ whose elements are easier to deal with.

According to the general theory of Lie algebras (see (A1.8)) the IR's of the double group $R^*(4)$ are characterized by a symbol (j_1j_2) with

$$\begin{aligned} j_1 &= 0, \tfrac{1}{2}, 1, \tfrac{3}{2}, \ldots \\ j_2 &= 0, \tfrac{1}{2}, 1, \tfrac{3}{2}, \ldots \end{aligned} \tag{9.6.9}$$

The dimension of the IR (j_1j_2) is given by (see (A1.8.3))

$$n_{(j_1j_2)} = (2j_1+1)(2j_2+1) \tag{9.6.10}$$

By analogy with (9.5.5) we consider the special transformations

$$\begin{aligned} x' &= e^{i\phi_1/2}x \qquad & z' &= e^{i\phi_2/2}z \\ y' &= e^{-i\phi_1/2}y \qquad & t' &= e^{-i\phi_2/2}t \end{aligned} \tag{9.6.11}$$

and then the elements of $SU(2)\otimes SU(2)$ are equivalent to $\mathbf{U}(\phi_1, \phi_2) = \text{diag}\,[e^{i\phi_1/2}, e^{-i\phi_1/2}]\otimes\text{diag}\,[e^{i\phi_2/2}, e^{-i\phi_2/2}]$. If we write

$$\psi = \phi_1+\phi_2$$

$$\theta = \phi_1-\phi_2$$

then

$$\mathbf{U}(\psi, \theta) = \text{diag}\,[e^{i\psi/2}, e^{-i\psi/2}, e^{i\theta/2}, e^{-i\theta/2}] \tag{9.6.12}$$

In terms of the parameters ψ, θ it can then be shown (see (A1.8.4)) that the character of $U(\psi, \theta)$ in the IR (j_1, j_2) is given by

$$\chi^{(j_1j_2)}(\psi, \theta) = \frac{\sin[(j_1+\frac{1}{2})\psi]\sin[(j_2+\frac{1}{2})\theta]}{\sin(\frac{1}{2}\psi)\sin(\frac{1}{2}\theta)} \tag{9.6.13}$$

In the group $R(4)$ the parameters ψ, θ correspond to angles and we then require

$$\chi^{(j_1j_2)}(\psi+2\pi, \theta+2\pi) = \chi^{(j_1j_2)}(\psi, \theta)$$

from which it follows that the IR's of $R(4)$ are given by (j_1j_2) with the condition that both $p = j_1+j_2$ and $q = j_1-j_2$ are integers.

Consider next the branching rule for the reduction $R(4)\to R(3)$. They are contained in the branching rule for the reduction $SU(2)\otimes SU(2)\to SU(2)$. We require $U(\psi, \theta)\to U(\phi)$ and this is achieved by setting $\phi_2 = 0$; $\phi_1 = \phi$. According to (9.6.13) with $\psi = \theta = \phi$ we find the branching rule

$$(j_1j_2)\to(j_1+j_2)\oplus(j_1+j_2-1)\oplus\ldots\oplus(j_1-j_2) \tag{9.6.14}$$

The result (9.6.13) can also be used to obtain the decomposition of the direct product of two IR's of $R^*(4)$. Thus we find

$$(j_1j_2)\otimes(j_1'j_2') = \sum_{J_1,J_2}^{\oplus}(J_1J_2) \tag{9.6.15}$$

with

$$\begin{aligned} J_1 &= j_1 + j_1', j_1 + j_1' - 1, \ldots, j_1 - j_1' \\ J_2 &= j_2 + j_2', j_2 + j_2' - 1, \ldots, j_2 - j_2' \end{aligned} \tag{9.6.16}$$

We now introduce for $R(4)$ operators which are analogous to the square of the angular momentum operator in $R(3)$. For the Lie algebra D_2 of $R(4)$ there exists two operators $\hat{F}_1$ and $\hat{F}_2$ which commute with all the IO's of D_2. From (9.6.6) and (9.4.16) it can be seen that we may take

$$\begin{aligned} \hat{F}_1 &= \hat{H}_1^2 + \hat{E}_\alpha \hat{E}_{-\alpha} + \hat{E}_{-\alpha}\hat{E}_\alpha \\ \hat{F}_2 &= \hat{H}_2^2 + \hat{E}_\beta \hat{E}_{-\beta} + \hat{E}_{-\beta}\hat{E}_\beta \end{aligned} \tag{9.6.17}$$

By comparison with (9.5.17) we have

$$\begin{aligned} \hat{F}_1 \,|j_1, j_2, m_1, m_2\rangle &= \tfrac{1}{2} j_1(j_1+1)\,|j_1, j_2, m_1, m_2\rangle \\ \hat{F}_2 \,|j_1, j_2, m_1, m_2\rangle &= \tfrac{1}{2} j_2(j_2+1)\,|j_1, j_2, m_1, m_2\rangle \end{aligned} \tag{9.6.18}$$

where $|j_1, j_2, m_1, m_2\rangle$ is a basis ket for the IR$(j_1 j_2)$. According to the general theory of Lie algebras (see (A1.10.14)) the Casimir operator for D_2 is simply

$$\hat{C} = \hat{F}_1 + \hat{F}_2$$

and we have

$$\hat{C}\,|j_1, j_2, m_1, m_2\rangle = \tfrac{1}{2}[j_1(j_1+1) + j_2(j_2+1)]\,|j_1, j_2, m_1, m_2\rangle \tag{9.6.19}$$

If we consider (9.6.4) and (9.6.7) we find

$$-2\hat{F}_1 = \hat{J}_1^2 + \hat{J}_2^2 + \hat{J}_3^2 = \tfrac{1}{4}\{(\hat{A}_1 + \hat{B}_1)^2 + (\hat{A}_2 + \hat{B}_2)^2 + (\hat{A}_3 + \hat{B}_3)^2\}$$

and it follows from (9.6.3) that

$$-2(\hat{F}_1 - \hat{F}_2) = \hat{A}_1\hat{B}_1 + \hat{A}_2\hat{B}_2 + \hat{A}_3\hat{B}_3$$

However using (9.6.2) we readily verify that

$$\hat{A}_1\hat{B}_1 + \hat{A}_2\hat{B}_2 + \hat{A}_3\hat{B}_3 = 0$$

We now have

$$\hat{F}_1 - \hat{F}_2 = 0$$

and it follows that the eigenvalues λ of the Casimir operator $\hat{C}$ are given by

$$\lambda = \tfrac{1}{4}[(2j+1)^2 - 1] = \tfrac{1}{4}(n^2 - 1) \tag{9.6.20}$$

where

$$j_1 = j_2 = j \qquad \text{and} \qquad n = 1, 2, 3, \ldots \tag{9.6.21}$$

9.7 The unitary unimodular group in three dimensions $SU(3)$

As we shall see in the next chapter this group plays an important role in the theory of the three-dimensional isotropic harmonic oscillator. The group

$SU(3)$ consists of all 3-dimensional unitary unimodular transformations

$$x'_i = \sum_{j=1}^{3} a_{ij}x_j \qquad i = 1, 2, 3 \tag{9.7.1}$$

The matrix $[a_{ij}]$ is unitary and has unit determinant. The infinitesimal transformations are given by

$$x_i + dx_i = \sum_{j=1}^{3} (\delta_{ij} + \alpha_{ij})x_j \qquad i = 1, 2, 3$$

where the parameters α_{ij} are infinitesimal. If we denote the matrix $[\delta_{ij} + \alpha_{ij}]$ by $\mathbf{A}$ the unitary condition is

$$\mathbf{A}^{\dagger} = \mathbf{A}^{-1} \tag{9.7.2}$$

Using the unimodularity condition and the fact that the α_{ij} are infinitesimal we find

$$\mathbf{A}^{-1} = \begin{bmatrix} 1+\alpha_{22}+\alpha_{33} & -\alpha_{12} & -\alpha_{13} \\ -\alpha_{21} & 1+\alpha_{11}+\alpha_{33} & -\alpha_{23} \\ -\alpha_{31} & -\alpha_{32} & 1+\alpha_{11}+\alpha_{22} \end{bmatrix}$$

and the condition (9.7.2) becomes

$$\begin{bmatrix} 1+\alpha^*_{11} & \alpha^*_{21} & \alpha^*_{31} \\ \alpha^*_{12} & 1+\alpha^*_{22} & \alpha^*_{32} \\ \alpha^*_{13} & \alpha^*_{23} & 1+\alpha^*_{33} \end{bmatrix} = \begin{bmatrix} 1+\alpha_{22}+\alpha_{33} & -\alpha_{12} & -\alpha_{13} \\ -\alpha_{21} & 1+\alpha_{11}+\alpha_{33} & -\alpha_{23} \\ -\alpha_{31} & -\alpha_{32} & 1+\alpha_{11}+\alpha_{22} \end{bmatrix} \tag{9.7.3}$$

If we write $\alpha_{ij} = p_{ij} + iq_{ij}$ then from (9.7.3) we readily find the conditions $p_{ij} = -p_{ji}$, $q_{ij} = q_{ji}$, $q_{33} = -(q_{11} + q_{22})$. There are thus only eight independent real parameters p_{12}, p_{13}, p_{23}, q_{11}, q_{12}, q_{13}, q_{22} amd q_{23}. The group $SU(3)$ is an eight parameter Lie group. If we write

$$x = x_1 \qquad y = x_2 \qquad z = x_3$$

we have

$$\begin{aligned} dx &= iq_{11}x + (p_{12} + iq_{12})y + (p_{13} + iq_{13})z \\ dy &= (-p_{12} + iq_{12})x + iq_{22}y + (p_{23} + iq_{23})z \\ dz &= (-p_{13} + iq_{13})x + (-p_{23} + iq_{23})y + i(-q_{11} - q_{22})z \end{aligned}$$

Now let $F(x, y, z)$ be an arbitrary function of x, y, z. We have

$$\begin{aligned} dF &= \frac{\partial F}{\partial x}dx + \frac{\partial F}{\partial y}dy + \frac{\partial F}{\partial z}dz \\ &= (q_{11}\hat{X}_1 + q_{22}\hat{X}_2 + p_{12}\hat{X}_3 + p_{13}\hat{X}_4 + p_{23}\hat{X}_5 + q_{12}\hat{X}_6 + q_{13}\hat{X}_7 + q_{23}\hat{X}_8)F \end{aligned} \tag{9.7.4}$$

and the natural basis for the Lie algebra of $SU(3)$ is given by the IO's

$$\begin{aligned} \hat{X}_1 &= i\left(x\frac{\partial}{\partial x} - z\frac{\partial}{\partial z}\right) & \hat{X}_2 &= i\left(y\frac{\partial}{\partial y} - z\frac{\partial}{\partial z}\right) \\ \hat{X}_3 &= \left(y\frac{\partial}{\partial x} - x\frac{\partial}{\partial y}\right) & \hat{X}_4 &= \left(z\frac{\partial}{\partial x} - x\frac{\partial}{\partial z}\right) \\ \hat{X}_5 &= \left(z\frac{\partial}{\partial y} - y\frac{\partial}{\partial z}\right) & \hat{X}_6 &= i\left(y\frac{\partial}{\partial x} + x\frac{\partial}{\partial y}\right) \\ \hat{X}_7 &= i\left(z\frac{\partial}{\partial x} + x\frac{\partial}{\partial z}\right) & \hat{X}_8 &= i\left(z\frac{\partial}{\partial y} + y\frac{\partial}{\partial z}\right) \end{aligned} \tag{9.7.5}$$

This Lie algebra is denoted by A_2 (see Rowlatt, 1966). A standard basis can be obtained as follows. If we examine the natural IO's (9.7.5) we immediately see that $\hat{X}_1$ and $\hat{X}_2$ are different in form from the other IO's. Furthermore they commute. We are therefore led to consider a new real basis given by the linear combinations

$$\begin{aligned} \hat{H}_1 &= \lambda(i\hat{X}_1) + \mu(i\hat{X}_2) \\ \hat{H}_2 &= \kappa(i\hat{X}_1) + \nu(i\hat{X}_2) \\ \hat{E}_{\pm\alpha} &= N_\alpha(\hat{X}_3 \mp i\hat{X}_6) \\ \hat{E}_{\pm\beta} &= N_\beta(\hat{X}_5 \mp i\hat{X}_8) \\ \hat{E}_{\pm\gamma} &= N_\gamma(\hat{X}_4 \mp i\hat{X}_7) \end{aligned} \tag{9.7.6}$$

The above IO's can be shown to give canonical CR's and a particular set of canonical CR's is obtained by an appropriate choice of λ, μ, κ, ν, N_α, N_β, N_γ (see (A1.4)).

We can turn our attention now to the IR's of $SU(3)$. It can be shown (see Hausner and Schwartz, 1968) that the covering group of the Lie algebra A_2 is just the group $SU(3)$ itself. Thus from the IR's of A_2 we obtain those of $SU(3)$. Using the techniques of the general theory of Lie algebras (see A1.9) it can be shown that the IR's of $SU(3)$ are labelled by a symbol (nm) where n and m are non-negative integers. Thus the IR's of $SU(3)$ are (00), (10), (01), (20), (02), (11), ... Again referring to the general theory it is possible to determine the dimension $n_{(nm)}$ of the IR (nm). It is found to be given by (see (A1.9.3))

$$n_{(nm)} = \tfrac{1}{2}(n+1)(m+1)(n+m+2) \tag{9.7.7}$$

The dimensions of the first few IR's are shown in Table 27.

TABLE 27

IR (nm)	(00)	(10)	(01)	(20)	(02)	(11)
Dimension	1	3	3	6	6	8

Comparison of Table 27 with Table 17 shows the connection between the Lie algebra notation and the permutational symmetry notation.

Finally we need the Casimir operator for $SU(3)$. This is given from the general theory (see (A1.10.15)) by

$$\begin{aligned}\hat{C} = \tfrac{1}{3}(\hat{H}_1^2 - \hat{H}_1\hat{H}_2 + \hat{H}_2^2) + \hat{E}_\alpha\hat{E}_{-\alpha} + \hat{E}_{-\alpha}\hat{E}_\alpha \\ + \hat{E}_\beta\hat{E}_{-\beta} + \hat{E}_{-\beta}\hat{E}_\beta + \hat{E}_\gamma\hat{E}_{-\gamma} + \hat{E}_{-\gamma}\hat{E}_\gamma\end{aligned} \tag{9.7.8}$$

Furthermore the eigenvalues of this operator are (see (A1.10.16))

$$\lambda = \tfrac{1}{9}(n^2 + m^2 + nm) + \tfrac{1}{3}(n + m) \tag{9.7.9}$$

References

Boerner, H. (1963) "Representations of Groups," North-Holland.

Eyring, H., Walter, J. and Kimball, G. E. (1944) "Quantum Chemistry," Wiley.

Griffith, J. S. (1964) "The Theory of Transition Metal Ions," Cambridge, Chapter 6.

Hammermesh, M. (1962) "Group Theory and its Application to Physical Problems," Addison-Wesley, Chapters 8 and 9.

Hausner, M. and Schwartz, J. T. (1968) "Lie Groups; Lie Algebras," Gordon & Breach.

Judd, B. R. (1963) "Operator Techniques in Atomic Spectroscopy," McGraw-Hill, Chapter 2.

Rowlatt, P. A. (1966) "Group Theory and Elementary Particles," Longmans.

CHAPTER 10

The Quantum Mechanics of Simple Systems

10.1 Harmonic oscillators

Using an appropriate system of units the Hamiltonian for the linear harmonic oscillator can be written as

$$\hat{H} = \tfrac{1}{2}(\hat{p}^2 + \hat{q}^2) \tag{10.1.1}$$

where p is the linear momentum and q the displacement coordinate. The operators $\hat{p}$, $\hat{q}$ and $\hat{1}$ form a three-dimensional Lie algebra (called the Hiesenberg algebra) since

$$[\hat{p}, \hat{1}] = \hat{0} \qquad [\hat{q}, \hat{1}] = \hat{0} \qquad [\hat{p}, \hat{q}] = -i$$

If we introduce the operators

$$\hat{a}^\dagger = \frac{1}{\sqrt{2}}(\hat{p} + i\hat{q}) \qquad \hat{a} = \frac{1}{\sqrt{2}}(\hat{p} - i\hat{q}) \tag{10.1.2}$$

the Hamiltonian can be written as

$$\hat{H} = (\hat{a}^\dagger \hat{a} + \tfrac{1}{2}) = (\hat{a}\hat{a}^\dagger - \tfrac{1}{2}) \tag{10.1.3}$$

The operators $\hat{1}$, $\hat{H}$, $\hat{a}$, $\hat{a}^\dagger$ span a Lie algebra since

$$[\hat{a}, \hat{a}^\dagger] = 1 \qquad [\hat{H}, \hat{a}^\dagger] = \hat{a}^\dagger \qquad [\hat{H}, \hat{a}] = -\hat{a}$$

For a representation of this algebra we consider the eigenvalue problem

$$\hat{H}\,|m\rangle = m\,|m\rangle$$

From the CR's it follows that $\hat{a}^\dagger\,|m\rangle$ is an eigenket of $\hat{H}$ with eigenvalue $(m+1)$ and $\hat{a}\,|m\rangle$ is an eigenket with eigenvalue $(m-1)$. If there exists a lowest eigenvalue m_0 then $(m_0 - 1)$ is not an eigenvalue and $\hat{a}\,|m_0\rangle = 0$. In this case

$$[\hat{a}, \hat{a}^\dagger]\,|m_0\rangle = \hat{a}\hat{a}^\dagger\,|m_0\rangle = |m_0\rangle$$

and the lowest eigenvalue of $\hat{H}$ is $\tfrac{1}{2}$. It follows that the energy levels are given by

$$E_n = (n + \tfrac{1}{2}) \qquad n = 0, 1, 2, \ldots \tag{10.1.4}$$

These energy levels are non-degenerate in accordance with the fact that $\hat{H}$ is invariant under the group $SU(1)$ which has only one-dimensional IR's.

We consider next the two-dimensional isotropic harmonic oscillator with Hamiltonian

$$\hat{H} = \tfrac{1}{2}(\hat{p}_1^2 + \hat{p}_2^2 + \hat{q}_1^2 + \hat{q}_2^2) \tag{10.1.5}$$

The angular momentum which is proportional to the operator

$$\hat{L} = \frac{1}{2i}(\hat{q}_1\hat{p}_2 - \hat{q}_2\hat{p}_1) \tag{10.1.6}$$

is a constant of the motion since $[\hat{H}, \hat{L}] = 0$. In general it is extremely difficult to find any other constants of the motion which might exist. In this case however, due to the very special nature of the potential function, we can obtain further constants of the motion. Let

$$\hat{P} = \hat{p}_1 + \hat{p}_2 \qquad \hat{Q} = \hat{q}_1 + \hat{q}_2 \tag{10.1.7}$$

The Hamiltonian can now be written in the form

$$\hat{H} = \tfrac{1}{2}(\hat{P}^2 + \hat{Q}^2) - (\hat{p}_1\hat{p}_2 + \hat{q}_1\hat{q}_2) \tag{10.1.8}$$

and the new operator $(\hat{p}_1\hat{p}_2 + \hat{q}_1\hat{q}_2)$ which appears is a constant of the motion. If we write

$$\hat{K} = \frac{1}{2i}(\hat{p}_1\hat{p}_2 + \hat{q}_1\hat{q}_2) \tag{10.1.9}$$

then it can be seen that $[\hat{K}, \hat{L}] = \hat{D}$ where

$$\hat{D} = \frac{1}{4i}\{(\hat{p}_1^2 + \hat{q}_1^2) - (\hat{p}_2^2 + \hat{q}_2^2)\} \tag{10.1.10}$$

is another constant of the motion. We now have the CR's

$$[\hat{L}, \hat{D}] = \hat{K} \qquad [\hat{K}, \hat{L}] = \hat{D} \qquad [\hat{D}, \hat{K}] = \hat{L} \tag{10.1.11}$$

which are characteristic of the group $SU(2)$. Thus the symmetry group of the 2-dimensional isotropic harmonic oscillator is not simply $R(2)$ but instead the higher group $SU(2)$. According to Section 9.5 the Casimir operator for $SU(2)$ can be written in the form

$$\hat{C} = -\tfrac{1}{2}(\hat{K}^2 + \hat{L}^2 + \hat{D}^2) \tag{10.1.12}$$

and from (9.5.17) the eigenvalues of $\hat{C}$ are $\tfrac{1}{2}j(j+1)$. Using the definitions of $\hat{L}$, $\hat{K}$ and $\hat{D}$ we find

$$\hat{C} = \tfrac{1}{8}(\hat{H}^2 - 1) \tag{10.1.13}$$

from which it follows that the energy levels are given by

$$\tfrac{1}{2}j(j+1) = \tfrac{1}{8}(E^2 - 1)$$

This equation yields the familiar result

$$E_{n_1,n_2} = (n_1 + n_2 + 1) = (N+1) \tag{10.1.14}$$

where $N = 2j$ is a non-negative integer. The $(N+1)$-fold degeneracy of the energy levels is now a consequence of the $SU(2)$ symmetry where the IR's are of dimension $(N+1)$.

We see that the extra constants of motion can be incorporated into a group theoretical framework which then solves the physical problem with much greater depth than is obtained by a direct solution of the Schrödinger equation. It is interesting to note that the constants of motion are either symmetric or antisymmetric with respect to a permutation of the coordinates.

The three-dimensional isotropic harmonic oscillator can be treated in a similar fashion. The Hamiltonian is

$$\hat{H} = \tfrac{1}{2}(\hat{p}_1^2 + \hat{q}_1^2 + \hat{p}_2^2 + \hat{q}_2^2 + \hat{p}_3^2 + \hat{q}_3^2) \tag{10.1.15}$$

By analogy with the two-dimensional case we introduce the following constants of motion

$$\begin{aligned}
\hat{X}_1 &= \frac{i}{2}(\hat{p}_1^2 + \hat{q}_1^2 - \hat{p}_2^2 - \hat{q}_2^2) & \hat{X}_4 &= i(\hat{q}_1\hat{p}_2 - \hat{q}_2\hat{p}_1) \\
\hat{X}_7 &= -i(\hat{p}_1\hat{p}_2 + \hat{q}_1\hat{q}_2) & \hat{X}_2 &= \frac{i}{2}(\hat{p}_2^2 + \hat{q}_2^2 - \hat{p}_3^2 - \hat{q}_3^2) \\
\hat{X}_3 &= i(\hat{q}_3\hat{p}_1 - \hat{q}_1\hat{p}_3) & \hat{X}_5 &= i(\hat{q}_2\hat{p}_3 - \hat{q}_3\hat{p}_2) \\
\hat{X}_6 &= -i(\hat{p}_3\hat{p}_1 + \hat{q}_3\hat{q}_1) & \hat{X}_8 &= -i(\hat{p}_2\hat{p}_3 + \hat{q}_2\hat{q}_3)
\end{aligned} \tag{10.1.16}$$

The above notation has been introduced by comparison with Section 9.7. It is not hard to verify that these constants of motion satisfy the CR's for the Lie algebra of $SU(3)$. Thus the symmetry group of the 3-dimensional isotropic harmonic oscillator is not merely $R(3)$ but the higher group $SU(3)$. Using (9.7.6) the Casimir operator for $SU(3)$ can be written in the form

$$-9\hat{C} = \hat{X}_1^2 - \hat{X}_1\hat{X}_2 + \hat{X}_2^2 + \tfrac{3}{4}(\hat{X}_3^2 + \hat{X}_4^2 + \hat{X}_5^2 + \hat{X}_6^2 + \hat{X}_7^2 + \hat{X}_8^2)$$

We now substitute (10.1.16) into this expression and collect up the terms. Using (10.1.15) we find the functional dependence of $\hat{C}$ on $\hat{H}$ to be

$$36\hat{C} = 4\hat{H}^2 - 9 \tag{10.1.17}$$

From (9.7.9) it follows that the energy levels are given by

$$4(n^2 + m^2 + nm) + 12(n+m) = 4E^2 - 9 \tag{10.1.18}$$

At this point we compare the Lie algebra description of the IR's with the irreducible tensor method and readily find that the IR labelled (nm) has associated with it the partition $[n+mm]$. Since the Hamiltonian is totally symmetric it follows in this case that the eigenfunctions are totally symmetric and we must discard all those representations (nm) for which $m \neq 0$. The IR $(n0)$ has permutational symmetry $[n]$ and functions spanning this IR are totally symmetric. From (10.1.18) the energy levels are now

$$E_{n_1,n_2,n_3} = (n_1 + n_2 + n_3 + \tfrac{3}{2}) = (n + \tfrac{3}{2}) \qquad (10.1.19)$$

where n is a non-negative integer. From (9.7.7) we find the degeneracy of the nth level to be

$$\omega_n = \tfrac{1}{2}(n+1)(n+2) \qquad (10.1.20)$$

10.2 The rigid rotator

It is very easy to obtain the energy levels and their degeneracy for this system. The Hamiltonian in appropriate units is

$$\hat{H} = \tfrac{1}{2}\hat{J}^2 \qquad (10.2.1)$$

where $\hat{J}^2$ is the square of the orbital angular momentum. From Section 9.5 we have

$$\hat{H} = \hat{C} \qquad (10.2.2)$$

and it follows using (9.5.17) that the energy levels are

$$E_J = \tfrac{1}{2}J(J+1) \qquad (10.2.3)$$

The degeneracy is simply $\omega_J = (2J+1)$. The symmetry group of $\hat{H}$ is clearly $R(3)$ and so J is restricted to non negative integer values.

10.3 The hydrogen atom

The Hamiltonian for the non-relativistic motion of the electron relative to the nucleus of charge Z is given, using atomic units, by

$$\hat{H} = \tfrac{1}{2}\hat{p}^2 - \frac{Z}{r} \qquad (10.3.1)$$

where $\hat{p}^2 = \hat{p}_x^2 + \hat{p}_y^2 + \hat{p}_z^2$.

The energy levels and their degeneracies can be obtained and explained without recourse to the Schrödinger equation by using the theory of Lie groups. In the quantum mechanical framework we consider the fundamental vector operators $\hat{\mathbf{r}}$ and $\hat{\mathbf{p}}$ where

$$\hat{\mathbf{p}} = -\mathrm{i}\nabla$$

The angular momentum operator is given by

$$\hat{\mathbf{l}} = \hat{\mathbf{r}}_\wedge \hat{\mathbf{p}} = -\mathrm{i}\hat{\mathbf{r}}_\wedge \nabla$$

and since $[\hat{H}, \hat{\mathbf{l}}]=0$ it follows that the angular momentum is a constant of the motion. In an attempt to find any further constants of the motion we are naturally led to consider, at least initially, operators of the form $\hat{\mathbf{p}}_\wedge\hat{\mathbf{l}}$ and $\hat{\mathbf{l}}_\wedge\hat{\mathbf{p}}$. By constructing the commutators of the components of these operators with $\hat{H}$ (a tedious calculation) it can be seen that the following quantity is a constant of the motion (see Hermann, 1966)

$$\hat{\mathbf{R}} = \tfrac{1}{2}(\hat{\mathbf{l}}_\wedge\hat{\mathbf{p}} - \hat{\mathbf{p}}_\wedge\hat{\mathbf{l}}) + \frac{Z\hat{\mathbf{r}}}{r} \tag{10.3.2}$$

This is called the *Runge-Lenz* vector. It can be rewritten in either of the forms

$$\hat{\mathbf{R}} = \frac{Z\hat{\mathbf{r}}}{r} - \hat{\mathbf{r}}\hat{p}^2 + \hat{\mathbf{p}}(\hat{\mathbf{r}}\,.\,\hat{\mathbf{p}})$$

or

$$\hat{\mathbf{R}} = \hat{\mathbf{p}}(\hat{\mathbf{r}}\,.\,\hat{\mathbf{p}}) - 2\hat{\mathbf{r}}\hat{H} - \frac{Z\hat{\mathbf{r}}}{r}$$

The operators $\hat{R}_x$, $\hat{R}_y$ and $\hat{R}_z$ satisfy the CR's of the form

$$[\hat{R}_x, \hat{R}_y] = -2i\hat{H}\hat{l}_z \qquad [\hat{R}_y, \hat{R}_z] = -2i\hat{H}\hat{l}_x \qquad [\hat{R}_z, \hat{R}_x] = -2i\hat{H}\hat{l}_y \tag{10.3.3}$$

If we are concerned with bound states then $\hat{H}$ is negative definite. Since $\hat{H}$ commutes with $\hat{\mathbf{R}}$ we can rewrite the above CR's in the form

$$\left[\frac{i\hat{R}_x}{\sqrt{-2\hat{H}}}, \frac{i\hat{R}_y}{\sqrt{-2\hat{H}}}\right] = -i\hat{l}_z \qquad \left[\frac{i\hat{R}_y}{\sqrt{-2\hat{H}}}, \frac{i\hat{R}_z}{\sqrt{-2\hat{H}}}\right] = -i\hat{l}_x$$
$$\left[\frac{i\hat{R}_z}{\sqrt{-2\hat{H}}}, \frac{i\hat{R}_x}{\sqrt{-2\hat{H}}}\right] = -i\hat{l}_y \tag{10.3.4}$$

We now write

$$\hat{A}_1 = -i\hat{l}_x \qquad \hat{A}_2 = -i\hat{l}_y \qquad \hat{A}_3 = -i\hat{l}_z$$
$$\hat{B}_1 = \frac{i\hat{R}_x}{\sqrt{-2\hat{H}}} \qquad \hat{B}_2 = \frac{i\hat{R}_y}{\sqrt{-2\hat{H}}} \qquad \hat{B}_3 = \frac{i\hat{R}_z}{\sqrt{-2\hat{H}}} \tag{10.3.5}$$

The CR's satisfied by these six operators are identical with (9.6.3) and it follows that the symmetry group of the hydrogen atom is not $R(3)$ but instead the higher group $R(4)$. From Section 9.6 the Casimir operator for $R(4)$ can be written as

$$\hat{C} = -\tfrac{1}{4}(\hat{A}_1^2 + \hat{A}_2^2 + \hat{A}_3^2 + \hat{B}_1^2 + \hat{B}_2^2 + \hat{B}_3^2)$$

and from (10.3.5) this becomes

$$\hat{C} = \frac{1}{4}\left(\hat{l}^2 - \frac{\hat{R}^2}{2\hat{H}}\right) \tag{10.3.6}$$

Now it can be shown by direct calculation that

$$\hat{R}^2 = 2\hat{H}\hat{l}^2 + 2\hat{H} + Z^2$$

and the Casimir operator becomes

$$\hat{C} = -\frac{Z^2}{8\hat{H}} - \frac{1}{4} \qquad (10.3.7)$$

Using (9.6.20) we obtain the energy levels

$$E_n = -\frac{Z^2}{2n^2} \qquad (10.3.8)$$

The degeneracy follows from (9.6.10) with $j_1 = j_2 = j$

$$\omega_n = (2j+1)^2 = n^2 \qquad (10.3.9)$$

We now see that the eigenfunctions of the hydrogen atom span IR's of $R(4)$. These IR's are of the form (jj) where j is a non-negative integer or half-integer. If we use the subgroup $R(3)$ to label the energy levels then the eigenfunctions span reducible representations. From (9.6.14) we have

$$(jj) \to (0) \oplus (1) \oplus (2) \oplus \ldots \oplus (2j) \qquad (10.3.10)$$

This result can be used to construct Table 28. We have now explained the so-called accidental degeneracy in the H atom and we have given group theoretical significance to the principal quantum number n. The IR's of $R(4)$ and $R(3)$ can be used to completely specify the eigenfunctions. Thus the eigenfunction $\psi[(jj)(J)M]$ spans the IR (jj) of $R(4)$ and belongs to the Mth row of the IR (J) of $R(3)$. The principal quantum number is $n = 2j+1$, the azimuthal quantum number is J and the magnetic quantum number is M.

TABLE 28

IR of $R(4)$	IR's of $R(3)$
(00)	s
$(\frac{1}{2}\frac{1}{2})$	$s \oplus p$
(11)	$s \oplus p \oplus d$
$(\frac{3}{2}\frac{3}{2})$	$s \oplus p \oplus d \oplus f$

Reference

Hermann, R. (1966) "Lie Groups for Physicists," Benjamin, Chapter 16.

CHAPTER 11

Spinors and Double Groups

11.1 The concept of a spinor

In many applications of group theory to atomic and molecular systems it becomes necessary to take into account explicitly the spin functions α and β. According to (9.5.11) we see that α and β can be taken as basis for the 2-dimensional IR $(\frac{1}{2})$ of the group $SU(2)$. Since $SU(2)$ is isomorphic with the rotation double group $R^*(3)$ we see that (α, β) spans the IR $(\frac{1}{2})$ of $R^*(3)$. Now $(\frac{1}{2})$ is not a representation of $R(3)$ and it follows that the spin functions do not form a basis for any representation of $R(3)$. In this respect the basis (α, β) behaves quite differently from the basis (x, y, z) transforming according to the IR (1) of $R^*(3)$ which is also a representation of $R(3)$. Because of this it is customary to call the entity (α, β) an *elementary spinor* or a spinor of rank one in 3-dimensional Euclidean space.

In general when j is one half an odd integer the $(2j+1)$ basis functions for the IR (j) of $R^*(3)$ are said to constitute the components of a spinor of rank $2j$ in 3-dimensional Euclidean space. When $j = l$ is an integer the IR's (l) of $R^*(3)$ coincide with the IR's of $R(3)$ and the $(2l+1)$ basis functions constitute the components of a tensor of rank l with respect to $R(3)$.

If we are concerned with the properties of a free atom or ion and we wish to incorporate the effects of electron spin it is now clear that we must replace the symmetry group $R(3)$ by the double group $R^*(3)$. Likewise if we are concerned with a molecule whose symmetry group is $\mathcal{G}$ and we wish to include spin effects then we must use the subgroup $\mathcal{G}^*$ of $R^*(3)$ in place of $\mathcal{G}$. The group $\mathcal{G}^*$ is called the molecular symmetry *double* group and it is constructed from $\mathcal{G}$ in exactly the same way as $R^*(3)$ is constructed from $R(3)$ (see Section 9.5). Now $R(3)$ consists entirely of proper rotations and it follows that if $\mathcal{G}$ is a finite subgroup of $R(3)$ given by $\{R_1, R_2, \ldots, R_g\}$ then $\mathcal{G}^*$ has elements $\{R_1, R_2, \ldots, R_g, QR_1, QR_2, \ldots, QR_g\}$ where Q is a rotation about any axis by 2π. The double group $\mathcal{G}^*$ thus has twice as many elements as $\mathcal{G}$ and we return to the identity only after a rotation about an axis by 4π.

The molecular symmetry groups which qualify for extension to double groups by the above procedure are the groups of proper rotations $\mathcal{C}_n$, $\mathcal{D}_n$ and $\mathcal{O}$. However if we use the results contained in Tables 2 and 3 it is clearly possible to obtain all the molecular symmetry double groups.

11.2 The double groups $\mathscr{C}_n^*$

The group $\mathscr{C}_n^*$ is given by

$$\mathscr{C}_n^* = \{E, C_n, C_n^2, \ldots, C_n^{n-1}, Q, QC_n, \ldots, QC_n^{n-1}\}$$

where $Q = C_n^n$. It is a cyclic group of order $2n$ and it follows that there exists $2n$ 1-dimensional IR's. Since $C_n^{2n} = E$ the character $\chi(C_n)$ of the element C_n is given by

$$\chi(C_n) = e^{m\pi i/n} \qquad m = 1, 2, \ldots, 2n \qquad (11.2.1)$$

For the group $\mathscr{C}_2^*$ we readily find from (11.2.1) the following character table. The representations Γ_1 and Γ_2 are of course also representations of $\mathscr{C}_2$ and as such they can be labelled A and B respectively. The complex conjugate representations Γ_3 and Γ_4 can be combined to form a single 2-dimensional representation Γ

$$\Gamma = \Gamma_3 \oplus \Gamma_4 \qquad (11.2.2)$$

$\mathscr{C}_2^*$	E	C_2	Q	QC_2
Γ_1	1	1	1	1
Γ_2	1	-1	1	-1
Γ_3	1	i	-1	$-i$
Γ_4	1	$-i$	-1	i

The labelling of IR's of the double group $\mathscr{G}^*$ which are not also representations of $\mathscr{G}$ is to a large extent still not standardized. One way in which this can be done is to retain the notation

$$\Gamma = A, B, E, T, U, V,$$

which describes the dimension of the IR. A subscript j is then attached to the label Γ to indicate the IR (j) of $SU(2)$ which first gives rise to the given IR of $\mathscr{G}^*$ upon making the transition

$$SU(2) \rightarrow \mathscr{G}^*$$

This allocation of j can be done by considering the character as given from (9.5.15) by

$$\chi^{(j)}(C_n^m) = \frac{\sin\left[(j+\frac{1}{2})\dfrac{2\pi m}{n}\right]}{\sin\left(\dfrac{\pi m}{n}\right)} \qquad (11.2.3)$$

Thus for the group $\mathscr{C}_2^*$ we have with $j = \frac{1}{2}$ the result

$$\chi^{(\frac{1}{2})}(C_2) = \frac{\sin \pi}{\sin \pi/2} = 0$$

$$\chi^{(\frac{1}{2})}(Q) = \chi^{(\frac{1}{2})}(C_2^2) = \frac{\sin 2\pi}{\sin \pi} = 2\cos \pi = -2$$

and it follows that we label the representation Γ of (11.2.2) by the symbol $E_{\frac{1}{2}}$. The character table can now be written in the form

$\mathscr{C}_2^*$	E	C_2	Q	QC_2
$A; z$	1	1	1	1
$B; x, y$	1	-1	1	-1
$E_{\frac{1}{2}}; (\alpha, \beta)$	2	0	-2	0

If we use (11.2.1) and combine complex conjugate IR's we readily find the character table for $\mathscr{C}_3^*$ to be

$\mathscr{C}_3^*$	E	C_3	C_3^2	Q	QC_3	QC_3^2
A	1	1	1	1	1	1
E	2	-1	-1	2	-1	-1
Γ_1	1	-1	1	-1	1	-1
Γ_2	2	1	-1	-2	-1	1

In order to label Γ_1 and Γ_2 we employ (11.2.3) and use (3.4.7). This yields the branching rules

$$(\tfrac{1}{2}) \to \Gamma_2$$

$$(\tfrac{3}{2}) \to 2\Gamma_1 \oplus \Gamma_2$$

and it follows that we label Γ_2 by $E_{\frac{1}{2}}$ and Γ_1 by $B_{\frac{3}{2}}$.

11.3 The double groups $\mathscr{D}_n^*$

The group $\mathscr{D}_2^*$ is a subgroup of $R^*(3)$ and consists of the eight elements

$$E, C_2(X), C_2(Y), C_2(Z), Q, QC_2(X), QC_2(Y), QC_2(Z)$$

These eight elements can be described by the eight unitary unimodular matrices

$$\begin{aligned}
\mathbf{E} &= \begin{bmatrix} 1 & 0 \\ 0 & 1 \end{bmatrix} & \mathbf{Q} &= \begin{bmatrix} -1 & 0 \\ 0 & -1 \end{bmatrix} \\
\mathbf{C}_2(X) &= \begin{bmatrix} 0 & i \\ i & 0 \end{bmatrix} & \mathbf{QC}_2(X) &= \begin{bmatrix} 0 & -i \\ -i & 0 \end{bmatrix} \\
\mathbf{C}_2(Y) &= \begin{bmatrix} 0 & 1 \\ -1 & 0 \end{bmatrix} & \mathbf{QC}_2(Y) &= \begin{bmatrix} 0 & -1 \\ 1 & 0 \end{bmatrix} \\
\mathbf{C}_2(Z) &= \begin{bmatrix} i & 0 \\ 0 & -i \end{bmatrix} & \mathbf{QC}_2(Z) &= \begin{bmatrix} -i & 0 \\ 0 & i \end{bmatrix}
\end{aligned} \tag{11.3.1}$$

which are obtained directly from (9.5.8) with $\omega = \pi$. From these matrices we readily verify that, unlike $\mathscr{D}_2$, the group $\mathscr{D}_2^*$ is not Abelian. Furthermore we see that

$$C_2^2(X) = C_2^2(Y) = C_2^2(Z) = Q \qquad Q^2 = E$$

and Q is thus regarded as a rotation about any axis by 2π. It follows that $\mathscr{D}_2^*$ can be regarded as a group of rotations in which it is necessary to rotate about any axis by 4π in order to return to the identity.

Using the matrices (11.3.1) it is a simple matter to show that in $\mathscr{D}_2^*$ there exists five classes of conjugate elements namely

$$\{E\}; \{Q\}; \{C_2(X), QC_2(X)\}; \{C_2(Y), QC_2(Y)\}; \{C_2(Z), QC_2(Z)\}$$

and consequently $\mathscr{D}_2^*$ admits of five IR's. We already know four of these IR's. They are the IR's of $\mathscr{D}_2$ in which E and Q have the same matrix. If we use (3.4.1) we see that it remains to find a single 2-dimensional IR. This of course is just the representation $E_{\frac{1}{2}}$ which is spanned by the elementary spinor (α, β). It is thus given by (11.3.1).

The character tables for $\mathscr{D}_3^*$, $\mathscr{D}_4^*$ and $\mathscr{D}_6^*$ are derived in Hamermesh (1962).

11.4 The octahedral double group $\mathscr{O}^*$

This important subgroup of $R^*(3)$ consists of the 48 elements $\{R, QR\}$ where R runs over the 24 elements of the octahedral group $\mathscr{O}$. Here Q is a rotation by 2π about any axis and $Q^2 = E$.

So far as the IR's of $\mathscr{O}^*$ are concerned we already know five of them. They are the IR's A_1, A_2, E, T_1 and T_2 of the group $\mathscr{O}$ in which E and Q are assigned the same matrix. In order to find those IR's of $\mathscr{O}^*$ which are not representations of $\mathscr{O}$ we can proceed as follows. In the discussion of the group $R(3)$ we noted that the character of a rotation in a given IR (l) is independent of the axis of rotation. When we go to a subgroup $\mathscr{G}$ of $R(3)$ the above assertion will in general only remain valid for those representations of $\mathscr{G}$ which are IR's of $R(3)$. Thus the IR (1) of $R(3)$ gives the IR T_1 of $\mathscr{O}$ and the character of a rotation in T_1 is dependent of the axis. Similarly the IR (2) of $R(3)$ gives the reducible representation $E \oplus T_2$ of $\mathscr{O}$ and the character of a rotation in $E \oplus T_2$ is independent of the axis. From (9.5.15) we see that the same kind of result holds for the double group $R^*(3)$ and its subgroups.

Consider now the representation Γ of $\mathscr{O}^*$ which is induced by the elementary spinor (α, β). The character $\chi^{(\Gamma)}$ of the element C_n^m is given from (9.5.8) by

$$\chi^{(\Gamma)}(C_n^m) = 2 \cos\left(\frac{m\pi}{n}\right)$$

Using this result the following set of characters is obtained

R	E	$8C_3$	$6C_2'$	$6C_4$	$3C_2$	Q	$8QC_3$	$6QC_2'$	$6QC_4$	$3QC_2$
$\chi^{(\Gamma)}(R)$	2	1	0	$\sqrt{2}$	0	-2	-1	0	$-\sqrt{2}$	0

Now

$$\sum_R |\chi^{(\Gamma)}(R)|^2 = 48$$

and it follows that Γ is a 2-dimensional IR of $\mathcal{O}^*$.

If we form the direct product

$$\Gamma' = A_2 \otimes \Gamma$$

we immediately obtain another 2-dimensional IR of $\mathcal{O}^*$ in which the characters are given by

R	E	$8C_3$	$6C_2'$	$6C_4$	$3C_2$	Q	$8QC_3$	$6QC_2'$	$6QC_4$	$3QC_2$
$\chi^{(\Gamma')}(R)$	2	1	0	$-\sqrt{2}$	0	-2	-1	0	$\sqrt{2}$	0

Thus far we have produced seven IR's of $\mathcal{O}^*$. They are

IR	A_1	A_2	E	T_1	T_2	Γ	Γ'
Dimension	1	1	2	3	3	2	2

According to (3.4.1) we have

$$1+1+4+9+9+4+4 = 32 = 48-(x_1^2+x_2^2+\ldots+x_p^2)$$

with

$$x_1^2+x_2^2+\ldots+x_p^2 = 16 \tag{11.4.1}$$

The equation (11.4.1) admits of several solutions in integers the simplest of which is

$$x_1 = 4 \qquad x_2 = x_3 = \ldots = x_p = 0$$

A 4-dimensional representation can be constructed by taking direct products such as $E \otimes \Gamma$. If we write $\Gamma'' = E \otimes \Gamma$ we readily find the set of characters

R	E	$8C_3$	$6C_2'$	$6C_4$	$3C_2'$	Q	$8QC_3$	$6QC_2'$	$6QC_4$	$3QC_2$
$\chi^{(\Gamma'')}(R)$	4	-1	0	0	0	-4	1	0	0	0

Since

$$\sum_R |\chi^{(\Gamma'')}(R)|^2 = 48$$

it follows that Γ'' is a 4-dimensional IR of $\mathcal{O}^*$ and we have now obtained all the IR's of this double group.

In order to label the IR's Γ, Γ' and Γ'' we use (11.2.3), (3.4.7) and the characters listed above. From this we readily deduce the branching rules

$$(\tfrac{1}{2}) \rightarrow \Gamma$$
$$(\tfrac{3}{2}) \rightarrow \Gamma''$$
$$(\tfrac{5}{2}) \rightarrow \Gamma' \oplus \Gamma''$$

and it follows that we label Γ by $E_{\frac{1}{2}}$, Γ' by $E_{\frac{5}{2}}$ and Γ'' by $U_{\frac{3}{2}}$.

Since there are eight IR's there must exist eight classes of conjugate elements. Now elements in a conjugate class have the same character in all IR's. From this we readily deduce that the eight classes in $\mathcal{O}^*$ are

$$\{E\}; \{Q\}; \{8C_3\}; \{8QC_3\}; \{6C_4\}; \{6QC_4\}; \{6C_2', 6QC_2'\}; \{3C_2, 3QC_2\}$$

The character table for $\mathcal{O}^*$ is shown in Table 29.

TABLE 29. Character table for $\mathcal{O}^*$

$\mathcal{O}^*$	E	$8C_3$	$\{3C_2, 3QC_2\}$	$\{6C_2', 6QC_2'\}$	$6C_4$	Q	$8QC_2$	$6QC_4$
A_1	1	1	1	1	1	1	1	1
A_2	1	1	1	−1	−1	1	1	−1
E	2	−1	2	0	0	2	−1	0
T_1	3	0	−1	−1	1	3	0	1
T_2	3	0	1	1	−1	3	0	1
$E_{\frac{1}{2}}$	2	1	0	0	$\sqrt{2}$	−2	−1	$-\sqrt{2}$
$E_{\frac{5}{2}}$	2	1	0	0	$-\sqrt{2}$	−2	−1	$\sqrt{2}$
$U_{\frac{3}{2}}$	4	−1	0	0	0	−4	1	0

Reference

Hamermesh, M. (1962) "Group Theory and its Application to Physical Problems," Addison-Wesley, Chapter 9.

CHAPTER 12

Direct Products and Coupling Coefficients

12.1 Introduction

One of the most powerful and elegant concepts in the application of group theoretical techniques is the direct product (Kronecker product) of two IR's. The direct product of two representations of a group $\mathcal{G}$ is defined in Section 5.2. The utility of the direct product in the derivation of selection rules is discussed in Section 5.3. In Section 6.7 we considered a particularly important direct product within the symmetric group $\mathcal{S}_n$. The method of irreducible tensors (see Chapter 8) depends entirely upon the use of the direct product.

The most famous direct product decomposition is the one connected with the group $SU(2)$ or $R^*(3)$. This decomposition is called the Clebsch-Gordan theorem and it is closely related to the theory of angular momentum in quantum mechanics.

12.2 Coupling of two angular momenta

It is well known in quantum mechanics that the components of orbital angular momentum $\hat{l}_x$, $\hat{l}_y$ and $\hat{l}_z$ satisfy the commutation relations

$$[\hat{l}_x, \hat{l}_y] = i\hat{l}_z \qquad [\hat{l}_y, \hat{l}_z] = i\hat{l}_x \qquad [\hat{l}_z, \hat{l}_x] = i\hat{l}_y \tag{12.2.1}$$

The operators $\hat{S}_x$, $\hat{S}_y$, $\hat{S}_z$ for the components of electron spin are not capable of being expressed analytically. In the Pauli theory of electron spin the hypothesis is made that electron spin is an intrinsic angular momentum of $\frac{1}{2}$ a.u. and it follows that $\hat{S}_x$, $\hat{S}_y$ and $\hat{S}_z$ can be defined by the commutation relations

$$[\hat{S}_x, \hat{S}_y] = i\hat{S}_z \qquad [\hat{S}_y, \hat{S}_z] = i\hat{S}_x \qquad [\hat{S}_z, \hat{S}_x] = i\hat{S}_y \tag{12.2.2}$$

We now generalize (12.2.1) and (12.2.2) by letting $j = (j_x, j_y, j_z)$ be an arbitrary quantum mechanical angular momentum defined by

$$[\hat{j}_x, \hat{j}_y] = i\hat{j}_z \qquad [\hat{j}_y, \hat{j}_z] = i\hat{j}_x \qquad [\hat{j}_z, \hat{j}_x] = i\hat{j}_y \tag{12.2.3}$$

The CR's (12.2.3) define the Lie algebra A_1 of $SU(2)$ (see Section 9.5) and it follows that the possible values j of a quantum mechanical angular momentum are given by

$$j = 0, \tfrac{1}{2}, 1, \tfrac{3}{2}, 2, \ldots$$

We have already seen that orbital angular momentum eigenstates $|l, m\rangle$, which correspond to integer values $j = l$, form bases for the IR's (l) of $R(3)$ or $R^*(3)$. According to Section 11.1 the angular momentum eigenstates $|j, m\rangle$ which correspond to $j = \frac{1}{2}, \frac{3}{2}, \frac{5}{2}, \ldots$ form bases for IR's (j) of $SU(2)$ or $R^*(3)$ but they do not form bases for any representations of $R(3)$.

Consider now two sets of eigenstates $|j_1, m_1\rangle$ and $|j_2, m_2\rangle$ which span respectively the IR's (j_1) and (j_2) of the group $R^*(3)$. The set of $(2j_1+1)\times(2j_2+1)$ states $|j_1, m_1\rangle\,|j_2, m_2\rangle$ forms a basis for the direct product representation $(j)=(j_1)\otimes(j_2)$. According to (9.5.16) we have the Clebsch-Gordan theorem

$$(j)=(j_1)\otimes(j_2)=(j_1+j_2)\oplus(j_1+j_2-1)\oplus\ldots\oplus(|j_1-j_2|) \qquad (12.2.4)$$

and it follows that the representation space of kets $|j_1, m_1\rangle\,|j_2, m_2\rangle$ is reducible into a direct sum of sub-spaces. Each of these subspaces is the representation space for an IR (j) of $R^*(3)$ with (j) given by (12.2.4). It is convenient to denote the kets which span the IR (j) by $|j_1, j_2, j, m\rangle$. In physical terms the above process is called coupling (or addition) of two angular momenta and the operator $\hat{j}$ defined by

$$\hat{j}=\hat{j}_1+\hat{j}_2$$

is called the resultant (or sum) of $\hat{j}_1$ and $\hat{j}_2$.

Since $|j_1, m_1\rangle\,|j_2, m_2\rangle$ and $|j_1, j_2, j, m\rangle$ are merely two different sets of basis functions for the representation $(j)=(j_1)\otimes(j_2)$ it follows that they are linearly related i.e.

$$|j_1, j_2, j, m\rangle=\sum_{m_1,m_2} c_{m_1m_2}\,|j_1, m_1\rangle\,|j_2, m_2\rangle \qquad (12.2.5)$$

The coefficients $c_{m_1m_2}$ are the elements of the matrix which reduces the representation space $|j_1, m_1\rangle\,|j_2, m_2\rangle$ (see Yutsis, Levison and Vanagas, 1962). According to Section 9.4 we have (with $\hat{j}_z=\sqrt{2}\,\hat{H}$)

$$\hat{j}_z\,|j, m\rangle=m\,|j, m\rangle$$

Now $\hat{j}_z=\hat{j}_{1z}+\hat{j}_{2z}$ and it readily follows from (12.2.5) that

$$m=m_1+m_2$$

The coefficients $c_{m_1m_2}$ are called *coupling coefficients.* For the group $R^*(3)$ these coefficients are usually referred to as Clebsch-Gordan coefficients and they are written out in full as

$$c_{m_1m_2}=(j_1m_1j_2m_2\,|\,jm)$$

We now have

$$|j_1, j_2, j, m\rangle=\sum_{m_1+m_2=m}(j_1m_1j_2m_2\,|\,jm)\,|j_1, m_1\rangle\,|j_2, m_2\rangle \qquad (12.2.6)$$

Since the sets $|j, m\rangle$ and $|j_1, m_1\rangle\ |j_2, m_2\rangle$ are both orthonormal the transformation (12.2.6) is unitary. By an appropriate choice of normalization (see Wigner, 1965) it can be shown that $(j_1m_1j_2m_2 \mid jm)$ may be taken to be real. The transformation which is the inverse of (12.2.6) can now be written as

$$|j_1, m_1\rangle|j_2, m_2\rangle = \sum_{j,m} (j_1m_1j_2m_2 \mid jm)\, |j_1, j_2, j, m\rangle \qquad (12.2.7)$$

In order to evaluate the coefficients $(j_1m_1j_2m_2 \mid jm)$ we use the standard basis relations for the Lie algebra A_1. Since A_1 and B_1 are isomorphic it follows from (9.4.21) and (9.4.18) that the standard basis relations in question are

$$\hat{j}_\pm\, |j, m\rangle = \sqrt{j(j+1) - m(m \pm 1)}\, |j, m \pm 1\rangle \qquad (12.2.8)$$

where $\hat{j}_\pm = 2\hat{E}_{\pm\alpha}$ (see (9.4.13)). As an example we consider the case

$$|1, 2, 2, 2\rangle = (1022 \mid 22)\, |1, 0\rangle\, |2, 2\rangle + (1121 \mid 22)\, |1, 1\rangle\, |2, 1\rangle$$

We have

$$\hat{j}_+\, |1, 2, 2, 2\rangle = 0 = (1022 \mid 22)[\hat{j}_{1+}\, |1, 0\rangle\, |2, 2\rangle + |1, 0\rangle\, (\hat{j}_{2+}\, |2, 2\rangle)]$$
$$+ (1121 \mid 22)[\hat{j}_{1+}\, |1, 1\rangle\, |2, 1\rangle + |1, 1\rangle\, (\hat{j}_{2+}\, |2, 1\rangle)]$$

Now $\hat{j}_{1+}\, |1, 1\rangle = \hat{j}_{2+}\, |2, 2\rangle = 0$. Using (12.2.8) we find

$$0 = \sqrt{2}[(1022 \mid 22) + \sqrt{2}(1121 \mid 22)]\, |1, 1\rangle\, |2, 2\rangle$$

and

$$(1022 \mid 22) = -\sqrt{2}\, (1121 \mid 22)$$

The normalization condition is

$$(1022 \mid 22)^2 + (1121 \mid 22)^2 = 1$$

and it follows that

$$(1121 \mid 22) = +\frac{1}{\sqrt{3}} \qquad (1022 \mid 22) = -\frac{\sqrt{2}}{\sqrt{3}}$$

The method illustrated in the above example can be carried through in the general case and the following formula for the Clebsch-Gordan coefficients is obtained (see Judd, 1963)

$$(j_1m_1j_2m_2 \mid jm)$$
$$= \delta_{m_1+m_2,m}\left[\frac{(2j+1)(j_1+j_2-j)!\,(j_1-m_1)!\,(j_2-m_2)!\,(j+m)!\,(j-m)!}{(j_1+j_2+j+1)!\,(j+j_1-j_2)!\,(j+j_2-j_1)!\,(j_1+m_1)!\,(j_2+m_2)!}\right]^{\frac{1}{2}}$$
$$\times \sum_x (-1)^{j_1-m_1-x} \frac{(j_1+m_1+x)!\,(j_2+j-m_1-x)!}{x!\,(j-m-x)!\,(j_1-m_1-x)!\,(j_2-j+m_1+x)!} \qquad (12.2.9)$$

An extremely useful result can be derived immediately using (12.2.9). It is

$$(j_2m_2j_1m_1 \mid jm) = (-1)^{j_1+j_2-j}(j_1m_1j_2m_2 \mid jm) \qquad (12.2.10)$$

Another useful result depends upon the unitarity of the matrix of coupling coefficients. We have

$$|j_1, j_2, j, m\rangle = \sum_{m_1,m_2} (j_1m_1j_2m_2 \mid jm)\, |j_1, m_1\rangle\, |j_2, m_2\rangle$$

and since the coupling coefficients may be taken to be real we also have

$$\langle j_1, j_2, j', m'| = \sum_{m_1',m_2'} (j_1m_1'j_2m_2' \mid j'm')\, \langle j_1, m_1'|\, \langle j_2, m_2'|$$

Now the angular momentum eigenstates are orthonormal and it follows that

$$\begin{aligned}\langle j_1, j_2, j', m' \mid j_1, j_2, j, m\rangle &= \delta_{jj'}\, \delta_{mm'} \\ &= \sum_{m_1,m_2} \sum_{m_1',m_2'} (j_1m_1'j_2m_2' \mid j'm') \\ &\quad \times (j_1m_1j_2m_2 \mid jm)\langle j_1, m_1' \mid j_1, m_1\rangle\langle j_2, m_2' \mid j_2, m_2\rangle\end{aligned}$$

This clearly reduces to

$$\sum_{m_1,m_2} (j_1m_1j_2m_2 \mid j'm')(j_1m_1j_2m_2 \mid jm) = \delta_{jj'}\, \delta_{mm'} \qquad (12.2.11)$$

For some purposes it is more convenient to replace the Clebsch-Gordan coefficient by the $3j$ symbol which is defined as

$$\begin{pmatrix} j_1 & j_2 & j_3 \\ m_1 & m_2 & m_3 \end{pmatrix} = \frac{(-1)^{j_1-j_2-m_3}}{\sqrt{2j_3+1}} (j_1m_1j_2m_2 \mid j_3 - m_3) \qquad (12.2.12)$$

It can be shown that even permutations of the columns leave the value of the $3j$ symbol unaltered while odd permutations of the columns introduce a phase factor $(-1)^{j_1+j_2+j_3}$ i.e.

$$\begin{pmatrix} j_2 & j_1 & j_3 \\ m_2 & m_1 & m_3 \end{pmatrix} = (-1)^{j_1+j_2+j_3} \begin{pmatrix} j_1 & j_2 & j_3 \\ m_1 & m_2 & m_3 \end{pmatrix} \qquad (12.2.13)$$

Further properties of the $3j$ symbol can be found in Judd (1963).

12.3 Coupling of representations

In this section we consider the analogue of the coupling of orbital angular momenta when we have a finite molecular symmetry group $\mathcal{G}$. In such a case the orbital angular momentum of an electron is "quenched". Thus the IR's of $\mathcal{G}$ which serve to label the various electronic states no longer have the direct physical significance of angular momentum such as occurs in spherical symmetry.

In what follows we shall restrict ourselves to those molecular symmetry groups $\mathscr{G}$ which, like $R^*(3)$, are simply reducible. By a *simply reducible* group we mean:

(a) if R is in $\mathscr{G}$ then for all R the elements R and R^{-1} belong to the same class.
(b) the direct product of any two IR's Γ_1 and Γ_2 of $\mathscr{G}$ contains in its decomposition each IR Γ of $\mathscr{G}$ no more than once. thus if

$$\Gamma_1 \otimes \Gamma_2 = \sum_i^{\oplus} c_i \Gamma_i$$

then for all i we have $c_i = 0$ or $c_i = 1$.

Many of the molecular symmetry groups $\mathscr{G}$ are simply reducible but most of the molecular symmetry double groups $\mathscr{G}^*$ are not. The restriction to simply reducible groups can be removed (see McWeeny, 1963) but when this is done much of the inherent symmetry amongst the coupling coefficients is lost.

It can be shown that if $\mathscr{G}$ is a finite group then its representations belong to one of three types. These are (see Wigner, 1965; Hamermesh, 1962):

Type I. the representation is equivalent to a real representation. Such a representation is called an *integer* representation.
Type II. the representation is not equivalent to a real representation but it is equivalent to its complex conjugate. Such a representation is called a *half-integer* representation.
Type III. the representation is not equivalent to its complex conjugate.

It can be further shown that simply reducible groups have no representations of Type III.

For a simply reducible group the analogue of (12.2.6) is written

$$|\Gamma_1, \Gamma_2, \Gamma, M\rangle = \sum_{M_1, M_2} (\Gamma_1 M_1 \Gamma_2 M_2 \mid \Gamma M) \, |\Gamma_1, M_1\rangle \, |\Gamma_2, M_2\rangle \quad (12.3.1)$$

The coefficients $(\Gamma_1 M_1 \Gamma_2 M_2 \mid \Gamma M)$ are called coupling coefficients for $\mathscr{G}$ and they are clearly non zero only when Γ is contained in $\Gamma_1 \otimes \Gamma_2$. In (12.3.1) the M's serve to label the rows of the standard IR's and the sum runs over all the standard basis functions. The effect of a set of generators on the standard basis functions serves to define a set of standard basis relations for the IR's of $\mathscr{G}$. This set of standard basis relations for $\mathscr{G}$ is the analogue of (12.2.8) in the atomic case. Whenever the group $\mathscr{G}$ is specified we can evaluate coupling coefficients by a consideration of the standard basis relations.

For the sake of simplicity we shall henceforth assume that the standard basis functions can be taken to be real and orthonormal. The standard representations are then real and we have only integer representations to consider. The coupling coefficients can now be taken to be real and the analogue of (12.2.11) is clearly

$$\sum_{M_1,M_2} (\Gamma_1 M_1 \Gamma_2 M_2 \mid \Gamma M)(\Gamma_1 M_1 \Gamma_2 M_2 \mid \Gamma' M') = \delta_{\Gamma\Gamma'}\,\delta_{MM'} \qquad (12.3.2)$$

As an example we now calculate from first principles a set of coupling coefficients for the group $\mathscr{D}_3$. With C_2 and C_3 as generators we have from (3.3.1) and (3.3.2) the following standard basis relations:

IR	Notation for standard basis function	Effect of $\hat{C}_2$	Effect of $\hat{C}_3$
A_1	a_1	a_1	a_1
A_2	a_2	$-a_2$	a_2
E	x	x	$-\frac{1}{2}x - \frac{\sqrt{3}}{2}y$
	y	$-y$	$\frac{\sqrt{3}}{2}x - \frac{1}{2}y$

As an illustration of the process of calculation we take

$$\begin{aligned}|E, E, A_2, a_2\rangle &= (ExEx \mid A_2a_2)\,|E, x\rangle\,|E, x\rangle + (ExEy \mid A_2a_2)\,|E, x\rangle\,|E, y\rangle \\ &\quad + (EyEx \mid A_2a_2)\,|E, y\rangle\,|E, x\rangle + (EyEy \mid A_2a_2)\,|E, y\rangle\,|E, y\rangle\end{aligned}$$

If we apply $\hat{C}_2$ to this equation we immediately deduce that

$$(ExEx \mid A_2a_2) = (EyEy \mid A_2a_2) = 0$$

Thus

$$|E, E, A_2, a_2\rangle = (ExEy \mid A_2a_2)\,|E, x\rangle\,|E, y\rangle + (EyEx \mid A_2a_2)\,|E, y\rangle\,|E, x\rangle$$

Application of $\hat{C}_3$ to this equation gives

$$\begin{aligned}&(ExEy \mid A_2a_2)\,|E, x\rangle\,|E, y\rangle + (EyEx \mid A_2a_2)\,|E, y\rangle\,|E, x\rangle \\ &\quad = (ExEy \mid A_2a_2)\left[-\tfrac{1}{2}|E, x\rangle - \frac{\sqrt{3}}{2}|E, y\rangle\right]\left[\frac{\sqrt{3}}{2}|E, x\rangle - \tfrac{1}{2}|E, y\rangle\right] \\ &\quad\quad + (EyEx \mid A_2a_2)\left[\frac{\sqrt{3}}{2}|E, x\rangle - \tfrac{1}{2}|E, y\rangle\right]\left[-\tfrac{1}{2}|E, x\rangle - \frac{\sqrt{3}}{2}|E, y\rangle\right]\end{aligned}$$

and we deduce that

$$(ExEy \mid A_2a_2) = -(EyEx \mid A_2a_2)$$

We now have

$$|E, E, A_2, a_2\rangle = (ExEyA_2a_2)[|E, x\rangle\,|E, y\rangle - |E, y\rangle\,|E, x\rangle]$$

The normalization condition is

$$(ExEy \mid A_2a_2)^2 = \tfrac{1}{2}$$

and we have

$$(EyEx \mid A_2a_2) = -(ExEy \mid A_2a_2) = \pm\frac{1}{\sqrt{2}}$$

In a similar way we readily find the other non-zero coupling coefficients to be

$$(A_1a_1A_1a_1 \mid A_1a_1) = \pm 1$$

$$(A_1a_1A_2a_2 \mid A_2a_2) = \pm 1$$

$$(A_1a_1Ex \mid Ex) = (A_1a_1Ey \mid Ey) = \pm 1$$

$$(A_2a_2A_1a_1 \mid A_2a_2) = \pm 1$$

$$(A_2a_2A_2a_2 \mid A_1a_1) = \pm 1$$

$$(ExA_1a_1 \mid Ex) = (EyA_1a_1 \mid Ey) = \pm 1$$

$$(ExA_2a_2 \mid Ey) = -(EyA_2a_2 \mid Ex) = \pm 1$$

$$(ExEx \mid A_1a_1) = (EyEy \mid A_1a_1) = \pm\frac{1}{\sqrt{2}}$$

$$(ExEy \mid A_2a_2) = -(EyEx \mid A_2a_2) = \pm\frac{1}{\sqrt{2}}$$

$$(ExEx \mid Ex) = -(EyEy \mid Ex) = -(ExEy \mid Ey) = -(EyEx \mid Ey) = \pm\frac{1}{\sqrt{2}}$$

The choice of sign in any given set of related coupling coefficients is of course arbitrary. Before making a definite choice for the group $\mathscr{D}_3$ we return for the moment to the general case.

For simply reducible groups it is convenient to define a 3Γ symbol by

$$\begin{pmatrix} \Gamma_1 & \Gamma_2 & \Gamma_3 \\ M_1 & M_2 & M_3 \end{pmatrix} = \frac{1}{\sqrt{n_{\Gamma_3}}}(\Gamma_1M_1\Gamma_2M_2 \mid \Gamma_3M_3) \qquad (12.3.3)$$

Here n_{Γ_3} is the dimension of the IR Γ_3. It is clear that the definitions (12.2.12) and (12.3.3) are of similar nature.

In order to maintain as close an analogy as possible with $R(3)$ we introduce the following phase convention. If Γ is an integer IR of a simply reducible molecular symmetry group $\mathscr{G}$ then we define a quantity $(-1)^{\Gamma}$ as follows

(a) $(-1)^{\Gamma_1+\Gamma_2}=(-1)^{\Gamma_1}(-1)^{\Gamma_2}$
(b) $(-1)^{2\Gamma}=1$
(c) $(-1)^{\Gamma}=1$ if Γ is contained in the reduction of any symmetric representation $(\Gamma_i\otimes\Gamma_i)^+$ (see (5.2.6))
(d) $(-1)^{\Gamma}=-1$ if Γ is contained in the reduction of any antisymmetric representation $(\Gamma_i\otimes\Gamma_i)^-$ (see (5.2.6))
(e) $(-1)^{\Gamma}=1$ if Γ is not contained in the reduction of any symmetric or antisymmetric representation.

As far as the author can ascertain the above phase convention for the 3Γ symbols is equivalent to that used for the V coefficients of Griffith (1962). As we shall see the results for the group $\mathscr{D}_3$ are certainly the same. However this whole question of phase is much more complicated than for the group $SU(2)$. The reader who intends to use, in actual computations, coupling coefficients for any molecular symmetry group or molecular symmetry double group should consult the recent work by Konig and Kremer (1973) who claim to have completely standardized the phase.

When $(-1)^{\Gamma}$ is defined as above it can be shown (see Griffith, 1962) that the 3Γ symbol is unchanged by an even permutation of its columns and is multiplied by $(-1)^{\Gamma_1+\Gamma_2+\Gamma_3}$ for an odd permutation of its columns. Thus we have

$$\begin{pmatrix}\Gamma_2 & \Gamma_1 & \Gamma_3\\ M_2 & M_1 & M_3\end{pmatrix}=(-1)^{\Gamma_1+\Gamma_2+\Gamma_3}\begin{pmatrix}\Gamma_1 & \Gamma_2 & \Gamma_3\\ M_1 & M_2 & M_3\end{pmatrix} \tag{12.3.4}$$

This result should be compared with (12.2.13). Further properties of the 3Γ symbol can be found in Griffith (1962).

TABLE 30. 3Γ symbols for $\mathscr{D}_3$

A_1	A_1	A_1	3Γ
a_1	a_1	a_1	1

A_1	A_2	A_2	3Γ
a_1	a_2	a_2	1

A_1	E	E	3Γ
a_1	x	x	$1/\sqrt{2}$
a_1	y	y	$1/\sqrt{2}$

A_2	E	E	3Γ
a_2	x	y	$1/\sqrt{2}$
a_2	y	x	$-1/\sqrt{2}$

E	E	E	3Γ
x	x	x	$-\frac{1}{2}$
x	y	y	$\frac{1}{2}$
y	x	y	$\frac{1}{2}$
y	y	x	$\frac{1}{2}$

If we now return to the group $\mathscr{D}_3$ we see that it is indeed possible to choose the signs of the coupling coefficients so that the above mentioned permutational symmetry is present in the 3Γ symbols. The independent non-vanishing 3Γ symbols for the group $\mathscr{D}_3$ are given in Table 30. According to (5.2.8) we have

$$(-1)^{A_1} = +1 \qquad (-1)^{A_2} = -1 \qquad (-1)^{E} = +1$$

By using (12.3.4) all the 3Γ symbols for the group $\mathscr{D}_3$ can be obtained from Table 30. As we shall presently see these tables of 3Γ symbols are useful when we require to obtain wave functions for spectroscopic states.

12.4 Recoupling of angular momenta

We consider three sets of eigenstates $|j_1, m_1\rangle$; $|j_2, m_2\rangle$ and $|j_3, m_3\rangle$ which form bases for the IR's (j_1), (j_2) and (j_3) of $R^*(3)$ respectively. The set of $(2j_1+1)(2j_2+1)(2j_3+1)$ states $|j_1, m_1\rangle\,|j_2, m_2\rangle\,|j_3, m_3\rangle$ forms a basis for the direct product representation

$$(j) = (j_1) \otimes (j_2) \otimes (j_3)$$

There are two distinct ways in which this direct product can be decomposed. Each of these ways corresponds to what is called a *coupling scheme.* The two coupling schemes in question are

$$[(j_1) \otimes (j_2)] \otimes (j_3) \qquad \text{and} \qquad (j_1) \otimes [(j_2) \otimes (j_3)]$$

The relation (12.2.10) shows that there are no other non-trivial coupling schemes of three angular momenta. If we write $(j_{12}) = (j_1) \otimes (j_2)$ and $(j_{23}) = (j_2) \otimes (j_3)$ so that $(j) = (j_{12}) \otimes (j_3) = (j_1) \otimes (j_{23})$ then we can apply (12.2.6) for each of the coupling schemes. We have

$$\begin{aligned}|(j_1, j_2), j_{12}, j_3, j, m\rangle &= \sum_{m_{12}, m_3} (j_{12}m_{12}j_3m_3\,|\,jm)\,|(j_1, j_2), j_{12}, m_{12}\rangle\,|j_3, m_3\rangle \\ &= \sum_{m_1, m_2, m_3, m_{12}} (j_{12}m_{12}j_3m_3\,|\,jm)(j_1m_1j_2m_2\,|\,j_{12}m_{12})\,|j_1, m_1\rangle\,|j_2, m_2\rangle\,|j_3, m_3\rangle\end{aligned} \tag{12.4.1}$$

and

$$\begin{aligned}&|j_1, (j_2, j_3), j_{23}, j, m\rangle \\ &\quad= \sum_{m_1, m_{23}} (j_1m_1j_{23}m_{23}\,|\,jm)\,|j_1, m_1\rangle\,|(j_2, j_3), j_{23}, m_{23}\rangle \\ &\quad= \sum_{m_1, m_2, m_3, m_{23}} (j_1m_1j_{23}m_{23}\,|\,jm)(j_2m_2j_3m_3\,|\,j_{23}m_{23})\,|j_1, m_1\rangle\,|j_2, m_2\rangle\,|j_3, m_3\rangle\end{aligned} \tag{12.4.2}$$

Now the eigenstates in the two coupling schemes are simply two different sets of basis functions for the representation (j) and as such they are linearly related. This relation may be written as

$$|j_1, (j_2, j_3), j_{23}, j, m\rangle = \sum_{j_{12}} a_{j_{12}} |(j_1, j_2), j_{12}, j_3, j, m\rangle \tag{12.4.3}$$

and $a_{j_{12}}$ is called the *transformation coefficient*. If we use the orthonormality of the eigenstates and (12.2.11) we find

$$\begin{aligned}
&\langle (j_1, j_2), j'_{12}, j_3, j', m' \mid j_1, (j_2, j_3), j_{23}, j, m\rangle \\
&\quad = \sum_{j_{12}} a_{j_{12}} \langle (j_1, j_2), j'_{12}, j_3, j', m' \mid (j_1, j_2), j_{12}, j_3, j, m\rangle \\
&\quad = \sum_{j_{12}} a_{j_{12}} \sum_{m'_{12}, m_{12}, m_3} (j'_{12}m'_{12}j_3m_3 \mid j'm')(j_{12}m_{12}j_3m_3 \mid jm) \\
&\qquad \times \langle (j_1, j_2), j'_{12}, m'_{12} \mid (j_1, j_2), j_{12}, m_{12}\rangle \langle j_3, m_3 \mid j_3, m_3\rangle \\
&\quad = \sum_{j_{12}} a_{j_{12}} \sum_{m_{12}, m_3} (j_{12}m_{12}j_3m_3 \mid j'm')(j_{12}m_{12}j_3m_3 \mid jm)\, \delta_{j_{12}j'_{12}}\, \delta_{m_{12}m'_{12}} \\
&\quad = \sum_{j_{12}} a_{j_{12}}\, \delta_{jj'}\, \delta_{mm'}\, \delta_{j_{12}j'_{12}}\, \delta_{m_{12}m'_{12}} = a_{j'_{12}}
\end{aligned}$$

It follows that the transformation coefficient is given by

$$a_{j_{12}} = \langle (j_1, j_2), j_{12}, j_3, j, m \mid j_1, (j_2, j_3), j_{23}, j, m\rangle \tag{12.4.4}$$

If we apply the operator $\hat{j}_+$ to (12.4.3) we find that the transformation coefficient is independent of m and we now write it as

$$\begin{aligned}
&\langle (j_1, j_2), j_{12}, j_3, j', m' \mid j_1, (j_2, j_3), j_{23}, j, m\rangle \\
&\qquad = \delta_{jj'}\, \delta_{mm'} ((j_1j_2)j_{12}j_3j \mid j_1(j_2j_3)j_{23}j)
\end{aligned} \tag{12.4.5}$$

If we substitute (12.4.1) and (12.4.2) in (12.4.5) we have

$$\begin{aligned}
&((j_1j_2)j_{12}j_3j \mid j_1(j_2j_3)j_{23}j)\, \delta_{jj'}\, \delta_{mm'} \\
&\quad = \sum_{\substack{m_1, m_2, m_3 \\ m_{12}, m_{23}}} (j_{12}m_{12}j_3m_3 \mid j'm')(j_1m_1j_2m_2 \mid j_{12}m_{12}) \\
&\qquad \times (j_1m_1j_{23}m_{23} \mid jm)(j_2m_2j_3m_3 \mid j_{23}m_{23})
\end{aligned} \tag{12.4.6}$$

At this stage we introduce the $6j$ symbol $\begin{Bmatrix} j_1 & j_2 & j_{12} \\ j_3 & j & j_{23} \end{Bmatrix}$ by the equation

$$((j_1j_2)j_{12}j_3j \mid j_1(j_2j_3)j_{23}j) = (-1)^{j_1+j_2+j_3+j} \sqrt{(2j_{12}+1)(2j_{23}+1)} \begin{Bmatrix} j_1 & j_2 & j_{12} \\ j_3 & j & j_{23} \end{Bmatrix} \tag{12.4.7}$$

The $6j$ symbol is related to the Racah coefficient W as follows

$$\begin{Bmatrix} j_1 & j_2 & j_3 \\ j_4 & j_5 & j_6 \end{Bmatrix} = (-1)^{j_1+j_2+j_4+j_5} W(j_1j_2j_5j_4; j_3j_6)$$

We now introduce the $3j$ and $6j$ symbols in order to cast (12.4.6) into the form

$$\begin{Bmatrix} j_1 & j_2 & j_{12} \\ j_3 & j & j_{23} \end{Bmatrix} \frac{\delta_{jj'}\,\delta_{mm'}}{(2j+1)} = \sum_{\substack{m_1,m_2,m_3 \\ m_{12},m_{23}}} (-1)^{j_3-j-3j_1+j_{23}-j_{12}-j_2-m'-m-m_{12}-m_{23}}$$

$$\times \begin{pmatrix} j_{12} & j_3 & j' \\ m_{12} & m_3 & -m' \end{pmatrix} \begin{pmatrix} j_1 & j_2 & j_{12} \\ m_1 & m_2 & -m_{12} \end{pmatrix} \begin{pmatrix} j_1 & j_{23} & j \\ m_1 & m_{23} & -m \end{pmatrix} \begin{pmatrix} j_2 & j_3 & j_{23} \\ m_2 & m_3 & -m_{23} \end{pmatrix} \tag{12.4.8}$$

If we set $j = j'$; $m = m'$ and sum over m we obtain

$$\begin{Bmatrix} j_1 & j_2 & j_{12} \\ j_3 & j & j_{23} \end{Bmatrix} = \sum_{\substack{m_1,m_2,m_3, \\ m_{12},m_{23},m}} (-1)^{j_3-j-3j_1+j_{23}-j_{12}-j_2-2m-m_{12}-m_{23}}$$

$$\times \begin{pmatrix} j_{12} & j_3 & j \\ m_{12} & m_3 & -m \end{pmatrix} \begin{pmatrix} j_1 & j_2 & j_{12} \\ m_1 & m_2 & -m_{12} \end{pmatrix} \begin{pmatrix} j_1 & j_{23} & j \\ m_1 & m_{23} & -m \end{pmatrix} \begin{pmatrix} j_2 & j_3 & j_{23} \\ m_2 & m_3 & -m_{23} \end{pmatrix} \tag{12.4.9}$$

The $6j$ symbol affords the simplest example of what is called a *recoupling* coefficient. Although by no means trivial it is possible to reduce (12.4.9) to a single summation (see Racah, 1942). The result is

$$\begin{Bmatrix} j_1 & j_2 & j_3 \\ j_4 & j_5 & j_6 \end{Bmatrix} = \Delta(j_1j_2j_3)\,\Delta(j_1j_5j_6)\,\Delta(j_4j_2j_6)\,\Delta(j_4j_5j_3)$$

$$\times \sum_x \frac{(-1)^x(x+1)!}{(x-j_1-j_2-j_3)!\,(x-j_1-j_5-j_6)!\,(x-j_4-j_2-j_6)!\,(x-j_4-j_5-j_3)!}$$

$$\times \frac{1}{(j_1+j_2+j_4+j_5-x)!\,(j_2+j_3+j_5+j_6-x)!\,(j_3+j_1+j_6+j_4-x)!} \tag{12.4.10}$$

where

$$\Delta(\alpha\beta\gamma) = \left[\frac{(\alpha+\beta-\gamma)!\,(\alpha-\beta+\gamma)!\,(\beta+\gamma-\alpha)!}{(\alpha+\beta+\gamma+1)!}\right]^{\frac{1}{2}} \tag{12.4.11}$$

It is evident that the $6j$ symbol is invariant under any permutation of its columns. It is also invariant when the upper and lower arguments in any two of its columns are interchanged.

Further properties of the $6j$ symbol can be found in Judd (1963).

We consider next the recoupling of four angular momenta. In particular we shall be concerned with the recoupling which is expressed by the transformation

$$|(j_1, j_3), j_{13}, (j_2, j_4), j_{24}, j, m\rangle$$

$$= \sum_{j_{12},j_{34}} ((j_1j_2)j_{12}(j_3j_4)j_{34}j \mid (j_1j_3)j_{13}(j_2j_4)j_{24}j)|(j_1, j_2), j_{12}, (j_3, j_4), j_{34}, j, m\rangle \tag{12.4.12}$$

The transformation coefficient which occurs in (12.4.12) can readily be expressed in terms of transformation coefficients for recouplings of three angular momenta. Thus we find

$$((j_1j_2)j_{12}(j_3j_4)j_{34}j \mid (j_1j_3)j_{13}(j_2j_4)j_{24}j)$$

$$= \sum_{j'} ((j_1j_2)j_{12}j_{34}j \mid j_1(j_2j_{34})j'j)(j_2(j_3j_4)j_{34}j' \mid j_3(j_2j_4)j_{24}j')(j_1(j_3j_{24})j'j \mid (j_1j_3)j_{13}j_{24}j)$$

This can be re-arranged using (12.2.10) to give

$$((j_1j_2)j_{12}(j_3j_4)j_{34}j \mid (j_1j_3)j_{13}(j_2j_4)j_{24}j)$$

$$= \sum_{j'} (-1)^{2j_1+2j_2+2j_3+j_4+j_{24}+j_{34}-2j-j'}((j_1j_2)j_{12}j_{34}j \mid j_1(j_2j_{34})j'j)$$
$$\times ((j_3j_4)j_{34}j_2j' \mid j_3(j_4j_2)j_{24}j')((j_{24}j_3)j'j_1j \mid j_{24}(j_3j_1)j_{13}j) \quad (12.4.13)$$

We are now in a position to substitute (12.4.9) into (12.4.13). This yields

$$((j_1j_2)j_{12}(j_3j_4)j_{34}j \mid (j_1j_3)j_{13}(j_2j_4)j_{24}j)$$

$$= \sum_{j'} (-1)^{p}(2j'+1)\sqrt{(2j_{12}+1)(2j_{13}+1)(2j_{24}+1)(2j_{34}+1)}$$
$$\times \begin{Bmatrix} j_1 & j_2 & j_{12} \\ j_{34} & j & j' \end{Bmatrix}\begin{Bmatrix} j_3 & j_4 & j_{34} \\ j_2 & j' & j_{24} \end{Bmatrix}\begin{Bmatrix} j_1 & j_3 & j_{13} \\ j_{24} & j & j' \end{Bmatrix} \quad (12.4.14)$$

where

$$(-1)^{p} = (-1)^{4j_1+4j_2+4j_3+2j_4+2j_{24}+2j_{34}}$$

Now j_i is either integral or half-integral and it follows that $(-1)^{4j_i} = 1$. We also have $j' = j_2 + j_{34}$ and it follows that $j_2 + j_{34} - j'$ is an integer. Thus

$$2j_2 + 2j_{34} - 2j' = \text{even integer}$$

Similarly $2j_2 + 2j_4 = 2j_{24} = $ even integer. The phase factor now reduces to $(-1)^{2j'}$.

If we define the $9j$ symbol by

$$\begin{Bmatrix} j_1 & j_2 & j_{12} \\ j_3 & j_4 & j_{34} \\ j_{13} & j_{24} & j \end{Bmatrix}$$

$$= \sum_{j'} (-1)^{2j'}(2j'+1)\begin{Bmatrix} j_1 & j_2 & j_{12} \\ j_{34} & j & j' \end{Bmatrix}\begin{Bmatrix} j_3 & j_4 & j_{34} \\ j_2 & j' & j_{24} \end{Bmatrix}\begin{Bmatrix} j_1 & j_3 & j_{13} \\ j_{24} & j & j' \end{Bmatrix} \quad (12.4.15)$$

then we can write (12.4.14) in the form

$$((j_1j_2)j_{12}(j_3j_4)j_{34}j \mid (j_1j_3)j_{13}(j_2j_4)j_{24}j)$$

$$= \sqrt{(2j_{12}+1)(2j_{13}+1)(2j_{24}+1)(2j_{34}+1)}\begin{Bmatrix} j_1 & j_2 & j_{12} \\ j_3 & j_4 & j_{34} \\ j_{13} & j_{24} & j \end{Bmatrix} \quad (12.4.6)$$

An even permutation of the rows or columns leaves the $9j$ symbol unaltered whilst an odd permutation multiplies the $9j$ symbol by $(-1)^J$ where J is the sum of the nine arguments. The $9j$ symbol is invariant upon transposition. Further properties of the $9j$ symbol can be found in Judd (1963).

12.5 Recoupling of representations

Let $|\Gamma_1, M_1\rangle$; $|\Gamma_2, M_2\rangle$ and $|\Gamma_3, M_3\rangle$ be real standard basis functions for integer IR's Γ_1; Γ_2 and Γ_3 respectively of a simply reducible molecular symmetry group. It is clear that the analogue of (12.4.3) is

$$|\Gamma_1, (\Gamma_2, \Gamma_3), \Gamma_{23}, \Gamma, M\rangle = \sum_{\Gamma_{12}} a_{\Gamma_{12}} |(\Gamma_1, \Gamma_2), \Gamma_{12}, \Gamma_3, \Gamma, M\rangle \quad (12.5.1)$$

where the transformation coefficient $a_{\Gamma_{12}}$ is given by

$$a_{\Gamma_{12}} = \langle(\Gamma_1, \Gamma_2), \Gamma_{12}, \Gamma_3, \Gamma, M \mid \Gamma_1, (\Gamma_2, \Gamma_3), \Gamma_{23}, \Gamma, M\rangle$$

If we apply an element R of $\mathcal{G}$ to (12.5.1) and use (3.4.2) it can be seen that $a_{\Gamma_{12}}$ is independent of M and we write it as

$$a_{\Gamma_{12}} = ((\Gamma_1\Gamma_2)\Gamma_{12}\Gamma_3\Gamma \mid \Gamma_1(\Gamma_2\Gamma_3)\Gamma_{23}\Gamma)$$

This transformation coefficient can be expressed in terms of coupling coefficients and we obtain the analogue of (12.4.5). Thus

$$\begin{aligned}((\Gamma_1\Gamma_2)\Gamma_{12}\Gamma_3\Gamma \mid \Gamma_1(\Gamma_2\Gamma_3)\Gamma_{23}\Gamma) \\ = \sum_{\substack{M_1,M_2,M_3\\M_{12},M_{23}}} (\Gamma_{12}M_{12}\Gamma_3M_3 \mid \Gamma M)(\Gamma_1M_1\Gamma_2M_2 \mid \Gamma_{12}M_{12}) \\ \times (\Gamma_1M_1\Gamma_{23}M_{23} \mid \Gamma M)(\Gamma_2M_2\Gamma_3M_3 \mid \Gamma_{23}M_{23}) \quad (12.5.2)\end{aligned}$$

Since the transformation coefficient is independent of M we obtain from (12.5.2) the following result which is expressed in terms of 3Γ symbols.

$$\begin{aligned}\sum_{M} ((\Gamma_1\Gamma_2)\Gamma_{12}\Gamma_3\Gamma \mid \Gamma_1(\Gamma_2\Gamma_3)\Gamma_{23}\Gamma) \\ = n_\Gamma \,.\, ((\Gamma_1\Gamma_2)\Gamma_{12}\Gamma_3\Gamma \mid \Gamma_1(\Gamma_2\Gamma_3)\Gamma_{23}\Gamma) \\ = \sum_{\text{all } M\text{'s}} (-1)^{\Gamma_1+\Gamma_2+\Gamma_3+\Gamma+2\Gamma_{23}} n_\Gamma \,.\, \sqrt{n_{\Gamma_{12}} \,.\, n_{\Gamma_{23}}} \\ \times \begin{pmatrix}\Gamma_1 & \Gamma_2 & \Gamma_{12}\\ M_1 & M_2 & M_{12}\end{pmatrix}\begin{pmatrix}\Gamma_1 & \Gamma & \Gamma_{23}\\ M_1 & M & M_{23}\end{pmatrix}\begin{pmatrix}\Gamma_2 & \Gamma_{23} & \Gamma_3\\ M_2 & M_{23} & M_3\end{pmatrix}\begin{pmatrix}\Gamma_{12} & \Gamma_3 & \Gamma\\ M_{12} & M_3 & M\end{pmatrix} \\ (12.5.3)\end{aligned}$$

If we define the 6Γ symbol by

$$\begin{Bmatrix} \Gamma_1 & \Gamma_2 & \Gamma_{12} \\ \Gamma_3 & \Gamma & \Gamma_{23} \end{Bmatrix} = \sum_{\text{all } M\text{'s}} \begin{pmatrix} \Gamma_1 & \Gamma_2 & \Gamma_{12} \\ M_1 & M_2 & M_{12} \end{pmatrix} \begin{pmatrix} \Gamma_1 & \Gamma & \Gamma_{23} \\ M_1 & M & M_{23} \end{pmatrix} \begin{pmatrix} \Gamma_2 & \Gamma_{23} & \Gamma_3 \\ M_2 & M_{23} & M_3 \end{pmatrix} \begin{pmatrix} \Gamma_{12} & \Gamma_3 & \Gamma \\ M_{12} & M_3 & M \end{pmatrix} \tag{12.5.4}$$

then (12.5.3) can be written as

$$((\Gamma_1\Gamma_2)\Gamma_{12}\Gamma_3\Gamma \mid \Gamma_1(\Gamma_2\Gamma_3)\Gamma_{23}\Gamma) = (-1)^{\Gamma_1+\Gamma_2+\Gamma_3+\Gamma}\sqrt{n_{\Gamma_{12}} \cdot n_{\Gamma_{23}}} \begin{Bmatrix} \Gamma_1 & \Gamma_2 & \Gamma_{12} \\ \Gamma_3 & \Gamma & \Gamma_{23} \end{Bmatrix} \tag{12.5.5}$$

since for integer representations $(-1)^{2\Gamma_{23}} = +1$. The equation (12.5.5) should be compared with (12.4.9).

It is evident from the definition that the 6Γ symbol is invariant under any permutation of its columns. It is also invariant when the upper and lower arguments of any two of its columns are interchanged. Further properties of the 6Γ symbol can be found in Griffith (1962).

In the group $\mathscr{D}_3$ the only non-necessarily zero non-trivial 6Γ symbols are

$$\begin{Bmatrix} A_2 & E & E \\ A_2 & E & E \end{Bmatrix} \qquad \begin{Bmatrix} A_2 & E & E \\ E & E & E \end{Bmatrix} \qquad \begin{Bmatrix} E & E & E \\ E & E & E \end{Bmatrix}$$

As an example of their calculation we take

$$\begin{aligned} \begin{Bmatrix} A_2 & E & E \\ A_2 & E & E \end{Bmatrix} &= \sum_{i,j,k,l} \begin{pmatrix} A_2 & E & E \\ a_2 & i & j \end{pmatrix} \begin{pmatrix} A_2 & E & E \\ a_2 & k & l \end{pmatrix} \begin{pmatrix} E & E & A_2 \\ i & l & a_2 \end{pmatrix} \begin{pmatrix} E & A_2 & E \\ j & a_2 & k \end{pmatrix} \\ &= \begin{pmatrix} A_2 & E & E \\ a_2 & y & x \end{pmatrix} \begin{pmatrix} A_2 & E & E \\ a_2 & y & x \end{pmatrix} \begin{pmatrix} E & E & A_2 \\ y & x & a_2 \end{pmatrix} \begin{pmatrix} E & A_2 & E \\ x & a_2 & y \end{pmatrix} \\ &\quad + \begin{pmatrix} A_2 & E & E \\ A_2 & x & y \end{pmatrix} \begin{pmatrix} A_2 & E & E \\ a_2 & x & y \end{pmatrix} \begin{pmatrix} E & E & A_2 \\ x & y & a_2 \end{pmatrix} \begin{pmatrix} E & A_2 & E \\ y & a_2 & x \end{pmatrix} \\ &= 2 \cdot \left(\frac{1}{\sqrt{2}}\right)\left(\frac{1}{\sqrt{2}}\right)\left(\frac{1}{\sqrt{2}}\right)\left(\frac{1}{\sqrt{2}}\right) = \frac{1}{2} \end{aligned}$$

In a similar way we find

$$\begin{Bmatrix} A_2 & E & E \\ E & E & E \end{Bmatrix} = \tfrac{1}{2} \quad \text{and} \quad \begin{Bmatrix} E & E & E \\ E & E & E \end{Bmatrix} = 0$$

When we consider the recoupling of four integer representations we define the 9Γ symbol by analogy with (12.4.15) as

$$\begin{Bmatrix} \Gamma_1 & \Gamma_2 & \Gamma_{12} \\ \Gamma_3 & \Gamma_4 & \Gamma_{34} \\ \Gamma_{13} & \Gamma_{24} & \Gamma \end{Bmatrix} = \sum_{\Gamma'} n_{\Gamma'} \begin{Bmatrix} \Gamma_1 & \Gamma_2 & \Gamma_{12} \\ \Gamma_{34} & \Gamma & \Gamma' \end{Bmatrix} \begin{Bmatrix} \Gamma_3 & \Gamma_4 & \Gamma_{34} \\ \Gamma_2 & \Gamma' & \Gamma_{24} \end{Bmatrix} \begin{Bmatrix} \Gamma_1 & \Gamma_3 & \Gamma_{13} \\ \Gamma_{24} & \Gamma & \Gamma' \end{Bmatrix}$$

Some of the properties of the 9Γ symbol can be found in Griffith (1962).

For the group $\mathcal{D}_3$ the only non-necessarily zero non-trivial 9Γ symbols are

$$\begin{Bmatrix} E & E & E \\ E & E & E \\ E & E & E \end{Bmatrix} \quad \begin{Bmatrix} A_2 & E & E \\ E & E & E \\ E & E & E \end{Bmatrix} \quad \begin{Bmatrix} A_2 & E & E \\ E & A_2 & E \\ E & E & E \end{Bmatrix} \quad \begin{Bmatrix} A_2 & E & E \\ E & A_2 & E \\ E & E & A_2 \end{Bmatrix}$$

As an example of their calculation we have

$$\begin{Bmatrix} A_2 & E & E \\ E & A_2 & E \\ E & E & E \end{Bmatrix} = \sum_{\Gamma'} n_{\Gamma'} \begin{Bmatrix} A_2 & E & E \\ E & E & \Gamma' \end{Bmatrix}\begin{Bmatrix} E & A_2 & E \\ E & \Gamma' & E \end{Bmatrix}\begin{Bmatrix} A_2 & E & E \\ E & E & \Gamma' \end{Bmatrix}$$

$$= 2 . \begin{Bmatrix} A_2 & E & E \\ E & E & E \end{Bmatrix}^3 = 2 . \tfrac{1}{8} = \tfrac{1}{4}$$

In a similar way we find

$$\begin{Bmatrix} E & E & E \\ E & E & E \\ E & E & E \end{Bmatrix} = -\begin{Bmatrix} A_2 & E & E \\ E & A_2 & E \\ E & E & A_2 \end{Bmatrix} = \tfrac{1}{4}$$

and

$$\begin{Bmatrix} A_2 & E & E \\ E & E & E \\ E & E & E \end{Bmatrix} = 0$$

For applications of the nj and $n\Gamma$ symbols the reader is referred to Judd (1963) and Griffith (1962) respectively.

References

Griffith, J. S. (1962) "The irreducible tensor method for molecular symmetry groups," Prentice Hall.

Hamermesh, M. (1962) "Group Theory and its application to Physical Problems," Addison-Wesley, Chapter 5.

Judd, B. R. (1963) "Operator Techniques in Atomic Spectroscopy," McGraw-Hill, Chapters 1 and 3.

Konig, E. and Kremer, S. (1973) *Theoret. Chim. Acta (Berl.)*, **32,** 27.

McWeeny, R. (1963) "Symmetry," Pergamon, Chapter 8.

Racah, G. (1942) *Phys. Rev.*, **62,** 438.

Wigner, E. P. (1965) "Quantum Theory of Angular Momentum," Ed. L. C. Biedenharn and H. Van Dam, Academic Press, p. 87.

Yutsis, A. P., Levinson, I. B. and Vanagas, V. V. (1962) "Theory of Angular Momentum," Israel Program for Scientific Translations, Jerusalem.

CHAPTER 13

Subgroups and Branching Rules

13.1 Introduction

In the application of group representation theory to atomic and molecular systems we often require a knowledge of the connections between the IR's of a group $\mathcal{G}$ and those of a subgroup $\mathcal{H}$ of $\mathcal{G}$. The best known example occurs in crystal field theory where we require the branching rules for the reduction $R^+(3) \to \mathcal{O}_h$. Here $R^+(3)$ is the symmetry group of a free atom or ion and $\mathcal{O}_h$ is the symmetry group of a regular octahedral complex.

There are however branching rules of a much more general nature which are of great importance in the classification of many electron states particularly for atomic systems. It is with such branching rules that the present chapter is mainly concerned.

We have already had occasion to consider certain branching rules in previous chapters. Thus in Section 5.4 we constructed branching rules for the reduction $\mathcal{O} \to \mathcal{D}_4$. Branching rules for the reduction $\mathcal{S}_n \to \mathcal{S}_{n-1}$ are obtained by a method which is discussed in Section 6.4. In Table 17 we have branching rules for the reduction $GL(3) \to SU(3)$ and (9.6.14) gives branching rules for the reduction $R(4) \to R(3)$. Finally in Chapter 11 we used branching rules for the reduction $SU(2) \to \mathcal{G}^*$ in order to label certain IR's of double groups.

13.2 Branching rules for the reduction $R^+(3) \to \mathcal{G}$

As a first example we consider $R^+(3) \to \mathcal{O}_h$. We start by considering the simpler case $R(3) \to \mathcal{O}$ in which there are only proper rotations. From (9.4.30) we have

$$\chi^{(l)}(C_2) = \frac{\sin (l+\frac{1}{2})\pi}{\sin \pi/2} = (-1)^l$$

$$\begin{aligned} \chi^{(l)}(C_3) = \frac{\sin (2l+1)\pi/3}{\sin \pi/3} &= 1 \qquad \text{for} \quad l = 0, 3, 6, \ldots \\ &= 0 \qquad \text{for} \quad l = 1, 4, 7, \ldots \\ &= -1 \qquad \text{for} \quad l = 2, 5, 8, \ldots \end{aligned}$$

$$\chi^{(l)}(C_4) = \frac{\sin(l+\frac{1}{2})\pi/2}{\sin \pi/4} = 1 \qquad \text{for} \quad l = 0, 4, 8, \ldots$$

$$= 1 \qquad \text{for} \quad l = 1, 5, 9, \ldots$$

$$= -1 \qquad \text{for} \quad l = 2, 6, 10, \ldots$$

$$= -1 \qquad \text{for} \quad l = 3, 7, 11, \ldots$$

and in terms of the character table for $\mathcal{O}$ we find

$\mathcal{O}$	E	$8C_3$	$3C_2$	$6C_2'$	$6C_4$
A_1	1	1	1	1	1
A_2	1	1	1	−1	−1
E	2	−1	2	0	0
T_1	3	0	−1	−1	1
T_2	3	0	−1	1	−1
S	1	1	1	1	1
P	3	0	−1	−1	1
D	5	−1	1	1	−1
F	7	1	−1	−1	−1
G	9	0	1	1	1

It remains to apply (3.4.7) in order to obtain the following branching rules

IR of $R(3)$	IR's of $\mathcal{O}$
S	A_1
P	T_1
D	E, T_2
F	A_2, T_1, T_2
G	A_1, E, T_1, T_2

For the extended reduction $R^+(3) \to \mathcal{O}_h$ it is evident that the g, u property is preserved and we thus have branching rules such as

$$D_g \to E_g \oplus T_{2g}$$

$$D_u \to E_u \oplus T_{2u}$$

As a second example we consider $R^+(3) \to \mathscr{C}_{3v}$. Now $\mathscr{C}_{3v}$ is a subgroup of $\mathcal{O}_h$ and it can be seen from Figure 4 that corresponding elements in the two groups are

$\mathcal{O}_h$	E	C_3	IC_2'
$\mathscr{C}_{3v}$	E	C_3	σ_v

In terms of the character table for $\mathscr{C}_{3v}$ we now have

$\mathscr{C}_{3v}$	E	$2C_3$	$3\sigma_v$
A_1	1	1	1
A_2	1	1	−1
E	2	−1	0
A_{1g}	1	1	1
A_{1u}	1	1	−1
A_{2g}	1	1	−1
A_{2u}	1	1	1

$\mathscr{C}_{3v}$	E	$2C_3$	$3\sigma_v$
E_g	2	−1	0
E_u	2	−1	0
T_{1g}	3	0	−1
T_{1u}	3	0	1
T_{2g}	3	0	1
T_{2u}	3	0	−1

It remains to apply (3.4.7) in order to obtain the following branching rules for the reduction $\mathscr{O}_h \to \mathscr{C}_{3v}$:

IR of $\mathscr{O}_h$:	A_{1g}	A_{1u}	A_{2g}	A_{2u}	E_g	E_u	T_{1g}	T_{1u}	T_{2g}	T_{2u}
IR's of $\mathscr{C}_{3v}$:	A_1	A_2	A_2	A_1	E	E	A_2, E	A_2, E	A_1, E	A_2, E

In order to obtain branching rules for $R^+(3) \to \mathscr{C}_{3v}$ we simply combine the results for the two reductions $R^+(3) \to \mathscr{O}_h$ and $\mathscr{O}_h \to \mathscr{C}_{3v}$. The first few of these branching rules are given as follows

IR of $R^+(3)$:	S_g	S_u	P_g	P_u	D_g	D_u	F_g	F_u
IR's of $\mathscr{C}_{3v}$:	A_1	A_2	A_2, E	A_1, E	$A_1, 2E$	$A_2, 2E$	$A_1, 2A_2, 2E$	$2A_1, A_2, 2E$

A fairly comprehensive set of the above kind of branching rules can be found in Herzberg (1966).

13.3 Branching rules for the reduction $SU(M) \to R(M)$

In this section we shall be concerned with branching rules for the reduction $SU(2l+1) \to R(2l+1)$ where l is a positive integer. In particular we restrict ourselves to those IR's $[\lambda_1\lambda_2 \ldots \lambda_{2l}]$ of $SU(2l+1)$ for which $\lambda_i \leq 2$; $i = 1, 2, \ldots, 2l$. We make this restriction because Young shapes for spin functions can have at most two rows.

The complexity of the character formulae for IR's of $SU(2l+1)$ and $R(2l+1)$ makes the derivation of branching rules by their use very involved. It is fortunate however that their exists an alternative method for obtaining such branching rules. We shall state this method here without proof and give an example of how to apply it (see Littlewood, 1940).

Draw the shape of the IR $[\lambda_1\lambda_2 \ldots \lambda_{2l}]$ of $SU(2l+1)$ and perform successively the following operations

1. Leave the shape untouched.
2. Delete the cells at the foot of the two columns if possible

3. Take the new shape and delete the cells at the foot of its two columns if possible
4. Continue in the above manner until a shape with a single column is obtained or until no cell remains.

At this stage we have a set of shapes deriving from the initial shape $[\lambda_1\lambda_2 \dots \lambda_{2l}]$. We now associate with each such shape a partition symbol $(\lambda_1\lambda_2 \dots)$. When this is done one of the two types of symbol is obtained. Either the symbol contains at most l integers or else it contains more than l integers. If the symbol contains no more than l integers it is the label for an IR of $R(2l+1)$. If the symbol contains more than l integers it cannot be the label for an IR of $R(2l+1)$. However we can use a modification rule due to Murnaghan (1963) which enables us to replace the symbol by another which does label an IR of $R(2l+1)$. This modification rule is expressed by

$$(\underbrace{22\dots2}_{\leftarrow a\rightarrow}\ \underbrace{11\dots1}_{\leftarrow b\rightarrow}) \equiv (\underbrace{22\dots2}_{\leftarrow a\rightarrow}\ \underbrace{11\dots\dots\dots1}_{\leftarrow(2l+1-2a-b)\rightarrow}\underbrace{0\ 0\dots\dots0}_{\leftarrow(l-a-c)})$$

As an example we consider the IR $[2^2\,1]$ of $SU(5)$ and the reduction $SU(5)\rightarrow R(5)$. We have

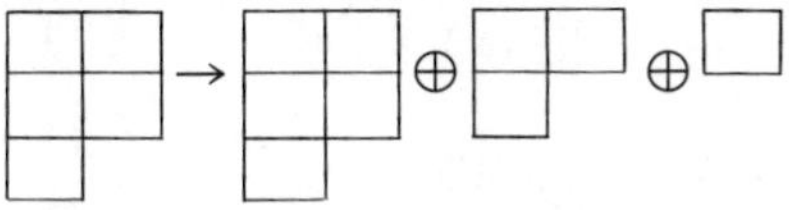

$$[2^2\,1] \rightarrow (221) \oplus (21) \oplus (10)$$

The modification rule gives $(221)\equiv(22)$ and we obtain the branching rule

$$[2^2 1]\rightarrow(22)\oplus(21)\oplus(10)$$

Branching rules for the cases $l=1$ and 2 are given in Tables 31 and 32 respectively. The branching rules for the case $l=3$ are given in Judd (1963).

TABLE 31. Branching rules for $SU(3)\rightarrow R(3)$

IR of $SU(3)$	IR's of $R(3)$
[00]	S
[10]	P
[20]	S, D
[11]	P
[21]	P, D
[22]	S, D

TABLE 32. Branching rules for $SU(5) \to R(5)$

IR of $SU(5)$	IR's of $R(5)$
[0]	(00)
[1]	(10)
[2]	(00), (20)
[11]	(11)
[21]	(10), (21)
[111]	(11)
[1111]	(10)
[211]	(11), (21)
[22]	(00), (20), (22)
[2111]	(11), (20)
[221]	(10), (21), (22)

13.4 Branching rules for the reduction $SU(M) \to R(3)$

The natural IO's for the Lie algebra A_{M-1} of the group $SU(M)$ are readily found to be (see Section 9.7)

$$\hat{I}_{rs} = x_r \frac{\partial}{\partial x_s}$$

with $\sum_r \hat{I}_{rr} = 0$. The CR's are clearly

$$[\hat{I}_{rs}, \hat{I}_{tu}] = \delta_{st}\hat{I}_{ru} - \delta_{ur}\hat{I}_{ts}$$

If we introduce a new basis by writing

$$\hat{J}_{rs} = \hat{I}_{rs} - \hat{I}_{sr}$$

then the CR's become

$$[\hat{J}_{rs}, \hat{J}_{tu}] = \delta_{st}\hat{J}_{ru} + \delta_{ur}\hat{J}_{st} + \delta_{tr}\hat{J}_{us} + \delta_{us}\hat{J}_{tr}$$

In particular for $s = t$, $u \neq r$, $t \neq r$, $u \neq s$ we have

$$[\hat{J}_{rs}, \hat{J}_{su}] = \hat{J}_{ru}$$

If we write

$$\hat{X}_1 = \hat{J}_{rs} \qquad \hat{X}_2 = \hat{J}_{su} \qquad \hat{X}_3 = \hat{J}_{ru}$$

then we obtain the CR's

$$[\hat{X}_1, \hat{X}_2] = \hat{X}_3 \qquad [\hat{X}_2, \hat{X}_3] = \hat{X}_1 \qquad [\hat{X}_3, \hat{X}_1] = \hat{X}_2$$

These CR's describe the Lie algebra B_1 of $R(3)$ (see (9.4.8)) and it follows that $R(3)$ is a subgroup of $SU(M)$.

It is clear that the $(2j+1)$ basis functions $|j, m\rangle$ which span the IR (j) of

$R^*(3)$ also serve as a basis for the IR [1] of $SU(2j+1)$ since they form a (complex) orthonormal set. Thus we have the branching rule

$$[1] \to (j) \tag{13.4.1}$$

The set of $(2j+1)^2$ functions $|j, m_1\rangle |j, m_2\rangle$ now form a basis not only for the direct product $(j) \otimes (j)$ of $R^*(3)$ but also for the direct product $[1] \otimes [1]$ of $SU(2j+1)$. According to (9.5.16) we have

$$(j) \otimes (j) = (2j) \oplus (2j-1) \oplus (2j-2) \oplus \ldots \oplus (1) \oplus (0)$$

and from Section 8.6 we have

$$[1] \otimes [1] = [2] \oplus [1^2]$$

It follows that

$$[2] \otimes [1^2] \to (2j) \oplus (2j-1) \oplus \ldots \oplus (0) = \sum_{J=0}^{2j \oplus} (J) \tag{13.4.2}$$

Now the functions which span [2] are symmetric whilst those which span $[1^2]$ are antisymmetric. According to (12.2.6) we have

$$|j, j, J, M\rangle = \sum_{m_1+m_2=M} (jm_1 jm_2 \mid JM) \, |j, m_1\rangle \, |j, m_2\rangle$$

which can be written in terms of coordinates as

$$\Psi_{J,M}(1, 2) = \sum_{m_1+m_2=M} (jm_1 jm_2 \mid JM) \psi_{j,m_1}(1) \psi_{j,m_2}(2)$$

Now

$$\begin{aligned} \Psi_{J,M}(2, 1) &= \sum_{m_1,m_2} (jm_1 jm_2 \mid JM) \psi_{j,m_1}(2) \psi_{j,m_2}(1) \\ &= \sum_{m_1,m_2} (jm_2 jm_1 \mid JM) \psi_{j,m_1}(1) \psi_{j,m_2}(2) \\ &= (-1)^{2j-J} \Psi_{J,M}(1, 2) \end{aligned}$$

where we have used the property (12.2.10). When $(2j-J)$ is even the function is symmetric and when $(2j-J)$ is odd the function is antisymmetric. It follows from (13.4.2) that

$$\left.\begin{aligned} [2] &\to (0) \oplus (2) \oplus (4) \oplus \ldots \oplus (2j) \\ [1^2] &\to (1) \oplus (3) \oplus (5) \oplus \ldots \oplus (2j-1) \end{aligned}\right\} j \text{ integral} \tag{13.4.3}$$

$$\left.\begin{aligned} [2] &\to (1) \oplus (3) \oplus \ldots \oplus (2j) \\ [1^2] &\to (0) \oplus (2) \oplus \ldots \oplus (2j-1) \end{aligned}\right\} j \text{ half integral} \tag{13.4.4}$$

As an example we shall now consider the reduction $SU(5) \to R(3)$ which corresponds to $j=2$. We have

$$[1] \to D$$
$$[2] \to S \oplus D \oplus G$$
$$[1^2] \to P \oplus F$$

and we can obtain further branching rules by using a recursive method based on the construction of direct products. The analysis is greatly simplified when an equivalence, which we now describe, is introduced.

We have seen that the IR's of $R(M)$ are self-contragredient. It follows that two contragredient representations of $SU(M)$ give rise to the same representations of the subgroup $R(M)$ (or $R(3)$). Thus according to (8.2.4) the IR's $[\lambda_1\lambda_2 \ldots \lambda_M]$ and $[\lambda_1-\lambda_M\lambda_1-\lambda_{M-1} \ldots \lambda_1-\lambda_2\ 0]$ of $SU(M)$ can be regarded as equivalent when they are considered as representations of the subgroup $R(M)$ (or $R(3)$). We write this equivalence as

$$[\lambda_1\lambda_2 \ldots \lambda_M] \stackrel{R}{=} [\lambda_1-\lambda_M\ \lambda_1-\lambda_{M-1} \ldots \lambda_1-\lambda_2\ 0] \qquad (13.4.5)$$

Returning to our example $SU(5) \to R(3)$ we have

$$[2] \otimes [1] = [21] \oplus [3]$$
$$\to (S \oplus D \oplus G) \otimes D = S \oplus P \oplus 3D \oplus 2F \oplus 2G \oplus H \oplus I \qquad (13.4.6)$$

and

$$[1^2] \otimes [1] = [21] \oplus [1^3]$$
$$\to (P \oplus F) \otimes D = 2P \oplus 2D \oplus 2F \oplus G \oplus H \qquad (13.4.7)$$

Now $[1^3] \stackrel{R}{=} [1^2] \to P \oplus F$ and it follows from (13.4.7) that

$$[21] \to P \oplus 2D \oplus F \oplus G \oplus H$$

When this result is used in (13.4.6) we find

$$[3] \to S \oplus D \oplus F \oplus G \oplus I$$

We can now consider

$$[21] \otimes [1] = [31] \oplus [2^2] \oplus [21^2]$$
$$\to (P \oplus 2D \oplus F \oplus G \oplus H) \otimes D$$
$$= 2S \oplus 4P \oplus 5D \oplus 6F \oplus 5G \oplus 3H \oplus 2I \oplus K$$

We also have

$$[2] \otimes [1^2] = [31] \oplus [21^2]$$
$$\to (S \oplus D \oplus G) \otimes (P \oplus F) = 4P \oplus 3D \oplus 5F \oplus 3G \oplus 3H \oplus I \oplus K \qquad (13.4.8)$$

and it follows that

$$[2^2] \to 2S \oplus 2D \oplus F \oplus 2G \oplus I$$

We take next

$$[1^2]\otimes[1^2]=[1^4]\oplus[21^2]\oplus[2^2]$$
$$\rightarrow(P\oplus F)\otimes(P\oplus F)=2S\oplus 2P\oplus 4D\oplus 3F\oplus 3G\oplus H\oplus I$$

Now $[1^4]\stackrel{R}{\equiv}[1]\rightarrow D$ and we already know $[2^2]$. Thus

$$[21^2]\rightarrow 2P\oplus D\oplus 2F\oplus G\oplus H$$

and we now obtain [31] from (13.4.8) as

$$[31]\rightarrow 2P\oplus 2D\oplus 3F\oplus 2G\oplus 2H\oplus I\oplus K$$

Finally we consider

$$[3]\otimes[1]=[4]\oplus[31]$$
$$\rightarrow(S\oplus D\oplus F\oplus G\oplus I)\otimes D$$
$$=S\oplus 2P\oplus 4D\oplus 3F\oplus 4G\oplus 3H\oplus 2I\oplus K\oplus L$$

from which we obtain

$$[4]\rightarrow S\oplus 2D\oplus 2G\oplus H\oplus I\oplus L$$

Branching rules for the reduction $SU(5)\rightarrow R(3)$ are given in Table 33. For branching rules $SU(7)\rightarrow R(3)$ the reader is referred to Judd (1963).

TABLE 33. Branching rules for $SU(5)\rightarrow R(3)$

IR [λ] of $SU(5)$	IR's (J) of R(3) J = 0 S	1 P	2 D	3 F	4 G	5 H	6 I	7 K	8 L	9 M
[0]	1									
[1]	.	.	1							
[2]	1	.	1	.	1					
[1²]	.	1	.	1						
[3]	1	.	1	1	1	.	1			
[21]	.	1	2	1	1	1				
[4]	1	.	2	.	2	1	1	.	1	
[31]	.	2	2	3	2	2	1	1		
[2²]	2	.	2	1	2	.	1			
[21²]	.	2	1	2	1	1				
[41]	1	2	3	4	4	3	3	2	1	1
[32]	1	2	4	3	4	3	2	1	1	
[31²]	.	3	2	4	2	3	1	1		
[2²1]	1	1	3	2	2	1	1			
[21⁴]	.	1	1	1	1					

13.5 Branching rules for the reduction $R(M)\rightarrow R(3)$

By examining the CR's for the Lie algebra of $R(M)$ it is not hard to see that $R(3)$ is a subgroup of $R(M)$. Branching rules for the reduction

$R(M) \rightarrow R(3)$ can be obtained by a consideration of those for the reductions $SU(M) \rightarrow R(M)$ and $SU(M) \rightarrow R(3)$.

As an example we take $R(5) \rightarrow R(3)$. From Table 32 and Table 33 we have

$$[0] \begin{array}{l} \nearrow (00) \\ \searrow S \end{array} \qquad\qquad [1] \begin{array}{l} \nearrow (10) \\ \searrow D \end{array}$$

$$[2] \begin{array}{l} \nearrow (00) \oplus (20) \\ \searrow S \oplus D \oplus G \end{array} \qquad\qquad [11] \begin{array}{l} \nearrow (11) \\ \searrow P \oplus F \end{array}$$

$$[21] \begin{array}{l} \nearrow (10) \oplus (21) \\ \searrow P \oplus 2D \oplus F \oplus G \oplus H \end{array}$$

and it immediately follows that

$$(00) \rightarrow S$$
$$(10) \rightarrow D$$
$$(11) \rightarrow P \oplus F$$
$$(20) \rightarrow D \oplus G$$
$$(21) \rightarrow P \oplus D \oplus F \oplus G \oplus H$$

Similarly

$$[22] \begin{array}{l} \nearrow (00) \oplus (20) \oplus (22) \\ \searrow 2S \oplus 2D \oplus F \oplus 2G \oplus I \end{array}$$

and we have

$$(22) \rightarrow S \oplus D \oplus F \oplus G \oplus I$$

These branching rules are collected together in Table 34 and branching rules for the reduction $R(7) \rightarrow R(3)$ are given in Judd (1963). When we consider f orbitals the exceptional Lie group corresponding to the Lie algebra G_2 (see Section (A1.3)) comes into play. This is because the new group (also denoted by G_2) is a subgroup of $R(7)$ and further $R(3)$ is a subgroup of G_2. Branching rules for the reductions $R(7) \rightarrow G_2 \rightarrow R(3)$ can be found in Judd (1963).

TABLE 34. Branching rules for $R(5) \rightarrow R(3)$

IR of $R(5)$	IR's of $R(3)$
(00)	S
(10)	D
(20)	D, G
(11)	P, F
(21)	P, D, F, G, H
(22)	S, D, F, G, I

References

Herzberg, G. (1966) "Molecular Spectra and Molecular Structure," Van Nostrand, Vol. III, Appendix 4.

Judd, B. R. (1963) "Operator Techniques in Atomic Spectroscopy," McGraw-Hill, Chapter 5.

Littlewood, D. E. (1940) "Theory of Group Characters," Oxford.

Murnaghan, F. D. (1963) "The Theory of Group Representations," Dover.

CHAPTER 14

Classification of Many-Electron States

14.1 Coupling schemes for atomic systems

The Hamiltonian for the non-relativistic motion of N electrons in the field of a nucleus of charge Z is

$$\hat{H}' = \sum_{i=1}^{N} \left(-\tfrac{1}{2}\nabla_i^2 - \frac{Z}{r_i} \right) + \sum_{i>j=1}^{N}\sum \frac{1}{r_{ij}} \tag{14.1.1}$$

As Z increases relativistic effects become important and the spin-orbit interaction must be taken into account. Let l_i denote the orbital angular momentum of the ith electron and let s_i denote the spin of the ith electron. The Hamiltonian with spin-orbit interaction included can be reduced to the form

$$\hat{H} = \hat{H}' + \sum_{i=1}^{N} \xi(r_i)\hat{\mathbf{l}}_i \cdot \hat{\mathbf{s}}_i \tag{14.1.2}$$

where H' is given by (14.1.1) and where ξ is a radial function which depends upon the potential field.

When the spin-orbit interaction is negligible by comparison with the electrostatic repulsions the spin-free Hamiltonian H' is appropriate. This is the situation in which we can speak of Russell-Saunders or LS coupling. At the other extreme when the electrostatic repulsions are negligible by comparison with the spin-orbit interaction the appropriate Hamiltonian is

$$\hat{H}'' = \sum_{i=1}^{N} \left(-\tfrac{1}{2}\nabla_i^2 - \frac{Z}{r_i} + \xi(r_i)\hat{\mathbf{l}}_i \cdot \hat{\mathbf{s}}_i \right) \tag{14.1.3}$$

and in this case "jj coupling" is appropriate. In the case of LS coupling the spin-orbit interaction is regarded as a perturbation, while in the case of jj coupling the electrostatic repulsions are regarded as a perturbation. Both these extremes are of course approximations and for intermediate strengths of the spin-orbit interaction the full Hamiltonian $\hat{H}$ must be used.

14.2 Terms in *LS* coupling

We begin by writing the Hamiltonian (14.1.1) in the form

$$\hat{H}' = \hat{H}_0 + \hat{H}_1 \tag{14.2.1}$$

with

$$\hat{H}_0 = \sum_{i=1}^{N} \left(-\tfrac{1}{2}\nabla_i^2 - \frac{Z}{r_i}\right) \tag{14.2.2}$$

and

$$\hat{H}_1 = \sum_{i>j=1}^{N}\sum \frac{1}{r_{ij}} \tag{14.2.3}$$

The Schrödinger equation

$$\hat{H}'\Phi = E\Phi \tag{14.2.4}$$

can now be solved approximately by using perturbation theory. We introduce the perturbation expansions

$$\Phi = \Phi_0 + \Phi_1 + \Phi_2 + \ldots \tag{14.2.5}$$

$$E = E_0 + E_1 + E_2 + \ldots \tag{14.2.6}$$

and we have the well-known perturbation formulae

$$\hat{H}_0\Phi_0 = E_0\Phi_0 \tag{14.2.7}$$

$$(\hat{H}_0 - E_0)\Phi_1 + (\hat{H}_1 - E_1)\Phi_0 = 0 \tag{14.2.8}$$

$$E_0 = \langle\Phi_0|\, \hat{H}_0\, |\Phi_0\rangle \tag{14.2.9}$$

$$E_1 = \langle\Phi_0|\, \hat{H}_1\, |\Phi_0\rangle \tag{14.2.10}$$

$$E_2 = \langle\Phi_0|\, \hat{H}_1\, |\Phi_1\rangle \tag{14.2.11}$$

The zeroth-order equation (14.2.7) clearly has solutions of the form

$$\Phi_0(1, 2, \ldots, N) = \phi_{n_1,l_1,m_1}(1)\phi_{n_2,l_2,m_2}(2) \ldots \phi_{n_N,l_N,m_N}(N) \tag{14.2.12}$$

$$E_0 = \sum_{i=1}^{N} \varepsilon_{n_i,l_i,m_i} \tag{14.2.13}$$

where $\phi_{n,l,m}$ is a hydrogen-like atomic orbital with energy $\varepsilon_{n,l,m}$. It is customary to associate with Φ_0 an electron configuration which we write as

$$\begin{gathered}(n_1l_1)^{x_1}(n_2l_2)^{x_2} \ldots (n_tl_t)^{x_t} \\ x_1 + x_2 + \ldots + x_t = N\end{gathered} \tag{14.2.14}$$

As an example the ground state electron configuration for the nitrogen atom is

$$(1s)^{x_1}(2s)^{x_2}(2p)^{x_3}$$

with $x_1 = 2$; $x_2 = 2$ and $x_3 = 3$.

We consider for the moment the case $N = 2$. The zeroth order eigenfunctions are clearly simultaneous eigenfunctions of $\hat{H}$, $\hat{l}_1^2$, $\hat{l}_2^2$, $\hat{l}_{1z}$ and $\hat{l}_{2z}$. For given n_1l_1, n_2l_2 the set of $(2l_1+1)(2l_2+1)$ eigenfunctions $\phi_{n_1,l_1,m_1}(1)$ $\phi_{n_2,l_2,m_2}(2)$ all have the same energy. They also form a basis for the direct product $(l_1) \otimes (l_2)$ of $R(3)$. When the interaction between the electrons is included it is now clear that we must use degenerate perturbation theory and solve a $(2l_1+1)(2l_2+1)\times(2l_1+1)(2l_2+1)$ secular problem in order to find the first order contribution to the energy. In general the $(2l_1+1)(2l_2+1)$-fold degeneracy of the zeroth order level will be at least partially lifted by the perturbation. Thus, up to the first order, a set of energy levels will be obtained. The zeroth order wave functions which describe these energy levels are given as usual by linear combinations

$$\Phi_0 = \sum_{m_1,m_2} c_{m_1m_2}\phi_{n_1,l_1,m_1}(1)\phi_{n_2,l_2,m_2}(2)$$

of the unperturbed wave functions. Here $c_{m_1m_2}$ are the coefficients which occur in the secular equations. If we define the total orbital angular momentum by

$$\hat{L} = \hat{l}_1 + \hat{l}_2$$

then it is clear from (12.2.6) that the coefficients $c_{m_1m_2}$ can be taken to be the Clebsch Gordan coefficients and we have

$$\Phi_0(\alpha l_1l_2LM \mid 1, 2) = \sum_{m_1+m_2=M} (l_1m_1l_2m_2 \mid LM)\phi_{n_1,l_1,m_1}(1)\phi_{n_2,l_2,m_2}(2)$$

Here α denotes any additional quantum numbers which may be required to specify the states completely.

From a group theoretical point of view we see that when we allow for the interaction between the electrons the Hamiltonian is no longer invariant under separate rotations of the electron coordinates. It is only invariant under simultaneous rotations of both electrons coordinates. In other words when the electrons are coupled the eigenstates are only simultaneously eigenstates of $\hat{H}'$, $\hat{L}^2$ and $\hat{L}_z$.

We see now that up to first order the energy levels are described by an eigenvalue L of $\hat{L}^2$. Since $\hat{H}'$ does not depend on spin we can clearly define the total spin by

$$\hat{S} = \hat{s}_1 + \hat{s}_2$$

Now $\hat{H}'$ commutes with $\hat{S}^2$ and $\hat{S}_z$ and the energy levels up to first order can also be described by an eigenvalue S of $\hat{S}^2$. The wave functions which describe these energy levels are called terms and a given term is denoted by ${}^{2S+1}L$.

We now return to the general case and consider, from a group theoretical point of view, a configuration $(nl)^N$ in which all the electrons are equivalent. When interaction between the electrons is taken into account it is necessary to obtain the terms ${}^{2S+1}L$ which arise. We already know the possible spin values S (see Table 14). Since $\phi_{n,l,m}$ is a basis function for the

IR (l) of $R(3)$ the values of L are obtained by decomposing the direct product

$$\underbrace{(l)\otimes(l)\ldots\otimes(l)}_{N \text{ factors}}$$

The problem now is to find which L values are to be associated with a given S value. This can be done by associating with each L value a definite permutational symmetry in the spatial function Φ. Now the $(2l+1)$ functions $\phi_{n,l,m}$ not only provide a basis for the IR (l) of $R(3)$ but they also provide a basis for the $(2l+1)$-dimensional IR $[10\ldots0]$ of the group $SU(2l+1)$. Thus the product functions $(nl)^N$ span the direct product representation

$$\underbrace{\square\otimes\square\otimes\ldots\quad\ldots\otimes\square}_{N \text{ factors}}$$

of $SU(2l+1)$. This direct product can be decomposed into IR's of $SU(2l+1)$. These IR's are given in terms of Young shapes and thus the basis functions have a definite permutational symmetry. It now follows that the branching rules for the reduction

$$SU(2l+1)\rightarrow R(3)$$

provide a means of determining which L values occur with a given permutational symmetry. According to (6.7.4) the Pauli principle requires that the permutational symmetry of the spatial function be dual to that of the spin function. The above considerations enable us to pick out the allowed terms very quickly.

We begin by considering $(np)^N$. From Table 14 and Table 31 we can immediately obtain Table 35 which is displayed in Appendix 2.

For the configurations $(nd)^N$ we use Table 14 and Table 33 to obtain Table 36 which is displayed in Appendix 2. To find the terms arising from the configurations $(nf)^N$ we refer the reader to Judd (1963). In passing it should be noted that the configurations $(nl)^x$ and $(nl)^{4l+2-x}$ give rise to the same set of terms (see Judd, 1963). It follows that we need only consider configurations up to and including the half-filled shell.

14.3 Seniority

If we examine Table 36 we see that configurations with more than two equivalent electrons can give rise to several terms with the same L and S values. Thus L and S are insufficient to completely classify the energy levels. To overcome this difficulty we attempt to find a further quantum number α which has different values for terms with the same L and S. Because the eigenfunctions for a term are simultaneous eigenfunctions of

$\hat{H}'$, $\hat{L}^2$ and $\hat{S}^2$ we say that L and S are *good* quantum numbers. By this we mean that L and S can be used to classify the different stationary states of the atomic system. If another constant of the motion could be found then its eigenvalues would serve as an additional label for a term. Unfortunately this has not yet been found possible. However in addition to being good quantum numbers L and S also have group theoretical significance. If we can find a subgroup $\mathcal{G}$ of $SU(2l+1)$ which has $R(3)$ as one of its subgroups then it is clear that the branching rules

$$SU(2l+1) \rightarrow \mathcal{G} \rightarrow R(3)$$

serve to associate a definite IR α of $\mathcal{G}$ with a given S and L. Although it is not essential for the purpose of state classification nevertheless it would be extremely useful if $\mathcal{G}$ was a symmetry group of the Hamiltonian. In that case α would be a good quantum number and the classified states would be exact eigenstates of $\hat{H}'$. This is not possible since further constants of the motion have not yet been found. However even if the Hamiltonian commutes approximately with the transformations of $\mathcal{G}$ then the quantum number α will have some kind of physical significance.

In the previous chapter we found that the rotation group $R(2l+1)$ is a subgroup of $SU(2l+1)$ and that $R(3)$ is a subgroup of $R(2l+1)$. It follows that the IR's $(w_1 w_2 \ldots w_l)$ of $R(2l+1)$ can be used as an extra label with which to classify the terms arising from the configurations l^N. Since the Young shapes for spin functions are limited to at most two rows it follows that $w_i \leqslant 2$ for all i and we may suppose in general that

$$w_1 = w_2 = \ldots = w_a = 2 \qquad (14.3.1)$$

$$\left.\begin{aligned} w_{a+1} &= \ldots = w_{a+b} = 1 \\ w_{a+b+1} &= \ldots = w_l = 0 \end{aligned}\right. \qquad (14.3.2)$$

For a term with a given spin value S it is convenient to define the *seniority number* v of the term by

$$v = 2(a+S) \qquad (14.3.3)$$

where a is given by (14.3.1). It can be shown (see Judd, 1963) that

$$b = \min(2S, 2l+1-v) \qquad (14.3.4)$$

If we know (w) and S we can find v. Alternatively if we know S and v we can find (w). When seniority is included the terms are labelled $v^{2S+1}L$. The seniority number v is obtained from (14.3.3) and the branching rules for the reductions

$$SU(2l+1) \rightarrow R(2l+1) \rightarrow R(3)$$

which can be found in Chapter 13.

For configurations p^N seniority is not required. For configurations d^N the seniority classification is given in Appendix 2 Table 37. From this table we see that in all cases v is sufficient to distinguish between terms with the same L and S. No two terms with the same L and S have the same v.

For configurations f^N (see Judd, 1963) the seniority number is not sufficient to separate terms with the same L and S. However the group $R(7)$ has a subgroup which contains $R(3)$. This is the exceptional Lie group G_2. It follows that in addition to v we can use the branching rules

$$SU(7) \to R(7) \to G_2 \to R(3)$$

and the IR's of G_2 in order to distinguish between terms with the same vSL. Unfortunately even this extra group is not sufficient to completely distinguish between all terms with the same vSL (see Judd, 1963).

14.4 Terms from configurations involving inequivalent electrons

The terms arising from a configuration of the kind

$$(n_1l_1)^{x_1}(n_2l_2)^{x_2}\ldots(n_tl_t)^{x_t}$$
$$x_1+x_2+\ldots+x_t=N$$

can readily be found from those of each of the subconfigurations $(n_il_i)^{x_i}$. Since electrons in different subconfigurations are inequivalent the Pauli principle causes no difficulty and we can classify the terms according to a *coupling format*

$$(n_1l_1)^{x_1}S_1L_1(n_2l_2)^{x_2}S_2L_2S_{\Sigma 2}L_{\Sigma 2}\ldots(n_tl_t)^{x_t}S_tL_tSL$$

Here a symbol such as $I_{\Sigma 2}$ denotes the result of coupling the angular momenta I_1 and I_2 using the Clebsch-Gordan theorem. In this coupling format S_iL_i refer to the ith shell whereas $S_{\Sigma i}L_{\Sigma i}$ refer to coupled resultants for shells up to and including the ith.

As a simple example consider the configuration $1s2s2p$. The terms arising from each of the subconfigurations are ^2S, ^2S and ^2P respectively. We couple first ^2S with ^2S and obtain ^1S and ^3S. Each of these is now coupled with ^2P to give ^2P, ^4P, ^2P. The ^4P term is unique. The two ^2P terms can be distinguished by writing them as $1s2s(^1S)2p\,^2P$ and $1s2s(^3S)2p\,^2P$ respectively. The intermediate terms ^1S and ^3S take the place here of the label α which we used for configurations of equivalent orbitals. It should be noted that the Pauli principle precludes the possibility of using the above coupling method for obtaining terms in the case of a configuration involving only equivalent orbitals.

The group theoretical scheme for the classification of states arising from a configuration l^n can also be used to classify the states arising from certain mixed configurations. As we have just seen this is not necessary but it can be extremely useful.

It is clear that the set of $(2l_1+1)+(2l_2+1)+\ldots+(2l_t+1)$ functions $\{\phi_{n_1,l_1,m_1}, \phi_{n_2,l_2,m_2}, \ldots \phi_{n_t,l_t,m_t}\}$ provide a basis for the IR $[100\ldots0]$ of the group $SU(2\{l_1+l_2\ldots+l_t\}+t)$ whenever the values of $l_1, l_2, \ldots, l_t$ are all different. It follows that we can classify the states arising from configurations

$$(n_1l_1)^{x_1}(n_2l_2)^{x_2}\ldots(n_tl_t)^{x_t}$$

$$x_1+x_2+\ldots+x_t=N$$

$$l_1\neq l_2\neq\ldots\neq l_t$$

by a consideration of branching rules for the reduction

$$SU(2\{l_1+l_2+\ldots+l_t\}+t)\rightarrow\mathscr{G}\rightarrow R(3)$$

where $\mathscr{G}$ is some intermediate group (or groups).

As an example we take the configurations s^mp^{n-m}. The relevant chain of groups is

$$SU(4)\rightarrow R(4)\rightarrow R(3)$$

and the states arising from the configurations p^n, $p^{n-1}s$, $p^{n-2}s^2$ are uniquely classified by $|n, (j_1j_2), S, L, M_S, M_L\rangle$ where (j_1j_2) is the label for an IR of $R(4)$. The required branching rules can readily be found and we then obtain the classification scheme shown in Table 38 which is contained in Appendix 2.

14.5 Multiplet structure

When the spin-orbit interaction is taken into account as a further perturbation the Hamiltonian is no longer invariant under separate transformations of the spatial and spin coordinates. It is invariant however under simultaneous transformations of both space and spin coordinates. The transformations referred to are of course those of $SU(2)$ or $R^*(3)$. The eigenstates are now simultaneous eigenstates of $\hat{H}$, $\hat{J}^2$ and $\hat{J}_z$ where $\hat{J}$ is the operator for the total angular momentum

$$\hat{J}=\hat{L}+\hat{S}$$

If $\Phi_0(\alpha, L, M_L)$ denotes the spatial function and $\Xi(S, M_S)$ denotes the spin function we have

$$\Psi_0(\alpha, J, M_J)=\sum_{M_L+M_S=M_J}(LM_LSM_S\mid JM_J)\Phi_0(\alpha, L, M_L)\Xi(S, M_S)$$

It is customary to associate with the function $\Psi_0(\alpha, S, L, J, M_J)$ a multiplet and to denote it simply by ${}^{2S+1}L_J$. When the spin orbit interaction is included the multiplets which arise from a given term will in general have different energies.

As an example we consider p^3. The terms are ${}^4S$, ${}^2D$ and ${}^2P$. It follows immediately that the multiplets are ${}^4S_{\frac{3}{2}}$, ${}^2D_{\frac{5}{2}}$, ${}^2D_{\frac{3}{2}}$, ${}^2P_{\frac{3}{2}}$ and ${}^2P_{\frac{1}{2}}$.

14.6 *jj* coupling

In the case of *jj* coupling the Hamiltonian for the unperturbed system is taken to be

$$\hat{H}_0 = \sum_i \{\hat{h}(i) + \xi(r_i)\hat{\mathbf{l}}_i \cdot \hat{\mathbf{s}}_i\} \qquad (14.6.1)$$

where

$$\hat{h}(i) = -\tfrac{1}{2}\nabla_i^2 - \frac{Z}{r_i} \qquad (14.6.2)$$

and the perturbation is

$$\hat{H}_1 = \sum_{i>j=1}^{N}\sum \frac{1}{r_{ij}} \qquad (14.6.3)$$

The one-electron Hamiltonian

$$\hat{h} + \xi(r)\hat{\mathbf{l}} \cdot \hat{\mathbf{s}}$$

is clearly only invariant under a simultaneous transformation of both the space and spin coordinate and it follows that we should use spin-orbitals given by

$$\psi_{j,m} \equiv \psi_{n,l,s,j,m} = \sum_{m_l+m_s=m} (lm_l sm_s \mid jm)\phi_{n,l,m_l}(r)\xi_{s,m_s}(\sigma_z) \qquad (14.6.4)$$

In (14.6.4) j is given by

$$\hat{j} = \hat{l} + \hat{s} \qquad (14.6.5)$$

and $s = \frac{1}{2}$. The spin orbital is a simultaneous eigenfunction of $\hat{l}^2$, $\hat{s}^2$ and $\hat{j}^2$.

The zeroth order equation

$$\hat{H}_0\Psi_0 = E_0\Psi_0$$

now has solutions of the form

$$\Psi_0 = \psi_{j_1,m_1}(1)\psi_{j_2,m_2}(2)\ldots\psi_{j_N,m_N}(N) \qquad (14.6.6)$$

with $\psi_{j,m}$ given by (14.6.4). If we note that

$$\hat{\mathbf{l}} \cdot \hat{\mathbf{s}} = \tfrac{1}{2}(\hat{j}^2 - \hat{l}^2 - \hat{s}^2) \qquad (14.6.7)$$

we find that the zeroth order energy is given by

$$E_0 = \sum_i \varepsilon_i \qquad (14.6.8)$$

with

$$\varepsilon_i = \varepsilon_{n_i,l_i} + \tfrac{1}{2}\zeta_{n_i,l_i}[j_i(j_i+1) - l_i(l_i+1) - \tfrac{3}{4}] \tag{14.6.9}$$

Here ε_{n_i,l_i} is the energy of a hydrogen like atomic orbital and

$$\zeta_{n,l} = \int_0^\infty R_{nl}(r)\xi(r)R_{nl}(r)r^2\,dr \tag{14.6.10}$$

The form (14.6.6) belongs to the configuration which is written as

$$\begin{gathered}(j_1)^{x_1}(j_2)^{x_2}\ldots(j_t)^{x_t}\\ x_1+x_2+\ldots+x_t=N\end{gathered} \tag{14.6.11}$$

When the electrostatic repulsions are included as a perturbation the Hamiltonian is invariant only under simultaneous transformations of all the space-spin coordinates and the eigenfunctions are now simultaneous eigenfunctions of $\hat{H}$, $\hat{J}^2$ and $\hat{J}_z$ where

$$\hat{J} = \sum_{i=1}^{N} \hat{j}_i$$

In the case of two electrons the proper zeroth order eigenfunctions are clearly given by

$$\Psi_0(j_1, j_2, J, M \mid 1, 2) = \sum_{m_1+m_2=M} (j_1m_1j_2m_2 \mid JM)\psi_{j_1,m_1}(1)\psi_{j_2,m_2}(2)$$

These eigenfunctions correspond to multiplets anu they are often simply denoted by J.

We now consider, from a group theoretical point of view, a configuration $(j)^N$ in which all the electrons are equivalent. When interaction between the electrons is taken into account it is necessary to obtain the multiplets J which arise. Since $\psi_{j,m}$ is a basis function for the IR (j) of $R^*(3)$ the values of J are obtained by decomposing the direct product

$$\underbrace{(j)\otimes(j)\otimes\ldots\otimes(j)}_{N \text{ factors}}$$

However the Pauli principle excludes many of the J values obtained in this way. The wave function must be totally antisymmetric and it follows that allowed J values are associated with permutational symmetry $[1^r]$. Now the $(2j+1)$ functions $\psi_{j,m}$ not only provide a basis for the IR (j) of $R^*(3)$ but they also provide a basis for the $(2j+1)$-dimensional IR $[10\ldots0]$ of $SU(2j+1)$. Thus the product $(j)^N$ spans the direct product

$$\underbrace{[1]\otimes[1]\otimes\ldots\otimes[1]}_{N \text{ factors}}$$

of $SU(2j+1)$. It follows that branching rules for the reduction

$$SU(2j+1) \to R^*(3)$$

provide a means of determining which J values occur with the permutational symmetry $[1^r]$. When j is an integer branching rules such as given in Tables 31 and 33 can be used. When j is half an odd integer we require branching rules for reductions such as $SU(4) \to R^*(3)$, $SU(6) \to R^*(3)$ and $SU(8) \to R^*(3)$. These branching rules can be obtained by the same method as used previously in the case of integral j. Results for the first few configurations are given in Table 39 which can be found in Appendix 2.

Just as the configurations $(nl)^x$ and $(nl)^{4l+2-x}$ give rise to the same terms so here the configurations $(j)^x$ and $(j)^{2j+1-x}$ give rise to the same multiplets.

The configuration $(\frac{7}{2})^4$ gives rise to different multiplets with the same value of J. These multiplets can be distinguished by means of a seniority number. It can be shown that when j is half an odd integer the symplectic group $Sp(2j+1)$ is a subgroup of $SU(2j+1)$ and further that $R^*(3)$ is a subgroup of $Sp(2j+1)$. Thus branching rules for the reductions

$$SU(2j+1) \to Sp(2j+1) \to R^*(3)$$

can be used to provide an extra label for the multiplets. This label is the IR $(\sigma_1\sigma_2 \ldots \sigma_\mu)$ of $Sp(2j+1)$ (see Section 8.5). Because the IR's of $SU(2j+1)$ with which we are concerned here all have the form $[111\ldots10\ldots0]$ so the IR's of $Sp(2j+1)$ arising from the branching rules have the form $(111\ldots100\ldots0)$ and we can define a seniority number v by

$$v = \sum_i \sigma_i$$

The multiplets are now denoted by v^J.

As an example we have the following branching rules for $(\frac{7}{2})^4$.

$[\lambda]$	$(\sigma_1\sigma_2\sigma_3\sigma_4)$	v	J
[1111]	(0000)	0	0
	(1100)	2	2, 4, 6
	(1111)	4	2, 4, 5, 8

We see that the two multiplets $J=2$ and the two multiplets $J=4$ are separated by their different seniority.

The multiplets arising from a configuration

$$(j_1)^{x_1}(j_2)^{x_2} \ldots (j_t)^{x_t}$$
$$x_1 + x_2 + \ldots + x_t = N$$

can be found from those of each of the subconfigurations $(j_i)^{x_i}$. The multiplets can be classified according to a coupling format

$$(j_1)^{x_1}J_1(j_2)^{x_2}J_2J_{\Sigma 2} \ldots (j_t)^{x_t}J_tJ$$

As an illustration we consider the configuration p^3. If SL coupling is used the multiplets are ${}^4S_{\frac{3}{2}}$, ${}^2P_{\frac{3}{2}}$, ${}^2P_{\frac{1}{2}}$, ${}^2D_{\frac{5}{2}}$ and ${}^2D_{\frac{3}{2}}$ with $J=\frac{3}{2}, \frac{3}{2}, \frac{1}{2}, \frac{5}{2}$ and $\frac{3}{2}$ respectively. When jj coupling is appropriate we have $l=1$; $s=\frac{1}{2}$; $j=\frac{1}{2}, \frac{3}{2}$ and $N=3$. The multiplets thus arise from configurations $(\frac{1}{2})^2(\frac{3}{2})$, $(\frac{1}{2})(\frac{3}{2})^2$ and $(\frac{3}{2})^3$. From Table 39 we have

	J
$(\frac{1}{2})^2$	0
$(\frac{3}{2})^2$	0, 2
$(\frac{3}{2})^3$	$\frac{3}{2}$

and it follows that the multiplets are $(\frac{1}{2})^2\,0(\frac{3}{2})\frac{3}{2}$, $(\frac{1}{2})\frac{1}{2}(\frac{3}{2})^2\,0\frac{1}{2}$, $(\frac{1}{2})\frac{1}{2}(\frac{3}{2})^2\,2\frac{5}{2}$, $(\frac{1}{2})\frac{1}{2}(\frac{3}{2})^2\,2\frac{3}{2}$, $(\frac{3}{2})^3\,\frac{3}{2}$ with $J=\frac{3}{2}, \frac{1}{2}, \frac{5}{2}, \frac{3}{2}$ respectively.

14.7 Molecular terms

Consider a molecular system consisting of M nuclei and N electrons. Let the charge on the ath nucleus be Z_a. Within the Born-Oppenheimer approximation the Hamiltonian $\hat{H}$ for the non relativistic motion of the electrons relative to the fixed nuclear framework is given by

$$\hat{H}=\hat{H}_0+\hat{H}_1 \tag{14.7.1}$$

with

$$\hat{H}_0=-\tfrac{1}{2}\sum_{i=1}^{N}\nabla_i^2-\sum_{i=1}^{N}\sum_{a=1}^{M}\frac{Z_a}{r_{ia}} \tag{14.7.2}$$

and

$$\hat{H}_1=\sum_{i>j=1}^{N}\sum\frac{1}{r_{ij}} \tag{14.7.3}$$

In general $\hat{H}$ is invariant under a subgroup $\mathcal{G}$ of the rotation-reflection group $R^+(3)$. The zeroth order equation

$$\hat{H}_0\Phi_0=E_0\Phi_0 \tag{14.7.4}$$

has the solutions of the form

$$\Phi_0(1,2,\ldots,N)=\phi_{n_1,\gamma_1,m_1}(1)\phi_{n_2,\gamma_2,m_2}(2)\ldots\phi_{n_N,\gamma_N,m_N}(N) \tag{14.7.5}$$

In (14.7.5) n if a file number, γ is the label for an IR of $\mathcal{G}$ and m is the label for a standard basis function belonging to γ. Furthermore $\phi_{n,\gamma,m}$ is a molecular orbital satisfying the equation

$$\left\{-\tfrac{1}{2}\nabla^2-\sum_{a=1}\frac{Z_a}{r_a}-\varepsilon(n,\gamma,m)\right\}\phi_{n,\gamma,m}=0 \tag{14.7.6}$$

The solution (14.7.5) has associated with it an electron configuration

$$\begin{gathered}(n_1\gamma_1)^{x_1}(n_2\gamma_2)^{x_2}\ldots(n_t\gamma_t)^{x_t}\\ x_1+x_2+\ldots+x_t=N\end{gathered} \tag{14.7.7}$$

The zeroth order energy is given by

$$E_0 = \sum_{i=1}^{t} x_i \varepsilon(n_i, \gamma_i)$$

When $M > 2$ it is not possible to obtain an exact solution of (14.7.6) and it is then only possible to obtain approximate values for the zeroth order energies.

We focus our attention now upon a configuration $(n\gamma)^N$ in which all the electrons are equivalent. When the electrostatic interaction between the electrons is taken into account as a perturbation it is necessary to obtain the terms ${}^{2S+1}\Gamma$ which arise. This is the analogue of LS coupling for atomic systems. For a given N the values of S are contained in Table 14. The values of Γ are given by decomposing the direct product

$$\underbrace{\gamma \otimes \gamma \otimes \ldots \otimes \gamma}_{N \text{ factors}}$$

In order to determine which values of Γ occur with a given spin S we associate with each Γ a definite permutational symmetry in the spatial function. This is done by a consideration of the branching rules for the reduction

$$SU(n_\gamma) \to \mathcal{G}$$

where n_γ is the dimension of the IR γ.

As an illustration we consider octahedral symmetry $\mathcal{G} = \mathcal{O}$ and the terms which arise from the configurations e^2, t_1^2, t_2^2, t_1^3, t_2^3.

For e^2 we have

$$[1] \otimes [1] = [2] \oplus [1^2] \to e \otimes e = \mathrm{A}_1 \oplus \mathrm{A}_2 \oplus E$$

Now $(E \otimes E)^- = \mathrm{A}_2$ and it follows that

$$[1^2] \to \mathrm{A}_2 \qquad [2] \to \mathrm{A}_1 \oplus E$$

The terms are thus ${}^3\mathrm{A}_2$, ${}^1\mathrm{A}_1$ and ${}^1E$.

For t_1^2 we have

$$[1] \otimes [1] = [2] \oplus [1^2] \to t_1 \otimes t_1 = \mathrm{A}_1 \oplus E \oplus T_1 \oplus T_2$$

Now $(T_1 \otimes T_1)^- = T_1$ and it follows that

$$[1^2] \to T_1 \qquad [2] \to \mathrm{A}_1 \oplus E \oplus T_2$$

The terms are thus ${}^3T_1$, ${}^1\mathrm{A}_1$, ${}^1E$ and ${}^1T_2$.

For t_2^2 we have

$$[1] \otimes [1] = [2] \oplus [1^2] \to t_2 \otimes t_2 = \mathrm{A}_1 \oplus E \oplus T_1 \oplus T_2$$

Now $(T_2 \otimes T_2)^- = T_1$ and we have

$$[1^2] \to T_1 \qquad [2] \to A_1 \oplus E \oplus T_2$$

The terms are ${}^3T_1$, ${}^1A_1$, ${}^1E$ and ${}^1T_2$.

For t_1^3 we have

$$[1^2] \otimes [1] = [21] \oplus [1^3] \to T_1 \otimes t_1 = A_1 \oplus E \oplus T_1 \oplus T_2$$

Now $[1^3]$ is a 1-dimensional IR of $SU(3)$ and we have

$$[1^3] \to A_1$$

$$[21] \to E \oplus T_1 \oplus T_2$$

The terms are thus ${}^4A_1$, ${}^2E$, ${}^2T_1$, ${}^2T_2$.

For t_2^3 we have

$$[1^2] \otimes [1] = [21] \oplus [1^3] \to T_1 \otimes t_2 = A_2 \oplus E \oplus T_1 \oplus T_2$$

and the terms are ${}^4A_2$, ${}^2E$, ${}^2T_1$, ${}^2T_2$.

14.8 Multiplet structure and $\gamma\gamma$ coupling

When the Coulomb repulsions between the electrons in a molecule are small by comparison with the interaction of the spin of each electron with its orbital motion it becomes necessary to classify the states by a method which is equivalent to jj coupling in atoms. We call this situation $\gamma\gamma$ coupling.

If the appropriate molecular symmetry group is $\mathcal{G}$ the spin functions α, β form the basis for a two dimensional IR $E_{\frac{1}{2}}$ of the double group $\mathcal{G}^*$ and we can introduce molecular spin orbitals by analogy with (14.6.4) as

$$\psi_{\gamma,m} = \sum_{\mu,\nu} (\Gamma\mu E_{\frac{1}{2}}\nu \mid \gamma m)\phi_{n,\Gamma,\mu}\xi_{E_{\frac{1}{2}},\nu}$$

Here $(\Gamma\mu E_{\frac{1}{2}}\nu \mid \gamma m)$ are coupling coefficients for the double group $\mathcal{G}^*$.

As an example we consider the configuration t_2^2 in octahedral symmetry. When $\gamma\gamma$ coupling is appropriate we construct the configuration which arise from the spin orbitals spanning $t_2 \otimes e_{\frac{1}{2}}$. Now

$$t_2 \otimes e_{\frac{1}{2}} = e_{\frac{5}{2}} \oplus u_{\frac{3}{2}}$$

and the configurations in question are

$$(e_{\frac{5}{2}})^2, \quad (e_{\frac{5}{2}})(u_{\frac{3}{2}}), \quad (u_{\frac{3}{2}})^2$$

When the Coulomb interaction between the electrons is included these configurations give rise to Γ multiplets. By using Table 29 and the result

(3.4.7) we find

$$(e_{\frac{5}{2}} \otimes e_{\frac{5}{2}})^- = A_1 \qquad (u_{\frac{3}{2}} \otimes u_{\frac{3}{2}})^- = A_1 \oplus E \oplus T_2$$
$$e_{\frac{5}{2}} \otimes u_{\frac{3}{2}} = E \oplus T_1 \oplus T_2$$

It follows that the Γ multiplets are given by

Configuration	Γ Multiplet
$(e_{\frac{5}{2}})^2$	A_1
$(u_{\frac{3}{2}})^2$	A_1, E, T_2
$(e_{\frac{5}{2}})(u_{\frac{3}{2}})$	E, T_1, T_2

We have already seen that the configuration t_2^2 gives rise to the terms $^3T_1$, $^1A_1$, ^1E and $^1T_2$. Now

$$(e_{\frac{1}{2}} \otimes e_{\frac{1}{2}})^+ = T_1 \qquad (e_{\frac{1}{2}} \otimes e_{\frac{1}{2}})^- = A_1$$

and it follows that the singlet spin function spans A_1 whilst the triplet spin functions span T_1. When spin orbit interaction is included as a perturbation on the terms the multiplets which arise are now given by

Configuration	Term		Multiplet
t_2^2	$^3T_1$	$T_1 \otimes T_1$	A_1, E, T_1, T_2
	$^1A_1$	$A_1 \otimes A_1$	A_1
	^1E	$A_1 \otimes E$	E
	$^1T_2$	$A_1 \otimes T_2$	T_2

Reference

Judd, B. R. (1963) "Operator Techniques in Atomic Spectroscopy", McGraw-Hill, Chapter 6.

CHAPTER 15

Fractional Parentage

15.1 Introduction

In the previous chapter we have seen how it is possible to obtain and classify the terms and multiplets which arise from a given atomic or molecular electronic configuration. In this chapter we shall be concerned with the construction of wave functions which describe these terms or multiplets.

Without loss of generality we shall restrict ourselves initially to the case of LS coupling in atomic systems. The wave function which describes a given term is required to satisfy two fundamental properties. It must be a simultaneous eigenfunction of $\hat{L}^2$, $\hat{S}^2$, $\hat{L}_z$ and $\hat{S}_z$ and it must also be totally antisymmetric.

In the case of a 2-electron system the situation is quite straightforward. The spatial coupled function

$$\Phi_0(\alpha, l_1, l_2, L, M \mid 1, 2) = \sum_{m_1+m_2=M} (l_1 m_1 l_2 m_2 \mid LM)\phi_{n_1,l_1,m_1}(1)\phi_{n_2,l_2,m_2}(2) \tag{15.1.1}$$

is an eigenfunction of $\hat{L}^2$ and $\hat{L}_z$. Likewise the spin coupled function

$$\Xi(\tfrac{1}{2}, \tfrac{1}{2}, S, M \mid 1, 2) = \sum_{m_1+m_2=M} (\tfrac{1}{2}m_1\tfrac{1}{2}m_2 \mid SM)\xi_{m_1}(1)\xi_{m_2}(2)$$

is an eigenfunction of $\hat{S}^2$ and $\hat{S}_z$. Now there are only two possible values of S. When $S = 0$ we have the antisymmetric spin function

$$\Xi_{0,0} = \Xi(\tfrac{1}{2}, \tfrac{1}{2}, 0, 0 \mid 1, 2) = \frac{1}{\sqrt{2}}(\alpha(1)\beta(2) - \beta(1)\alpha(2))$$

When $S = 1$ we have three symmetric spin functions

$$\Xi_{1,1} = \Xi(\tfrac{1}{2}, \tfrac{1}{2}, 1, 1 \mid 1, 2) = \alpha(1)\alpha(2)$$

$$\Xi_{1,0} = \Xi(\tfrac{1}{2}, \tfrac{1}{2}, 1, 0 \mid 1, 2) = \frac{1}{\sqrt{2}}(\alpha(1)\beta(2) + \beta(1)\alpha(2))$$

$$\Xi_{1,-1} = \Xi(\tfrac{1}{2}, \tfrac{1}{2}, 1, -1 \mid 1, 2) = \beta(1)\beta(2)$$

When the electrons are not equivalent we can construct spatial functions from (15.1.1) which are either symmetric or antisymmetric. They are given by

$$\Phi_0(\alpha, l_1, l_2, L, M \mid 1, 2) \pm \Phi_0(\alpha, l_1, l_2, L, M \mid 2, 1)$$

Since the total wave function must be antisymmetric it follows that the singlets are described by

$$^1\Psi_0(1, 2) = [\Phi_0(1, 2) + \Phi_0(2, 1)]\Xi_{0,0}$$

while the triplets are described by

$$^3\Psi_0(1, 2) = [\Phi_0(1, 2) - \Phi_0(2, 1)]\begin{pmatrix} \Xi_{1,1} \\ \Xi_{1,0} \\ \Xi_{1,-1} \end{pmatrix}$$

When the electrons are equivalent we have from (12.2.10)

$$\Phi_0(\alpha, l, l, L, M \mid 1, 2) = (-1)^L \Phi_0(\alpha, l, l, L, M \mid 2, 1)$$

It follows that for even L we obtain singlets and for odd L we obtain triplets.

When the number of electrons exceeds two the situation becomes more involved. The best known method for constructing totally antisymmetric eigenfunctions of $\hat{L}^2$, $\hat{S}^2$, $\hat{L}_z$ and $\hat{S}_z$ makes use of determinants. A determinant of spin-orbitals is set up. This determinant is necessarily totally antisymmetric but it is not in general an eigenfunction of $\hat{L}^2$, $\hat{S}^2$, $\hat{L}_z$ and $\hat{S}_z$. However by taking an appropriate linear combination of such determinants the required eigenfunction can be obtained. We shall not make use of this method here since there exists an alternative method which offers great advantages particularly when we go beyond the usual orbital approximation. This alternative method is called *fractional parentage*. It is based on a generalization of the method discussed in this section in connection with 2-electron systems. Instead of starting from a totally antisymmetric function we set up eigenfunctions of $\hat{L}^2$, $\hat{S}^2$, $\hat{L}_z$ and $\hat{S}_z$ and then obtain an appropriate linear combination of these eigenfunctions which is totally antisymmetric.

15.2 Substantive coefficients of fractional parentage

The method of fractional parentage is best approached through an example. The LS terms which arise from the configurations p^2 and p^3 are ^1S, ^1D, ^3P and ^2P, ^2D, ^4S respectively. If we couple a third (equivalent) p orbital to the terms of p^2 we obtain both allowed and forbidden terms of p^3. The allowed terms arise as follows

$$p(^2P)p^2(^1S)^2P$$

$$p(^2P)p^2(^1D)^2P, {}^2D$$

$$p(^2P)p^2(^3P)^2P, {}^2D, {}^4S$$

Whereas the ^4S term arises only from ^3P the ^2D term arises from both ^3P and ^1D. We say that ^3P and ^1D are *parent terms* of ^2D. Likewise the ^2P term of p^3 has three parent terms namely ^1S, ^1D and ^3P.

We now define a *zeroth order vector coupled function* by

$$\psi_0(p(1); p(2), p(3)S', L'; S, L, M_S, M_L)$$

$$\equiv \sum_{\substack{M_{S'}+m_s=M_S\\ M_{L'}+m_l=M_L}} (\tfrac{1}{2}m_sS'M_{S'} \mid SM_S)(1m_lL' \mid LM_L)\phi_{n,p,m_l}(1)\xi_{m_s}(1)$$

$$\times \Phi_0(p, p, L', M_{L'} \mid 2, 3)\Xi(\tfrac{1}{2}, \tfrac{1}{2}, S', M_{S'} \mid 2, 3) \quad (15.2.1)$$

In ψ_0 and in all that follows we shall use a semi-colon when we wish to emphasize the division of a many-electron state into vector coupled parts. The function ψ_0 is assumed to be antisymmetric in coordinates 2 and 3. When LS coupling is appropriate it is convenient to introduce the following shorthand notation

$$I \equiv L, S$$

$$M \equiv M_L, M_S$$

$$\not{l} \equiv \tfrac{1}{2}, l$$

$$(\not{p}mI'M' \mid IM) \equiv (\tfrac{1}{2}m_sS'M_{S'} \mid SM_S)(1m_lL'M_{L'} \mid LM_L)$$

$$\psi(\not{p}(1), m) \equiv \phi_{n,p,m_l}(1)\xi_{m_s}(1)$$

$$\Psi_0(\not{p}(2), \not{p}(3), I', M') \equiv \Phi_0(p, p, L', M_{L'} \mid 2, 3)\Xi(\tfrac{1}{2}, \tfrac{1}{2}, S', M'_S \mid 2, 3)$$

The zeroth order vector coupled function (15.2.1) can now be written in the form

$$\psi_0(\not{p}(1); \not{p}(2), \not{p}(3), I'; I, M)$$

$$\equiv \sum_{m+M'=M} (\not{p}mI'M' \mid IM)\psi(\not{p}(1), m)\Psi_0(\not{p}(2), \not{p}(3)I', M')$$

The pair of coupled functions $\psi_0(\not{p}(1); \not{p}(2), \not{p}(3), {}^1D; {}^2D)$ and $\psi_0(p(1); \not{p}(2), \not{p}(3), {}^3P; {}^2D)$ which can be constructed from the parent terms ^1D and ^3P respectively are clearly both eigenfunctions for ^2D but neither of them is totally antisymmetric. It is possible however to construct a linear combination of these two eigenfunctions which is totally antisymmetric. Consider a linear combination

$$\Psi_0(p^3, {}^2D) = \alpha\psi_0(\not{p}(1); \not{p}(2), \not{p}(3), {}^1D; {}^2D) + \beta\psi_0(\not{p}(1); \not{p}(2), \not{p}(3), {}^3P; {}^2D)$$

It is required to find values of α and β which make $\Psi_0(p^3, {}^2D)$ a totally antisymmetric function. Since the electrons are equivalent and Ψ_0 is already antisymmetric in 2 and 3 it is sufficient to make Ψ_0 antisymmetric in

1 and 2 in order to achieve total antisymmetry. Thus we require

$$(12)\Psi_0 = -\Psi_0$$

and we have

$$\begin{aligned}\alpha\psi_0(p(2); p(1), p(3), {}^1D; {}^2D) + \beta\psi_0(p(2); p(1); p(3), {}^3P; {}^2D)\\ = -\alpha\psi_0(p(1); p(2), p(3), {}^1D; {}^2D) - \beta\psi_0(p(1); p(2), p(3), {}^3P; {}^2D) \quad (15.2.2)\end{aligned}$$

In order to compare the two sides of this equation it is necessary to have the coordinates on the left hand side in the same order (*dictionary order*) as those on the right hand side. This can be done by means of a recoupling transformation (see (12.4.7)). Thus we have

$$\begin{aligned}\psi_0(p(2); p(1), p(3), I; {}^2D) = \sum_{I''} (-1)^{I''+I'} \sqrt{(2I'+1)(2I''+1)} \begin{Bmatrix} p & p & I'' \\ p & {}^2D & I' \end{Bmatrix}\\ \times \psi_0(p(1); p(2), p(3), I''; {}^2D) \quad (15.2.3)\end{aligned}$$

where $(-1)^I \equiv (-1)^{S+L}$, $(2I+1) \equiv (2S+1)(2L+1)$ and

$$\begin{Bmatrix} p & p & I'' \\ p & {}^2D & I' \end{Bmatrix} \equiv \begin{Bmatrix} \frac{1}{2} & \frac{1}{2} & S'' \\ \frac{1}{2} & \frac{1}{2} & S' \end{Bmatrix} \begin{Bmatrix} 1 & 1 & L'' \\ 1 & 2 & L' \end{Bmatrix}$$

The sum over I'' clearly includes the forbidden terms ${}^3D$ and ${}^1P$ of p^2. These terms do not occur on the right hand side of (15.2.2). It follows that the coefficient of each of these forbidden terms on the left hand side must vanish when (15.2.3) is substituted. For ${}^1P$ we have from (15.2.3) the coefficients

$$\begin{aligned}(-1)^{{}^1P+{}^1D}\sqrt{(2\,{}^1P+1)(2\,{}^1D+1)} \begin{Bmatrix} p & p & {}^1D \\ p & {}^2D & {}^1D \end{Bmatrix}\\ = (-1)^3\sqrt{3.5} \begin{Bmatrix} 1 & 1 & 1 \\ 1 & 2 & 2 \end{Bmatrix} \begin{Bmatrix} \frac{1}{2} & \frac{1}{2} & 0 \\ \frac{1}{2} & \frac{1}{2} & 0 \end{Bmatrix} = -\frac{\sqrt{3}}{4}\end{aligned}$$

and

$$(-1)^{{}^1P+{}^3P}\sqrt{(2\,{}^1P+1)(2\,{}^3P+1)} \begin{Bmatrix} p & p & {}^1P \\ p & {}^2D & {}^3P \end{Bmatrix} = -\frac{\sqrt{3}}{4}$$

where we have used the values of the $6j$ symbols given by (12.4.10). From (15.2.2) we now have the equation

$$-\frac{\sqrt{3}}{4}\alpha - \frac{\sqrt{3}}{4}\beta = 0$$

The normalization condition is $\alpha^2 + \beta^2 = 1$ and we find

$$\alpha = -\beta = \frac{1}{\sqrt{2}}$$

We have now obtained the *fractional parentage* (FP) *expansion*

$$\Psi_0(p^3, {}^2D) = \frac{1}{\sqrt{2}} \psi_0(p; p^2, {}^1D; {}^2D) - \frac{1}{\sqrt{2}} \psi_0(p; p^2, {}^3P; {}^2D)$$

and Ψ_0 is totally antisymmetric.

The coefficients α and β are examples of what are called *coefficients of fractional parentage* (CFP). When we are concerned with configurations which comprise only equivalent orbitals the CFP have a number of special properties (see Racah, 1943) and for this reason we refer to them as *substantive* CFP. Just as the Clebsch Gordan coefficients are written in descriptive form so the CFP are written as, (see Racah, 1943)

$$\alpha = (p; p^2\ {}^1D \mid \} p^3\ {}^2D)$$
$$\beta = (p; p^2\ {}^3P \mid \} p^3\ {}^2D)$$

The *director* $\mid \}$ leads from the parent to the *offspring*.

Similar considerations to the above lead to the FP expansions

$$\Psi_0(p^3, {}^4S) = \psi_0(p; p^2, {}^3P; {}^4S)$$

$$\Psi_0(p^3, {}^2P) = \frac{\sqrt{2}}{\sqrt{9}} \psi_0(p; p^2, {}^1S; {}^2P) - \frac{\sqrt{5}}{\sqrt{18}} \psi_0(p; p^2, {}^1D; {}^2P) - \frac{1}{\sqrt{2}} \psi_0(p; p^2, {}^3P; {}^2P)$$

These results are usually displayed in a table of CFP.

TABLE 40. The CFP $(p; p^2 I' \mid \} p^3 I)$

	p^2		
p^3	${}^1S$	${}^3P$	${}^1D$
${}^4S$	0	1	0
${}^2P$	$\frac{\sqrt{2}}{\sqrt{9}}$	$-\frac{1}{\sqrt{2}}$	$-\frac{\sqrt{5}}{\sqrt{18}}$
${}^2D$	0	$-\frac{1}{\sqrt{2}}$	$\frac{1}{\sqrt{2}}$

Due to the existence of forbidden terms in the parent configuration the CFP (unlike the Clebsch-Gordan coefficients) do not form a unitary matrix but only a rectangular matrix which is part of a unitary matrix (see Racah, 1943).

We now turn our attention to the general case and define for equivalent orbitals the FP expansion

$$\Psi_0(l^N I) = \sum_{I', I''} (l^K I'; l^{N-K} I'' | \} l^N I) \psi_0(l^K, I'; l^{N-K}, I''; I) \quad (15.2.4)$$

The symbol I now stands for all the quantum numbers required to specify the term completely (e.g. seniority in the case of d orbitals). The equation (15.2.4) is called a K-*electron* fractional parentage expansion and K can assume values $1, 2, 3, \ldots$. The particular value of K chosen depends upon the type of operator whose matrix elements are to be evaluated. The coefficient $(l^K I'; l^{N-K} I'' | \} l^N I)$ is called a K-electron CFP and the FP expansion achieves total antisymmetry by sums over intermediate terms rather than by permutations of the coordinates. In (15.2.4) the configuration l^N has been split into two subconfigurations and for this reason we say that the *symmetry division* is two-fold. In a two-fold symmetry division the vector coupled function $\psi_0(l^K, I'; l^{N-K}, I''; I)$ is antisymmetric in coordinates $1, 2, \ldots, K$ and is antisymmetric in coordinates $K+1, K+2, \ldots, N$. Thus

$$\psi_0(I'; I''; I, M) = \sum_{M'+M''=M} (I'M'I''M'' | IM) \Psi_0(I', M') \Psi_0(I'', M'')$$

We shall refer to $\Psi_0(l^N I)$ as the *post-parent*. Of the functions $\Psi_0(l^K I')$ and $\Psi_0(l^{N-K} I'')$ one is called the *preparent* and the other is called the *co-parent*. In the evaluation of matrix elements of an operator it is convenient to identify as co-parent that wave function upon which the operator acts. In general the co-parent contains fewer electrons than the preparent. In what follows we make the convention of locating the co-parent first in the CFP and to place the preparent immediately behind the director.

The most useful CFP are the 1-electron CFP and the 2-electron CFP. We have already shown by an example how the 1-electron CFP can be evaluated. The 2-electron CFP can be obtained from 1-electron CFP by means of a formula which we now establish.

We have for the 2-electron FP expansion

$$\Psi_0(l^N, I) = \sum_{I_1, I_2} (l^2 I_1; l^{N-2} I_2 | \} l^N I) \psi_0(l^2, I_1; l^{N-2}, I_2; I) \quad (15.2.5)$$

Now

$$\Psi_0(l^N, I) = \sum_{I'} (l; l^{N-1} I' | \} l^N I) \psi_0(l; l^{N-1}, I'; I)$$

and

$$\Psi_0(l^{N-1}, I') = \sum_{I''} (l; l^{N-2} I'' | \} l^{N-1} I') \psi_0(l; l^{N-2}, I''; I')$$

It follows that

$$\Psi_0(l^N, I) = \sum_{I',I''} (l; l^{N-1}I' \mid \} l^N I)(l; l^{N-2}I'' \mid \} l^{N-1}I')$$

$$\times \psi_0(l; (l; l^{N-2}, I'')I'; I) \qquad (15.2.6)$$

The function ψ_0 in (15.2.6) is now recoupled by using (12.4.7) and we have

$$\Psi_0(l^N, I) = \sum_{I',I''} (l; l^{N-1}I \mid \} l^N I)(l; l^{N-2}I'' \mid \} l^{N-1}I')$$

$$\times \sum_{I_{12}} ((ll)I_{12}I''I \mid l(lI'')I'I)$$

$$\times \psi_0(l^2, I_{12}; l^{N-2}, I''; I) \qquad (15.2.7)$$

We write $I_{12} = I_1$; $I'' = I_2$ and compare (15.2.7) with (15.2.5) to obtain the following formula for the 2-electron CFP.

$$(l^2 I_1; l^{N-2} I_2 \mid \} l^N I) = \sum_{I'} (l; l^{N-1}I' \mid \} l^N I)(l; l^{N-2}I_2 \mid \} l^{N-1}I')$$

$$\times (-1)^{l+l+I_2+I} \sqrt{(2I_1+1)(2I'+1)} \begin{Bmatrix} l & l & I_1 \\ I_2 & I & I' \end{Bmatrix} \qquad (15.2.8)$$

15.3 CFP in the spin free formalism

The method of fractional parentage can be set up within the spin free formalism (see Section 7.3) by making use of the standard Young-Yamanouchi IR's of the symmetric group.

According to (6.7.4) we have

$$\Psi_0(I) = \frac{1}{\sqrt{n_\lambda}} \sum_{r=1}^{n_\lambda} (-1)^{R_r} \Phi^{[\lambda]}_{(r)}(L) \Theta^{[\tilde{\lambda}]}_{(\tilde{r})}(S) \qquad (15.3.1)$$

If we are required to calculate matrix elements of a 1-electron spin free operator we only require the spatial function Φ. We thus define a 1-electron spatial FP expansion by

$$\Phi^{[\lambda]}_{(r_n r_{n-1} \dots r_2 r_1)}(l^n L) = \sum_{L'} (l^{N-1}[\lambda']L'; l \mid \} l^n [\lambda] L)$$

$$\times \phi(l^{n-1}, [\lambda'], L'(r_{n-1} \dots r_1); l; L)$$

In a similar way we can define a 1-electron spin FP expansion by

$$\Theta^{[\tilde{\lambda}]}_{(\tilde{r}_n \tilde{r}_{n-1} \dots \tilde{r}_2 \tilde{r}_1)}(S) = \sum_{S'} (\tfrac{1}{2}^{n-1}[\tilde{\lambda}']S'; \tfrac{1}{2} \mid \} \tfrac{1}{2}^n [\tilde{\lambda}] S)$$

$$\times \theta(\tfrac{1}{2}^{n-1}, [\lambda'], S'(\tilde{r}_{n-1} \dots \tilde{r}_1); \tfrac{1}{2}; S) \qquad (15.3.2)$$

As an example we consider the $p^3\,{}^2P$ term. In this case (15.3.1) becomes

$$\Psi_0(p^3, {}^2P) = \frac{1}{\sqrt{2}}[\Phi^{[21]}_{(211)}(P)\Theta^{[\widetilde{21}]}_{(\widetilde{211})}(\tfrac{1}{2}) - \Phi^{[21]}_{(121)}(P)\Theta^{[\widetilde{21}]}_{(\widetilde{121})}(\tfrac{1}{2})] \qquad (15.3.3)$$

The spatial FP expansions are

$$\Phi_{(211)} = \alpha\phi(p^2, [2], S(11); p; P) + \beta\phi(p^2, [2], D(11); p; P) \qquad (15.3.4)$$

and

$$\Phi_{(121)} = \gamma\phi(p^2, [11], P(21); p; P) \qquad (15.3.5)$$

According to (6.4.5) we have

$$(23)\Phi_{(121)} = \frac{\sqrt{3}}{2}\Phi_{(211)} + \tfrac{1}{2}\Phi_{(121)} \qquad (15.3.6)$$

Now

$$(23)\Phi_{(121)} = \gamma\phi(p(1), p(3), [11], P(21); p(2); P)$$

This can be recoupled to give

$$(23)\Phi_{(121)} = \frac{1}{\sqrt{3}}\gamma\phi(p^2, [2], S(11); p; P) + \tfrac{1}{2}\gamma\phi(p^2, [11], P(21); p, P)$$
$$- \frac{\sqrt{15}}{6}\gamma\phi(p^2, [2], D(11); p; P) \qquad (15.3.7)$$

Comparison of (15.3.7) with (15.3.6) shows that

$$\alpha = \sqrt{\frac{4}{9}}\gamma \qquad \beta = -\sqrt{\frac{5}{9}}\gamma$$

The normalization conditions are $\alpha^2 + \beta^2 = 1$ and $\gamma^2 = 1$. We choose $\gamma = +1$. The spin FP expansions are

$$\Theta_{(\widetilde{211})} = a\theta(\tfrac{1}{2}^2, [\tilde{2}], 0(\widetilde{11}); \tfrac{1}{2}; \tfrac{1}{2}) \qquad (15.3.8)$$

and

$$\Theta_{(\widetilde{121})} = b\theta(\tfrac{1}{2}^2, [\widetilde{11}], 1(\widetilde{21}); \tfrac{1}{2}; \tfrac{1}{2}) \qquad (15.3.9)$$

According to (6.6.1) we have

$$(23)\Theta_{(\widetilde{211})} = \tfrac{1}{2}\Theta_{(\widetilde{211})} + \frac{\sqrt{3}}{2}\Theta_{(\widetilde{121})}$$

and it readily follows on recoupling that

$$a = b = 1$$

The total CFP can be obtained by substitution of (15.3.9), (15.3.8),

(15.3.5) and (15.3.4) into (15.3.3). Thus

$$\Psi_0(^2P) = \frac{1}{\sqrt{2}}\Big[\Big(\sqrt{\frac{4}{9}}\,\phi(p^2, [2], S(11); p; P\Big) - \sqrt{\frac{5}{9}}\,\phi(p^2, [2], D(11); p; P)\Big)$$
$$\times\, \theta(\tfrac{1}{2}^2, [\tilde{2}], 0(\widetilde{11}); \tfrac{1}{2}; \tfrac{1}{2}) - \phi(p^2, [11], P(21); p; P)\theta(\tfrac{1}{2}^2, [\widetilde{11}], 1(\widetilde{21}); \tfrac{1}{2}; \tfrac{1}{2})\Big]$$

By coupling the spatial and spin functions and using (15.3.1) we see that

$$\Psi_0(p^3, {}^2P) = \sqrt{\frac{2}{9}}\,\psi_0(p^2, {}^1S; p; {}^2P)$$
$$- \sqrt{\frac{5}{18}}\,\psi_0(p^2, {}^1D; p; {}^2P) - \frac{1}{\sqrt{2}}\,\psi_0(p^2, {}^3P; p; {}^2P)$$

which is in agreement with Table 40.

The theory of fractional parentage within the spin free formalism has been completely generalized (see, for example, Jahn and Van Wieringen, 1951; Hassitt, 1955; Kaplan, 1962a, 1962b; Horie, 1964).

15.4 Factorization of CFP

A theorem due to Racah (1949) shows that it is possible to use the group $R(2l+1)$ (and G_2 in the case of f orbitals) in order to express a CFP in terms of a product of several factors each of which is easier to calculate than the total CFP. This result is of great value in reducing the labour involved in the calculation of CFP for f^N configurations. The reader is referred to Judd (1963) for a detailed description of this elegant but somewhat involved factorization.

15.5 Explicit expressions for substantive CFP

There exists a method by which substantive CFP can be evaluated explicitly without recourse to the solution of a set of simultaneous equations. As an example we consider the FP expansion

$$\Psi_0(l^3, \alpha, I) = \sum_{I'} (l; l^2I' \mid \} l^3\alpha I)\psi_0(l; l^2, I'; \alpha, I) \qquad (15.5.1)$$

It is clear from the orthonormality of the basis functions that

$$(l; l^2I' \mid \} l^3\alpha I) = \langle \psi_0(l; l^2, I'; \alpha I) \mid \Psi_0(l^3, \alpha, I)\rangle \qquad (15.5.2)$$

Thus the CFP can be evaluated if we have an explicit expression for $\Psi_0(l^3, \alpha, I)$. Now we write

$$\Psi_0(l^3, \alpha, I) = N\hat{A}^{(3)}\psi_0(l; l^2, I_1; \alpha, I) \qquad (15.5.3)$$

where $\hat{A}^{(3)}$ is the anitsymmetrizer, N is a normalization constant and I_1 is called the *principal* parent. The normalization constant is given by

$$N^{-2}=\langle\psi_0(l;\, l^2,\, I_1;\, \alpha,\, I)|\, \hat{A}^{(3)}\, |\psi_0(l;\, l^2,\, I_1;\, \alpha,\, I)\rangle \qquad (15.5.4)$$

If we substitute (15.5.3) in (15.5.2) we find

$$(l;\, l^2I'\,|\,\}\, l^3\alpha I)=N\,\langle\psi_0(l;\, l^2,\, I';\, \alpha,\, I)|\, \hat{A}^{(3)}\, |\psi_0(l;\, l^2,\, I_1;\, \alpha,\, I)\rangle$$

According to the discussion in Section 6.9 the antisymmetrizer can be expressed in the form

$$\hat{A}^{(3)}=\frac{2!}{3!}\hat{A}^{(2)}[1-2(12)]\hat{A}^{(2)}$$

We now have

$$N^{-2}=\frac{2!}{3!}-\frac{2.2!}{3!}\langle\psi_0(l;\, l^2,\, I_1;\, \alpha,\, I)|\, \hat{A}^{(2)}\, |\psi_0(l(2);\, l(1),\, l(3),\, I_1;\, \alpha,\, I)\rangle$$

This can be recoupled to restore dictionary order in the coordinates and we find

$$N^{-2}=\frac{2!}{3!}-\sum_J\frac{2.2!}{3!}(-1)^{2l+I_1-J}\sqrt{(2J+1)(2I_1+1)}\begin{Bmatrix} l & l & I_1\\ l & I & J\end{Bmatrix}$$
$$\times\langle\psi_0(l;\, l^2,\, I_1;\, \alpha,\, I)|\, \hat{A}^{(2)}\, |\psi_0(l(1);\, l(2);\, l(3);\, J;\, \alpha,\, I)\rangle$$

Now

$$\Psi_0(l^2,\, J)=M\hat{A}^{(2)}\psi_0(l;\, l;\, J)$$

and

$$M^{-2}=\langle\psi_0(l;\, l;\, J)|\, \hat{A}^{(2)}\, |\psi_0(l;\, l;\, J)\rangle$$
$$=\frac{1}{2!}\langle\psi_0(l;\, l;\, J)|\, 1-(12)\, |\psi_0(l;\, l;\, J)\rangle=\tfrac{1}{2}[1-(-1)^{2l-J}]=1$$

It follows that

$$N^{-2}=\tfrac{1}{3}-\tfrac{2}{3}(-1)^{2l}(2I_1+1)\begin{Bmatrix} l & l & I_1\\ l & I & I_1\end{Bmatrix} \qquad (15.5.5)$$

In a similar way we find that

$$(l;\, l^2I'\,|\,\}\, l^2\alpha I)=\tfrac{1}{3}N\,\delta_{I'I_1}-\tfrac{2}{3}N(-1)^{2l+I_1-I'}\sqrt{(2I''+1)(2I_1+1)}\begin{Bmatrix} l & l & I_1\\ l & I & I'\end{Bmatrix} \qquad (15.5.6)$$

As a first example we consider the configuration p^3. When $I={}^2P$ we choose $I_1={}^1S$. When $I={}^2D$ we choose $I_1={}^1D$ and when $I={}^4S$ we choose $I_1={}^3P$. It is then quite straightforward to obtain Table 40 directly from (15.5.5) and (15.5.6).

As a second example we take the pair of ${}^2D$ terms which arise from the configuration d^3. We have

$$\Psi_0(d^3,\, \alpha,\, {}^2D)=\sum_{I'}(d;\, d^2I'\,|\,\}\, d^3\alpha,\, {}^2D)\psi_0(d;\, d^2,\, I';\, \alpha,\, {}^2D)$$

If we distinguish the terms not by an arbitrary label α but by the seniority number v we have

$$\Psi_0(d^3, v, {}^2D) = \sum_{v'I'} (d; d^2 v' I' \mid \} d^3 v\, {}^2D)\psi_0(d; d^2, v', I'; v, {}^2D)$$

This equation can be expressed in terms of the IR's $(w_1 w_2)$ of $R(5)$ as

$$\Psi_0(d^3, (w_1 w_2), {}^2D) = \sum_{(w_1' w_2')I'} (d; d^2(w_1' w_2')I' \mid \} d^3(w_1 w_2)^2D)$$
$$\times \psi_0(d; d^2, (w_1' w_2'), I'; (w_1 w_2), {}^2D)$$

The possible parent terms are

$$(00)^1S \qquad (20)^1D \qquad (20)^1G \qquad (11)^3P \qquad (11)^3F$$

whilst the relevant offspring are $(10)^2D$ and $(21)^2D$. By making use of the branching rules for the reduction $SU(5) \rightarrow R(5)$ together with direct product decompositions of IR's of $SU(5)$ we can readily establish the following results

$$(00) \otimes (10) = (10)$$
$$(20) \otimes (10) = (10) \oplus (30) \oplus (21)$$
$$(11) \otimes (10) = (10) \oplus (11) \oplus (21)$$

which pertain to the $R(5)$ group. We now see that the $(10)^2D$ term has contributions from all the parents while $(21)^2D$ is characterized by the fact that it does not have $(00)^1S$ as a parent. For $(10)^2D$ we select $(00)^1S$ as principal parent. Application of (15.5.5) and (15.5.6) now yields the FP expression

$$\Psi_0(d^3, {}^2_1D) = \sqrt{\frac{4}{15}}\,\psi_0(d; d^2, {}^1S, {}^2_1D) - \frac{1}{\sqrt{12}}\,\psi_0(d; d^2, {}^1D; {}^2_1D)$$
$$- \sqrt{\frac{9}{60}}\,\psi_0(d; d^2, {}^1G; {}^2_1D) - \sqrt{\frac{9}{60}}\,\psi_0(d; d^2, {}^3P; {}^2_1D)$$
$$- \sqrt{\frac{21}{60}}\,\psi_0(d; d^2, {}^3F; {}^2_1D) \qquad (15.5.7)$$

We consider next $(11)^3P$ as principal parent. This yields a FP expansion for some ^2D state which is a mixture of ${}^2_1D$ and ${}^2_3D$. Thus

$$\Psi_0(d^3, \alpha, {}^2D) = -\frac{\sqrt{2}}{5}\,\psi_0(d; d^2, {}^1S; \alpha, {}^2D)$$
$$- \frac{1}{\sqrt{10}}\,\psi_0(d; d^2, {}^1D; \alpha, {}^2D) + \frac{2\sqrt{2}}{5}\,\psi_0(d; d^2, {}^1G; \alpha, {}^2D)$$
$$+ \frac{1}{\sqrt{2}}\,\psi_0(d; d^2, {}^3P; \alpha, {}^2D)$$

If we write

$$\Psi_0(d^3, {}^2_3D) = \lambda\Psi_0(d^3, {}^2_1D) + \mu\Psi_0(d^3, \alpha, {}^2D)$$

then the requirement that no ${}^1S$ parent be present shows that $\mu = \sqrt{\tfrac{10}{3}}\,\lambda$. The normalization condition gives $\lambda = \sqrt{\tfrac{3}{7}}$ and we have

$$\begin{aligned}\Psi_0(d^3, {}^2_3D) = &-\sqrt{\tfrac{45}{140}}\psi_0(d; d^2, {}^1D; {}^2_3D)\\ &+\sqrt{\tfrac{25}{140}}\psi_0(d; d^2, {}^1G; {}^2_3D) + \sqrt{\tfrac{49}{140}}\psi_0(d; d^2, {}^3P; {}^2_3D)\\ &-\sqrt{\tfrac{21}{140}}\psi_0(d; d^2, {}^3F; {}^2_3D)\end{aligned} \tag{15.5.8}$$

We now see that a careful choice of principal parent can force the label α to have the quality of a group theoretical description.

In passing it should be noted that the explicit form of a CFP can be derived very quickly by using the method of second quantization (see Judd, 1967).

15.6 Fractional parentage and molecular configurations

It should be clear that everything which we have established concerning fractional parentage in LS coupled atomic terms can be taken over directly to ΓS coupled molecular terms. Here Γ is an IR of the appropriate molecular symmetry group $\mathcal{G}$.

As an example we consider the group $\mathcal{D}_3$. The configurations e^2 and e^3 give rise to the terms ${}^3A_2$, ${}^1A_1$, ${}^1E$ and ${}^2E$ respectively. In the FP expansion

$$\Psi_0(e^3, {}^2E) = \alpha\psi_0(e; e^2, {}^3A_2; {}^2E) + \beta\psi_0(e; e^2, {}^1A_1; {}^2E) + \gamma\psi_0(e; e^2, {}^1E; {}^2E) \tag{15.6.1}$$

the coefficients α, β, γ are determined from the requirement

$$(12)\Psi_0 = -\Psi_0 \tag{15.6.2}$$

Now

$$\begin{aligned}(12)\psi_0(e(1); e(2), e(3), I; {}^2E) &= \psi_0(e(2); e(1), e(3), I; {}^2E)\\ &= \sum_{I'} (-1)^{2\not{e}+\Gamma'-S'}(-1)^{3\not{e}+{}^2E}\sqrt{(2I'+1)(2I+1)}\begin{Bmatrix}\not{e} & \not{e} & I'\\ \not{e} & {}^2E & I\end{Bmatrix}\\ &\qquad\times\psi_0(e(1), e(2), I'; e(3); {}^2E)\end{aligned} \tag{15.6.3}$$

This result is used to find the coefficients of the forbidden terms ${}^1A_2$, ${}^3A_1$, ${}^3E$ which occur on the left hand side of (15.6.2). By equating the coefficients of the forbidden terms to zero we obtain the equations

$$-\frac{1}{4}\alpha+\frac{\sqrt{3}}{4}\beta+\frac{\sqrt{6}}{4}\gamma=0$$

$$\frac{\sqrt{3}}{4}\alpha+\frac{1}{4}\beta-\frac{\sqrt{2}}{4}\gamma=0$$

$$\frac{\sqrt{2}}{4}\alpha+\frac{\sqrt{6}}{4}\beta=0$$

The normalization condition is $\alpha^2+\beta^2+\gamma^2=1$ and we have now the following CFP

$$\alpha=(e;e^{2\,3}\mathrm{A}_2|\}e^{3\,2}\mathrm{E})=\frac{1}{\sqrt{2}}$$

$$\beta=(e;e^{2\,1}\mathrm{A}_1|\}e^{3\,2}\mathrm{E})=-\frac{1}{\sqrt{6}}$$

$$\gamma=(e;e^{2\,1}\mathrm{E}|\}e^{3\,2}\mathrm{E})=\frac{1}{\sqrt{3}}$$

The result (15.2.8) can be used to obtain 2-electron CFP. Thus

$$\begin{aligned}(e^2I;e^2I|\}e^{3\,1}\mathrm{A}_1)\\ &=(e;e^{3\,2}\mathrm{E}|\}e^{4\,1}\mathrm{A}_1)(e;e^2I|\}e^{3\,2}\mathrm{E})\\ &\quad\times(-1)^{e+e+I+{}^1\mathrm{A}_1}\sqrt{(2I+1)(2^2\mathrm{E}+1)}\begin{Bmatrix}e & e & I\\ I & {}^1\mathrm{A}_1 & {}^2\mathrm{E}\end{Bmatrix}\\ &=(e;e^2I|\}e^{3\,2}\mathrm{E})(-1)^{1+I}\sqrt{4(2I+1)}\begin{Bmatrix}e & e & \Gamma\\ \Gamma & \mathrm{A}_1 & \mathrm{E}\end{Bmatrix}\begin{Bmatrix}\frac{1}{2} & \frac{1}{2} & S\\ S & 0 & \frac{1}{2}\end{Bmatrix}\\ &=(-1)^{2S}(e;e^2I|\}e^{3\,2}\mathrm{E})\end{aligned}$$

15.7 Adjective coefficients of fractional parentage

In this section we extend the concept of fractional parentage to deal with a configuration

$$(n_1l_1)^{x_1}(n_2l_2)^{x_2}\ldots(n_tl_t)^{x_t}$$

which is made up of several subconfigurations each of which contains a set of equivalent orbitals. As we shall presently see the CFP are obtained by considering substantive FP expansions for each of the subconfigurations. For this reason we call CFP which pertain to configurations involving more than a single set of equivalent orbitals *adjective* CFP.

The general derivation of adjective CFP is very cumbersome and we shall be content to illustrate how it works by means of two examples. As a first example we consider the calculation of 1-electron adjective CFP for the term $l_1 l_2^3(I_1)I$.

The 1-electron FP expansion takes the form

$$\Psi_0(l_1, l_2^3, I_1, I) = \sum_{I'} (l_1; l_2^3 I' \mid \} l_1 l_2^3 I_1 I)\psi_0(l_1; l_2^3, I, I'; I)$$

$$+ \sum_{I_2, I''} (l_2; l_1 l_2^2 I_2 I'' \mid \} l_1 l_2^3 I_1 I)\psi_0(l_2; l_1, l_2^2, I_2, I''; I)$$

from which it follows that

$$(l_1; l_2^3 I' \mid \} l_1 l_2^3 I_1 I) = \langle \psi_0(l_1; l_2^3, I'; I) \mid \Psi_0(l_1, l_2^3, I_1, I)\rangle$$

and

$$(l_2; l_1 l_2^2 I_2 I'' \mid \} l_1 l_2^3 I_1 I) = \langle \psi_0(l_2; l_1, l_2^2, I_2, I''; I) \mid \Psi_0(l_1, l_2^3, I_1, I)\rangle$$

Now we can write

$$\Psi_0(l_1, l_2^3, I_1, I) = N\hat{A}^{(4)}\psi_0(l_1; l_2^3, J; I)$$

where N is a normalization factor, $\hat{A}^{(4)}$ is the antisymmetrizer and J is a principal parent. According to (6.9.13) we have

$$\hat{A}^{(4)} = \frac{1!\,3!}{4!}\hat{A}^{(3)}[1 - 3(12)]\hat{A}^{(3)}$$

and it follows that

$$N^{-2} = \langle \psi_0(l_1; l_2^3, J; I)|\, \hat{A}^{(4)}\, |\psi_0(l_1; l_2^3, J; I)\rangle = \frac{1!\,3!}{4!}$$

We also have

$$(l_1; l_2^3 I' \mid \} l_1 l_2^3 I_1 I) = \langle \psi_0(l_1; l_2^3, I'; I)|\, N\hat{A}^{(4)}\, |\psi_0(l_1; l_2^3, J; I)\rangle$$

$$= N\frac{1!\,3!}{4!}\delta_{I'J} = \frac{1}{2}$$

and

$$(l_2; l_1 l_2^2 I_2 I'' \mid \} l_1 l_2^3 I_1 I) = \langle \psi_0(l_2; l_1, l_2^2, I_2, I''; I)|\, N\hat{A}^{(4)}\, |\psi_0(l_1; l_2^3, J; I)\rangle$$

$$= N\frac{1!\,3!}{4!}\langle \psi_0(l_2; l_1, l_2^2, I_2, I''; I)|\, \hat{A}^{(3)}$$

$$\times [1 - 3(12)]\hat{A}^{(3)}\, |\psi_0(l_1; l_2^3, J; I)\rangle$$

$$= (-1)N\,.\,3\,.\,\frac{1!\,3!}{4!}\langle \psi_0(l_2; l_1, l_2^2, I_2, I''; I)|\, \hat{A}^{(3)}$$

$$\times |\psi_0(l_1(2); l_2(1), l_2(3), l_2(4), J; I)\rangle$$

We now use a substantive FP expansion in order to detach $l_2(1)$. Thus

$$(l_2;\, l_1 l_2^2 I_2 I''\,|\,\}\, l_1 l_2^3 I_1 I) = (-1)N\frac{1!\,3!}{4!}\,3\sum_K (l_2;\, l_2^2 K\,|\,\}\, l_2^3 J) \times \langle \psi_0(l_2;\, l_1,\, l_2^2,\, I_2,\, I'';\, I)|\, \hat{A}^{(3)}\, |\psi_0(l_1(2);\, l_2(1);\, l_2^2,\, K;\, J;\, I)\rangle$$

The wavefunction on the right hand side is now recoupled in order to bring the coordinates into dictionary order. Thus

$$(l_2;\, l_1 l_2^2 I_2 I''\,|\,\}\, l_1 l_2^3 I_1 I) = -\tfrac{3}{4}N\sum_{K,A} (l_2;\, l_2^2 K\,|\,\}\, l_2^3 J) \times (-1)^{3l_1+l_2+J+2K-A}\sqrt{(2A+1)(2J+1)}\begin{Bmatrix} l_2 & K & J \\ l_1 & I & A \end{Bmatrix} \times \langle \psi_0(l_2;\, l_1,\, l_2^2,\, I_2,\, I'';\, I)|\, \hat{A}^{(3)} |\psi_0(l_2;\, l_1;\, l_2^2,\, K;\, A;\, I)\rangle$$

Now

$$\Psi_0(l_1,\, l_2^2,\, K,\, A) = M\hat{A}^{(3)}\psi_0(l_1;\, l_2^2,\, B;\, A)$$

where M is a normalization constant. If we use

$$\hat{A}^{(3)} = \frac{1!\,2!}{3!}\,\hat{A}^{(2)}[1-2(12)]\hat{A}^{(2)}$$

then it is readily verified that

$$M = \sqrt{\frac{3!}{1!\,2!}}$$

we now have

$$\langle \psi_0(l_2;\, l_1,\, l_2^2,\, I_2,\, I'';\, I)|\, \hat{A}^{(3)}\, |\psi_0(l_2;\, l_1;\, l_2^2,\, K;\, A;\, I)\rangle = \sqrt{\frac{1!\,2!}{3!}}\,\langle \psi_0(l_2;\, l_1,\, l_2^2,\, I_2,\, I'';\, I)\,|\,\psi_0(l_2;\, l_1,\, l_2^2,\, K;\, A;\, I)\rangle = \sqrt{\frac{1}{3}}\,\delta_{I_2K}\,\delta_{I''A}$$

and it follows that the adjective CFP is given by

$$(l_2;\, l_1 l_2^2 I_2 I''\,|\,\}\, l_1 l_2^3 I_1 I) = (-1)\sqrt{\frac{1!\,3!}{4!}}\sqrt{\frac{3!}{2!\,1!}}\,(l_2;\, l_2^2 I_2\,|\,\}\, l_2^3 J)(-1)^{3l_1+l_2+2I_2+J-I''} \times \sqrt{(2I''+1)(2J+1)}\begin{Bmatrix} l_2 & I_2 & J \\ l_1 & I & I'' \end{Bmatrix} \qquad (15.7.1)$$

As a second example we consider the 2-electron adjective FP expansion

$$\Psi_0(l_1^2, I_1, l_2^2, I_2, I) = \sum_{I',I''} (l_1^2 I'; l_2^2 I'' |\,\}\, l_1^2 I_1 l_2^2 I_2 I)$$

$$\times \psi_0(l_1^2, I'; l_2^2, I''; I) + \sum_{I',I''} (l_1 l_2 I'; l_1 l_2 I'' |\,\}\, l_1^2 I_1 l_2^2 I_2 I)$$

$$\times \psi_0(l_1, l_2, I'; l_1, l_2, I''; I)$$

$$+ \sum_{I',I''} (l_2^2 I'; l_1^2 I'' |\,\}\, l_1^2 I_1 l_2^2 I_2 I) \psi_0(l_2^2, I'; l_1^2, I''; I)$$

We have

$$(l_1^2 I'; l_2^2 I'' |\,\}\, l_1^2 I_1 l_2^2 I_2 I) = \langle \psi_0(l_1^2, I'; l_2^2, I''; I) \,|\, \Psi_0(l_1^2, I_1, l_2^2, I_2, I)\rangle$$

$$(l_1 l_2 I'; l_1 l_2 I'' |\,\}\, l_1^2 I_1 l_2^2 I_2 I) = \langle \psi_0(l_1, l_2, I'; l_1, l_2, I''; I) \,|\, \Psi_0(l_1^2, I_1, l_2^2, I_2, I)\rangle$$

$$(l_2^2 I'; l_1^2 I'' |\,\}\, l_1^2 I_1 l_2^2 I_2 I) = \langle \psi_0(l_2^2, I; l_1^2, I''; I) \,|\, \Psi_0(l_1^2, I_1, l_2^2, I_2, I)\rangle$$

and

$$\Psi_0(l_1^2, I_1, l_2^2, I_2, I) = N\hat{A}^{(4)}\psi_0(l_1^2, A; l_2^2, B; I)$$

If we use (6.9.12) we have

$$\hat{A}^{(4)} = \frac{2!\,2!}{4!} \hat{A}^{(2)}\hat{A}^{(2)}[1 - 4(23) + (14)(23)]\hat{A}^{(2)}\hat{A}^{(2)}$$

and it readily follows that

$$N = \sqrt{\frac{4!}{2!\,2!}}$$

We now consider

$$(l_1 l_2 I'; l_1 l_2 I'' |\,\}\, l_1^2 I_1 l_2^2 I_2 I) = -4N \frac{2!\,2!}{4!} \langle \psi_0(l_1, l_2, I'; l_1, l_2, I''; I) |\, \hat{A}^{(2)}\hat{A}^{(2)}$$

$$\times |\psi_0(l_1(1), l_1(3), A; l_2(2), l_2(4), B; I)\rangle$$

The wave function on the right hand side is recoupled and we find

$(l_1 l_2 I'; l_1 l_2 I'' |\,\}\, l_1^2 I_1 l_2^2 I_2 I)$

$$= -4N \frac{2!\,2!}{4!} \sum_{C,D} \sqrt{(2C+1)(2D+1)(2A+1)(2B+1)} \begin{Bmatrix} l_1 & l_1 & A \\ l_2 & l_2 & B \\ C & D & I \end{Bmatrix}$$

$$\times \langle \psi_0(l_1, l_2, I'; l_1, l_2, I''; I) |\, \hat{A}^{(2)}\hat{A}^{(2)}\, |\psi_0(l_1; l_2; C; l_1; l_2; D; I)\rangle$$

Now

$$\Psi_0(l_1, l_2, C) = M\hat{A}^{(2)}\psi_0(l_1; l_2; C)$$

and we readily find that $M = \sqrt{2}$. It follows that the adjective CFP is given by

$$(l_1l_2I'; l_1l_2I''| \} l_1^2I_1l_2^2I_2I)$$

$$= (-1) \sqrt{\frac{2!\,2!}{4!}} \sqrt{\frac{2!}{1!\,1!}} \sqrt{\frac{2!}{1!\,1!}} (l_1; l_1| \} l_1^2A)$$

$$\times (l_2; l_2| \} l_2^2B) \sqrt{(2I'+1)(2I''+1)(2A+1)(2B+1)} \begin{Bmatrix} l_1 & l_1 & A \\ l_2 & l_2 & B \\ I' & I'' & I \end{Bmatrix} \quad (15.7.2)$$

If we examine (15.7.1) and (15.7.2) we see that in general an adjective CFP can be partitioned. There are essentially two distinct parts, namely the substantive CFP and the "adjective". The "adjective" consists of a product of a transposition phase, several normalization factors, recoupling phases and a recoupling coefficient.

Tables of 1-electron and 2-electron adjective CFP for configurations $s^{\lambda}s'^{\mu}p^{q}$ can be found in Chisholm, Dalgarno and Innes (1969).

The concept of adjective fractional parentage can clearly be taken over directly to deal with molecular configurations such as

$$(n_1\gamma_1)^{x_1}(n_2\gamma_2)^{x_2}\ldots(n_t\gamma_t)^{x_t}$$

where γ_i are standard IR's of the appropriate molecular symmetry group.

References

Chisholm, C. D. H., Dalgarno, A. and Innes, F. R. (1969) *Adv. in Atomic and Molecular Phys.*, **5,** 297.

Hassitt, A. (1955) *Proc. Roy. Soc. (Lond.)*, **A229,** 110.

Horie, H. (1964) *J. Phys. Soc. Japan* **19,** 1783.

Jahn, H. A. and Van Wieringen, H. (1951) *Proc. Roy. Soc. (Lond.)*, **A209,** 502.

Judd, B. R. (1963) "Operator Techniques in Atomic Spectroscopy", McGraw-Hill.

Judd, B. R. (1967) "Second Quantization in Atomic Spectroscopy", Johns Hopkins.

Kaplan, I. G. (1962a) *J.E.T.P.*, **14,** 401.

Kaplan, I. G. (1962b) *J.E.T.P.*, **14,** 568.

Racah, G. (1943) *Phys. Rev.*, **63,** 367.

Racah, G. (1949) *Phys. Rev.*, **76,** 1352.

CHAPTER 16

Tensor Operator Analysis

16.1 Introduction

The calculation of a given atomic or molecular property depends upon the evaluation of a matrix element (or matrix elements) of the associated operator between a state (or states) of the system. We have already seen how the states can be classified according to IR's of certain groups. We have further seen how the state functions can be expressed in terms of fractional parentage expansions. The labour involved in the evaluation of the above mentioned matrix elements becomes very much reduced when the appropriate operator is written in a form which associates with it IR's of the same groups used to classify the states.

16.2 Irreducible tensor operators

We begin by considering the angular momentum vector operator $(\hat{J}_x, \hat{J}_y, \hat{J}_z)$. As we have seen this operator can be defined by the commutation relations

$$[\hat{J}_x, \hat{J}_y] = i\hat{J}_z \qquad [\hat{J}_y, \hat{J}_z] = i\hat{J}_x \qquad [\hat{J}_z, \hat{J}_x] = i\hat{J}_y$$

We now write $\hat{J}_\pm = \hat{J}_x \pm i\hat{J}_y$ and introduce the set of three operators $\hat{T}^{(1)}_q$, $q = 0, \pm 1$ where

$$\hat{T}^{(1)}_0 = \hat{J}_z \qquad \hat{T}^{(1)}_{\pm 1} = \mp \frac{1}{\sqrt{2}} \hat{J}_\pm$$

These operators are then defined by the commutation relations (see (9.4.21))

$$\begin{aligned} &[\hat{J}_z, \hat{T}^{(1)}_q] = q\hat{T}^{(1)}_q \\ &[\hat{J}_\pm, \hat{T}^{(1)}_q] = \sqrt{1(1+1) - q(q \pm 1)}\, \hat{T}^{(1)}_{q\pm 1} \end{aligned} \qquad (16.2.1)$$

The set of operators $\hat{T}^{(1)}_q$ thus forms a basis for the IR (1) of the group $R^*(3)$.

We can generalize (16.2.1) and introduce a set of $(2k+1)$ operators $\hat{T}^{(k)}_q$, $q = k, k-1, \ldots, -k+1, -k$ which are defined by the commutation relations

$$[\hat{J}_z, \hat{T}_q^{(k)}] = q\hat{T}_q^{(k)}$$

$$[\hat{J}_\pm, \hat{T}_q^{(k)}] = \sqrt{k(k+1)-q(q\pm 1)}\,\hat{T}_{q\pm 1}^{(k)}$$

These operators transform according to the IR (k) of the group $R^*(3)$ and for this reason we say that $\hat{T}_q^{(k)}$ form the components of an *irreducible tensor operator* $\mathbf{T}^{(k)}$ of *rank* k with respect to the group $R^*(3)$.

For a molecular symmetry group $\mathcal{G}$ there are no operators equivalent to the infinitesimal operators so the above indirect definition of a set spanning a representation must be replaced by the following direct definition. Let $|\Gamma, M\rangle$ be a set of n_Γ basis kets for the real standard IR Γ of $\mathcal{G}$. If R_i is a generator of $\mathcal{G}$ we have

$$\hat{R}_i\,|\Gamma, M\rangle = \sum_{M'} |\Gamma, M'\rangle\,\Gamma_{M'M}(R_i) \qquad i = 1, 2, \ldots$$

Any set of n_Γ operators $\hat{T}_M^{(\Gamma)}$ which transform according to

$$\hat{R}_i\hat{T}_M^{(\Gamma)} = \sum_{M'} \hat{T}_{M'}^{(\Gamma)}\,\Gamma_{M'M}(R_i) \qquad i = 1, 2, \ldots$$

is said to form an irreducible operator $\hat{\mathbf{T}}^{(\Gamma)}$ of *symmetry type* Γ with respect to $\mathcal{G}$.

16.3 The Wigner-Eckart theorem

Consider a matrix element

$$\mu = \langle \alpha, J, M|\,\hat{T}_q^{(k)}\,|\alpha', J', M'\rangle$$

where $\hat{T}_q^{(k)}$ is an irreducible tensor operator of rank k and $|\alpha, J, M\rangle$ is an atomic state which has been classified according to the IR (J) (or (L) and (S)) of $SU(2)$ and according to a set of quantum numbers α. As we have seen the quantum numbers α can often be taken as labels for IR's of higher groups. The operators $\hat{T}_q^{(k)}$ can also be further classified according to IR's of these higher groups. Such a classification greatly reduces the labour involved in calculating matrix elements when f orbital configurations are involved (see Judd, 1963). We shall not however pursue this further classification of operators.

According to (12.2.7) we can define a vector coupled ket as follows

$$\hat{T}_q^{(k)}\,|\alpha', J', M'\rangle = \sum_{K,Q} (kqJ'M' \mid KQ)\,|\hat{\mathbf{T}}^{(k)}; \alpha', J'; K, Q\rangle$$

$$\mu = \sum_{K,Q} (kqJ'M' \mid KQ)\,\langle \alpha, J, M \mid \hat{\mathbf{T}}^{(k)}; \alpha', J'; K, Q\rangle$$

$$= (kqJ'M' \mid JM)\,\langle \alpha, J, M \mid \hat{\mathbf{T}}^{(k)}; \alpha', J'; J, M\rangle$$

where we have used $\langle J, M \mid K, Q\rangle = \delta_{JK}\,\delta_{MQ}$. The complex conjugate transformation to (9.5.1) is

$$u' = a^*u + b^*v$$

$$v' = -bu + av$$

and this is obtained from (9.5.1) by making the substitution $u \to v$; $v \to -u$. It now follows from (9.5.11) that

$$\psi^*(j, m) = \frac{v^{j+m}(-u)^{j-m}}{\sqrt{(j+m)!\,(j-m)!}} = (-1)^{j-m}\psi(j, -m)$$

Using this result we can carry out a further coupling. Thus

$$\mu = (kqJ'M' \mid JM)(-1)^{J-M}\sum_{J'',M''}(J-MJM \mid J''M'') \,.\, \langle \alpha, J; \hat{\mathbf{T}}^{(k)}; \alpha', J'; J; J'', M''\rangle$$

We shall call $\langle \alpha, J; \hat{\mathbf{T}}^{(k)}; \alpha', J'; J; J'', M''\rangle$ a *fully coupled* matrix element. Now since μ is a definite integral and hence a scalar which spans the IR with $J''=0$ it follows that the only term in the summation which survives is $J''=M''=0$. Thus

$$\mu = (kqJ'M' \mid JM)(-1)^{J-M}(J-MJM \mid 00)\langle \alpha, J; \hat{\mathbf{T}}^{(k)}; \alpha', J'; J; 0, 0\rangle$$

and in terms of the $3j$ symbol we now have

$$\begin{aligned}\mu &= \langle \alpha, J, M|\, \hat{T}^{(k)}_q \,|\alpha', J', M'\rangle \\ &= (-1)^{2J+J'-k-M}\begin{pmatrix} J & k & J' \\ -M & q & M' \end{pmatrix}\langle \alpha, J; \hat{\mathbf{T}}^{(k)}; \alpha', J'; J; 0, 0\rangle\end{aligned} \tag{16.3.1}$$

The fully coupled matrix element is independent of M', q and M and the result (16.3.1) is referred to as the Wigner-Eckart theorem. This theorem is of great importance in the theory of tensor operators. The Wigner-Eckart theorem is more usually expressed in terms of the *reduced* matrix element $\langle \alpha, J\|\, \hat{\mathbf{T}}^{(k)} \,\|\alpha', J'\rangle$ which is defined through the relation

$$\begin{aligned}&\langle \alpha, J, M|\, \hat{T}^{(k)}_q \,|\alpha', J', M'\rangle \\ &\qquad = (-1)^{J-J'+k}\frac{(kqJ'M' \mid JM)}{\sqrt{2J+1}}\langle \alpha, J\|\, \hat{\mathbf{T}}^{(k)} \,\|\alpha', J'\rangle\end{aligned} \tag{16.3.2}$$

The relation between the fully coupled matrix element and the reduced matrix element is

$$\langle \alpha, J\|\, \hat{\mathbf{T}}^{(k)} \,\|\alpha', J'\rangle = (-1)^{J+J'-k}\langle \alpha, J; \hat{\mathbf{T}}^{(k)}; \alpha', J'; J; 0, 0\rangle \tag{16.3.3}$$

The double bar notation in the reduced matrix element makes it somewhat difficult to see how to carry out recoupling within it whereas the fully coupled matrix element clearly does not suffer from this disadvantage.

For a simply reducible molecular symmetry group $\mathcal{G}$ the Wigner-Eckart theorem can similarly be established. Thus if Γ, Γ' and Γ'' are standard IR's of $\mathcal{G}$ we have

$$\langle\Gamma', M'|\, \hat{T}^{(\Gamma)}_M \,|\Gamma'', M''\rangle = \frac{(\Gamma M \Gamma'' M'' \mid \Gamma' M')}{\sqrt{n_{\Gamma'}}} \cdot \langle\Gamma' \|\, \hat{\mathbf{T}}^{(\Gamma)} \,\|\Gamma''\rangle \tag{16.3.4}$$

16.4 Products of tensor operators

Let $\hat{U}^{(k_1)}$ and $\hat{V}^{(k_2)}$ be two irreducible tensor operators of ranks k_1 and k_2 respectively. According to (12.2.7) we have

$$\hat{U}^{(k_1)}_{q_1} \hat{V}^{(k_2)}_{q_2} = \sum_{k,q} (k_1 q_1 k_2 q_2 \mid kq)(\hat{\mathbf{U}}^{(k_1)} \otimes \hat{\mathbf{V}}^{(k_2)})^{(k)}_q$$

This can be inverted to give the components of the irreducible tensor operator $\hat{\mathbf{T}}^{(k)} = (\hat{\mathbf{U}}^{(k_1)} \otimes \hat{\mathbf{V}}^{(k_2)})^{(k)}$ of rank k. Thus

$$(\hat{\mathbf{U}}^{(k_1)} \otimes \hat{\mathbf{V}}^{(k_2)})^{(k)}_q = \sum_{q_1, q_2} (k_1 q_1 k_2 q_2 \mid kq)\, \hat{U}^{(k_1)}_{q_1} \hat{V}^{(k_2)}_{q_2}$$

When $k_1 = k_2 = k$ we have a scalar operator

$$(\hat{\mathbf{U}}^{(k)} \otimes \hat{\mathbf{V}}^{(k)})^{(0)}_0 = \sum_q \frac{(-1)^{k+q}}{\sqrt{2k+1}}\, \hat{U}^{(k)}_q \hat{V}^{(k)}_{-q}$$

The scalar product of $\hat{\mathbf{U}}^{(k)}$ and $\hat{\mathbf{V}}^{(k)}$ is by tradition defined as

$$(\mathbf{U}^{(k)} \cdot \mathbf{V}^{(k)}) = \sum_q (-1)^q \hat{U}^{(k)}_q \hat{V}^{(k)}_{-q} = (-1)^k \sqrt{2k+1}\, (\hat{\mathbf{U}}^{(k)} \otimes \hat{\mathbf{V}}^{(k)})^{(0)}_0 \tag{16.4.1}$$

In the case of a molecular symmetry group the analogous result is

$$(\hat{\mathbf{U}}^{(\Gamma)} \cdot \hat{\mathbf{V}}^{(\Gamma)}) = \sum_M \hat{U}^{(\Gamma)}_M V^{(\Gamma)}_M = \sqrt{n_\Gamma}\, (\hat{\mathbf{U}}^{(\Gamma)} \otimes \hat{\mathbf{V}}^{(\Gamma)})^{(A_1)}_{a_1} \tag{16.4.2}$$

An operator which occurs very frequently is the electrostatic repulsion $1/r_{12}$ between two electrons a distance r_{12} apart. This operator can be cast into tensorial form by using the well known expansion in spherical harmonics

$$\frac{1}{r_{12}} = \sum_{k=0}^{\infty} \sum_{q=-k}^{k} \left(\frac{4\pi}{2k+1}\right) \frac{r_<^k}{r_>^{k+1}}\, Y_{k,q}(\theta_1, \phi_1) Y^*_{k,q}(\theta_2, \phi_2) \tag{16.4.3}$$

If we define the spherical tensor $\mathbf{C}^{(k)}$ by

$$\hat{C}^{(k)}_q = \sqrt{\frac{4\pi}{2k+1}}\, Y_{k,q} \tag{16.4.4}$$

then we have

$$\frac{1}{r_{12}} = \sum_{k=0}^{\infty} \frac{r_<^k}{r_>^{k+1}} \sum_q (-1)^q \hat{C}_{-q}^{(k)}(1) \hat{C}_q^{(k)}(2)$$

and it follows from (16.4.1) that

$$\frac{1}{r_{12}} = \sum_{k=0}^{\infty} \frac{r_<^k}{r_>^{k+1}} (\hat{\mathbf{C}}^{(k)}(1) \cdot \hat{\mathbf{C}}^{(k)}(2)) \tag{16.4.5}$$

16.5 Simplification of matrix elements

We begin by considering a matrix element of the form

$$\langle \alpha_1, J_1, \alpha_2, J_2, J, M | (\hat{\mathbf{U}}^{(k_1)} \otimes \hat{\mathbf{V}}^{(k_2)})_q^{(k)} | \alpha_1', J_1', \alpha_2', J_2', J', M' \rangle$$

in which $\hat{\mathbf{U}}^{(k_1)}$ acts upon the coordinates included in $|J_1', M_1'\rangle$ and $\hat{\mathbf{V}}^{(k_2)}$ acts upon the coordinates included in $|J_2', M_2'\rangle$. Application of the Wigner-Eckart theorem gives

$$\langle \alpha_1, J_1, \alpha_2, J_2, J, M | (\hat{\mathbf{U}}^{(k_1)} \otimes \hat{\mathbf{V}}^{(k_2)})_q^{(k)} | \alpha_1', J_1', \alpha_2', J_2', J', M' \rangle$$
$$= (-1)^{J-J'+k} \frac{(kqJ'M' \mid JM)}{\sqrt{2J+1}} \langle \alpha_1, J_1, \alpha_2, J_2, J \| (\hat{\mathbf{U}}^{(k_1)} \otimes \hat{\mathbf{V}}^{(k_2)}) \| \alpha_1', J_1', \alpha_2', J_2', J' \rangle$$

Now

$$\langle \alpha_1, J_1, \alpha_2, J_2, J \| (\hat{\mathbf{U}}^{(k_1)} \otimes \hat{\mathbf{V}}^{(k_2)})^{(k)} \| \alpha_1', J_1', \alpha_2', J_2', J' \rangle$$
$$= (-1)^{J+J'-k} \langle \alpha_1, J_1, \alpha_2, J_2, J; (\hat{\mathbf{U}}^{(k_1)} \otimes \hat{\mathbf{V}}^{(k_2)})^{(k)}; \alpha_1', J_1', \alpha_2', J_2', J'; J; 0, 0 \rangle$$
$$= (-1)^{J+J'-k} \sqrt{(2k+1)(2J+1)(2J'+1)} \begin{Bmatrix} k_1 & k_2 & k \\ J_1' & J_2' & J' \\ J_1 & J_2 & J \end{Bmatrix}$$
$$\times \langle \alpha_1, J_1; \hat{\mathbf{U}}^{(k_1)}; \alpha_1', J_1'; J_1; 0, 0 \rangle \langle \alpha_2, J_2; \hat{\mathbf{V}}^{(k_2)}; \alpha_2', J_2'; J_2; 0, 0 \rangle$$
$$= \sqrt{(2k+1)(2J+1)(2J'+1)} \begin{Bmatrix} J_1 & J_1' & k_1 \\ J_2 & J_2' & k_2 \\ J & J' & k \end{Bmatrix} \langle \alpha_1, J_1 \| \hat{\mathbf{U}}^{(k_1)} \| \alpha_1', J_1' \rangle$$
$$\times \langle \alpha_2, J_2 \| \hat{\mathbf{V}}^{(k_2)} \| \alpha \quad J_2' \rangle$$

where we have used (16.. [illegible] nd (12.4.16). We have thus arrived at the result

$$\langle \alpha_1, J_1, \alpha_2, J_2, J, M | (\hat{\mathbf{U}}^{(k_1)} \otimes \hat{\mathbf{V}} \ [illegible] \ \alpha_2', J_2', J', M' \rangle$$

$$= (-1)^{J-M} \begin{pmatrix} k & J' & J \\ q & M' & -M \end{pmatrix} \begin{Bmatrix} J_1 & [illegible] \\ J_2 & [illegible] \\ J & J \ [illegible] \end{Bmatrix} [illegible] \ 2J+1)(2J'+1)$$
$$\times \langle \alpha_1, J_1 \| \mathbf{U} \ [illegible] \ _2', J_2' \rangle \tag{16.5.1}$$

If we take $\hat{\mathbf{V}}^{(k_2)}=1$ so that $k_2=0$ and $k_1=k$ we obtain a second useful result

$$\langle \alpha_1, J_1, \alpha_2, J_2, J, M|\, \hat{U}_q^{(k)}\, |\alpha_1', J_1', \alpha_2', J_2', J', M'\rangle$$

$$= (-1)^{J_1+J_2+J'+k+j-M} \sqrt{(2J+1)(2J'+1)}\, \delta(J_2, J_2') \begin{pmatrix} k & J' & J \\ q & M' & -M \end{pmatrix}$$

$$\times \begin{Bmatrix} J_1 & J & J_2 \\ J' & J_1' & k \end{Bmatrix} \langle \alpha_1, J_1 \|\, \hat{\mathbf{U}}^{(k)} \,\| \alpha_1', J_1' \rangle \quad (16.5.2)$$

In establishing this result we have used $\langle J, M|\, 1\, |J, M\rangle = 1$ which in conjunction with (16.3.2) shows that

$$\langle J \|\, 1 \,\| J \rangle = \sqrt{2J+1} \quad (16.5.3)$$

16.6 The Coulomb interaction

The results established in Section 16.5 are very powerful since they can be used, for example, to circumvent the tedious angular integrations which occur when we consider matrix elements of the Coulomb interaction $1/r_{12}$. The results can also be used to greatly simplify the calculation of matrix elements of other operators such as spin-orbit interaction, external magnetic field, magnetic hyperfine interaction, magnetic spin-spin interaction and many more. For details the reader is referred to Fano and Racah (1959), Edmonds (1960), Judd (1963). In this book we shall be content to consider just the Coulomb interaction $1/r_{12}$. The angular part of $1/r_{12}$ is given in (16.4.5). Using (16.4.1) and (16.5.1) we have

$$\langle \alpha_1, j_1, \alpha_2, j_2, j, m|\, (\hat{\mathbf{C}}^{(k)}(1) \cdot \hat{\mathbf{C}}^{(k)}(2))\, |\alpha_1', j_1', \alpha_2', j_2', j', m'\rangle$$

$$= (-1)^{j_1'+j_2+j}\, \delta(j, j')\, \delta(m, m') \begin{Bmatrix} j_1' & j_2' & j \\ j_2 & j_1 & k \end{Bmatrix}$$

$$\times \langle \alpha_1, j_1 \|\, \hat{\mathbf{C}}^{(k)} \,\| \alpha_1', j_1' \rangle \langle \alpha_2, j_2 \|\, \hat{\mathbf{C}}^{(k)} \,\| \alpha_2', j_2' \rangle$$

In the *LS* coupling scheme this result becomes

$$\langle \alpha_1, S_1, L_1, \alpha_2, S_2, L_2, S, L, M_S, M_L|\, (\hat{\mathbf{C}}^{(k)}(1) \cdot \hat{\mathbf{C}}^{(k)}(2))$$

$$\times |\alpha_1', S_1', L_1', \alpha_2', S_2', L_2', S', L', M_S', M_L'\rangle$$

$$= (-1)^{L_1'+L_2+L}\, \delta(L, L')\, \delta(S, S')\, \delta(M_L, M_L')\, \delta(M_S, M_S')\, \delta(S_1, S_1')\, \delta(S_2, S_2')$$

$$\times \begin{Bmatrix} L_1' & L_2' & L \\ L_2 & L_1 & k \end{Bmatrix} \langle \alpha_1, L_1 \|\, \hat{\mathbf{C}}^{(k)} \,\| \alpha_1', L_1' \rangle \langle \alpha_2, L_2 \|\, \hat{\mathbf{C}}^{(k)} \,\| \alpha_2', L_2' \rangle \quad (16.6.1)$$

When $|n, l, m\rangle$ is an atomic orbital the matrix element

$$\langle n_1, l_1, n_2, l_2, S, L, M_S, M_L|\, \frac{1}{r_{12}}\, |n_1', l_1', n_2', l_2', S', L', M_S', M_L'\rangle$$

can be written in the spin-free formalism as

$$\langle n_1, l_1, n_2, l_2, [\lambda], L, M_L | \frac{1}{r_{12}} | n_1', l_1', n_2', l_2', [\lambda'], L', M_L' \rangle$$

where $[\lambda]$ is an IR of $\mathscr{S}_2$. According to (16.6.1) the matrix element vanishes unless $L' = L$, $M_{L'} = M_L$ and $\lambda' = \lambda$. If $n_1 l_1$ and $n_2 l_2$ are inequivalent we have

$$|n_1, l_1, n_2, l_2, [\lambda], L, M_L\rangle = \frac{1}{\sqrt{2}} \{ \, | n_1, l_1, n_2, l_2, L, M_L\rangle \pm |n_2, l_2, n_1, l_1, L, M_L\rangle\}$$

and the matrix element becomes

$$\langle n_1, l_1, n_2, l_2, L, M_L | \frac{1}{r_{12}} | n_1', l_1', n_2', l_2', L, M_L\rangle \pm \langle n_2, l_2, n_1, l_1, L, M_L | \frac{1}{r_{12}} | n_1', l_1', n_2', l_2', L, M_L\rangle$$

Upon using (16.6.1) this can be written as

$$\sum_k (n_1 l_1 n_2 l_2 n_1' l_1' n_2' l_2')_k (-1)^{l_1'+l_2+L} \langle l_1 \| \hat{\mathbf{C}}^{(k)} \| l_1' \rangle \times \langle l_2 \| \hat{\mathbf{C}}^{(k)} \| l_2' \rangle \begin{Bmatrix} l_1' & l_2' & L \\ l_2 & l_1 & k \end{Bmatrix} \pm \sum_k (n_2 l_2 n_1 l_1 n_1' l_1' n_2' l_2')_k \times (-1)^{l_1'+l_1+L} \langle l_2 \| \hat{\mathbf{C}}^{(k)} \| l_1' \rangle \langle l_1 \| \hat{\mathbf{C}}^{(k)} \| l_2' \rangle \begin{Bmatrix} l_1' & l_2' & L \\ l_1 & l_2 & k \end{Bmatrix} \quad (16.6.2)$$

where $(n_1 l_1 n_2 l_2 n_1' l_1' n_2' l_2')_k$ is the radial integral

$$(n_1 l_1 n_2 l_2 n_1' l_1' n_2' l_2')_k = \int_0^\infty \int_0^\infty \frac{r_<^k}{r_{>k+1}} \times R_{n_1,l_1}(r_1) R_{n_1',l_1'}(r_1) R_{n_2,l_2}(r_2) R_{n_2',l_2'}(r_2) r_1^2 r_2^2 \, dr_1 \, dr_2$$

The reduced matrix elements $\langle l \| \hat{\mathbf{C}}^{(k)} \| l' \rangle$ can be obtained by using the Wigner-Eckart theorem. We have

$$\langle l, 0 | \hat{C}_0^{(k)} | l', 0\rangle = (-1)^l \begin{pmatrix} l & k & l' \\ 0 & 0 & 0 \end{pmatrix} \langle l \| \hat{\mathbf{C}}^{(k)} \| l' \rangle$$

Now

$$\langle l, 0 | \hat{C}_0^{(k)} | l', 0\rangle = \tfrac{1}{2}\sqrt{(2l+1)(2l'+1)} \int_{-1}^{1} P_l(\mu) P_k(\mu) P_{l'}(\mu) \, d\mu$$

where $P_l(\mu)$ is the Legendre polynomial. The above integral has a value which can be written as (see Judd, 1963)

$$\int_{-1}^{1} P_l(\mu) P_k(\mu) P_{l'}(\mu) \, d\mu = 2 \begin{pmatrix} l & k & l' \\ 0 & 0 & 0 \end{pmatrix}^2$$

and it follows that

$$\langle l \| \hat{\mathbf{C}}^{(k)} \| l' \rangle = (-1)^l \sqrt{(2l+1)(2l'+1)} \begin{pmatrix} l & k & l' \\ 0 & 0 & 0 \end{pmatrix} \tag{16.6.3}$$

When $n_1'l_1' \equiv n_1l_1$ and $n_2'l_2' \equiv n_2l_2$ we obtain from (16.6.2) the following expression for the first order energy E_1 of the state $n_1l_1n_2l_2SL$

$$E_1(n_1l_1n_2l_2SL)$$
$$= \langle n_1, l_1, n_2, l_2S, L, M_S, M_L | \frac{1}{r_{12}} | n_1, l_1, n_2, l_2, S, L, M_S, M_L \rangle$$
$$= \sum_k (-1)^L (2l_1+1)(2l_2+1) \begin{pmatrix} l_1 & k & l_1 \\ 0 & 0 & 0 \end{pmatrix} \begin{pmatrix} l_2 & k & l_2 \\ 0 & 0 & 0 \end{pmatrix} \begin{Bmatrix} l_1 & l_2 & L \\ l_2 & l_1 & k \end{Bmatrix} F_k(n_1l_1n_2l_2)$$
$$\pm \sum_k (-1)^{L+k} (2l_1+1)(2l_2+1) \begin{pmatrix} l_1 & k & l_2 \\ 0 & 0 & 0 \end{pmatrix}^2 \begin{Bmatrix} l_1 & l_2 & L \\ l_1 & l_2 & k \end{Bmatrix} G_k(n_1l_1n_2l_2) \tag{16.6.4}$$

where F_k and G_k are the radial integrals given by

$$F_k(n_1l_1n_2l_2) = \int_0^\infty \int_0^\infty \frac{r_<^k}{r_>^{k+1}} R_{n_1,l_1}^2(r_1) R_{n_2,l_2}^2(r_2) r_1^2 r_2^2 \, dr_1 \, dr_2$$

$$G_k(n_1l_1n_2l_2) = \int_0^\infty \int_0^\infty \frac{r_<^k}{r_>^{k+1}} R_{n_1,l_1}(r_1) R_{n_1,l_1}(r_2) R_{n_2,l_2}(r_1) R_{n_2,l_2}(r_2) r_1^2 r_2^2 \, dr_1 \, dr_2$$

When $n_2l_2 \equiv n_1l_1 = nl$ we find

$$E_1((nl)^2SL) = \sum_k (-1)^L (2l+1)^2 \begin{pmatrix} l & k & l \\ 0 & 0 & 0 \end{pmatrix}^2 \begin{Bmatrix} l & l & L \\ l & l & k \end{Bmatrix} F_k(nlnl) \tag{16.6.5}$$

Finally we consider the case when $n_2l_2 \equiv n_1l_1 = nl$ and $n_2'l_2' \equiv n_1'l_1' = n'l'$. This is important when configuration interaction calculations are being performed. Thus for example the two terms $2s^2\ ^1S$ and $2p^2\ ^1S$ have the same symmetry and also have the same zeroth order energy when hydrogenic orbitals are being considered. It follows that we have a two-fold degenerate problem and we must allow for configuration interaction by writing the zeroth order function as

$$\psi_0(^1S) = a\psi_0(2s^2\ ^1S) + b\psi_0(2p^2\ ^1S)$$

where a and b are coefficients to be determined from the secular equations. The off-diagonal element in the secular determinant is clearly

$$\langle \psi_0(2s^2\ ^1S) | \frac{1}{r_{12}} | \psi_0(2p^2\ ^1S) \rangle$$

which is of the form we are considering. In the general case we readily establish the result

$$\langle (nl)^2, S, L, M_S, M_L | \frac{1}{r_{12}} | (n'l')^2, S, L, M_S, M_L \rangle$$

$$= \sum_k (-1)^{l+l'+L}(2l+1)(2l'+1) \begin{pmatrix} l & k & l' \\ 0 & 0 & 0 \end{pmatrix}^2 \begin{Bmatrix} l' & l' & L \\ l & l & k \end{Bmatrix} G_k(nln'l') \quad (16.6.6)$$

When using the results (16.6.4), (16.6.5) and (16.6.6) the $3j$ and $6j$ symbols which occur can be found in standard tables (see for example Rotenberg *et al.*, 1959). Alternatively when small arguments occur in the nj symbols useful expressions can be found in Judd (1963).

16.7 First order energies for N-electron atomic states

We consider an N-electron atomic system in which we choose the unit of length to be Z Bohr and the unit of energy to be Z^2 Hartree. Here Z is the nuclear charge. The Hamiltonian for the non-relativistic motion of the electrons now takes the form

$$\hat{H} = \hat{H}_0 + \frac{1}{Z}\hat{H}_1 \quad (16.7.1)$$

with

$$\hat{H}_0 = \sum_{i=1}^{N} \hat{h}(i) \quad (16.7.2)$$

$$\hat{h}(i) = -\tfrac{1}{2}\nabla_i^2 - \frac{1}{r_i} \quad (16.7.3)$$

and

$$\hat{H}_1 = \sum_{i>j=1}^{N}\sum \frac{1}{r_{ij}} \quad (16.7.4)$$

If we regard $\hat{H}_1$ as a perturbation the solutions of the Schrödinger equation

$$\hat{H}\Psi = E\Psi$$

take the form of perturbation series

$$\Psi = \Psi_0 + \frac{1}{Z}\Psi_1 + \ldots \quad (16.7.5)$$

$$E = E_0 + \frac{1}{Z}E_1 + \frac{1}{Z^2}E_2 + \ldots \quad (16.7.6)$$

The zeroth order energy E_0 is given by (14.2.14) while the zeroth order function Ψ_0 is associated with an electronic configuration (14.2.4). If we assume LS coupling the zeroth order function for a term $I = L, S$ is

$\Psi_0(l_1^{x_1}, I_1; l_2^{x_2}, I_2; I_{\Sigma 2} \ldots l_t^{x_t}, I_t; I) \equiv \Psi_0(I)$. Provided that no other configurations have the same zeroth order energy then the first order energy for the term I is given by

$$E_1(I) = \langle \Psi_0(I) | \sum_{i>j=1} \sum \frac{1}{r_{ij}} | \Psi_0(I) \rangle \tag{16.7.7}$$

Since Ψ_0 is totally antisymmetric we can write

$$E_1(I) = \tfrac{1}{2} N(N-1) \langle \Psi_0(I) | \frac{1}{r_{12}} | \Psi_0(I) \rangle \tag{16.7.8}$$

It is at this point that we recognize the value of adopting a fractional parentage expansion for Ψ_0. The 2-electron FP expansion is

$$\Psi_0 = \sum_{i \leq j=1}^{t} \sum_{I_2, I^{(2)}} (l_i l_j I_2; I^{(2)} | \} I) \, \psi_0(l_i(1), l_j(2), I_2; I^{(2)}; I) \tag{16.7.9}$$

where $I^{(2)}$ comes from an $(N-2)$ electron configuration involving the coordinates $3, 4, \ldots, N$. We substitute (16.7.9) in (16.7.8) to obtain

$$\begin{aligned} E_1(I) = \tfrac{1}{2} N(N-1) & \sum_{i \leq j=1}^{t} \sum_{i' \leq j'=1}^{t} \sum_{I_2, I^{(2)}} \sum_{I_2', I^{(2)'}} \\ & \times (l_i l_j I_2; I^{(2)} | \} I)(l_{i'} l_{j'} I_2'; I^{(2)'} | \} I) \\ & \times \langle \psi_0(l_i, l_j, I_2; I^{(2)}; I) | \frac{1}{r_{12}} | \psi_0(l_{i'}, l_{j'}, I_2'; I^{(2)'}; I) \rangle \end{aligned}$$

Now $\hat{X}_0^{(0)} \equiv 1/r_{12}$ is a scalar operator acting only on the coordinates 1 and 2. It follows from (16.5.2), the orthogonality of the orbitals, and the Wigner-Eckart theorem, that

$$E_1(I) = \tfrac{1}{2} N(N-1) \sum_{i \leq j=1}^{t} \sum_{I_2, I^{(2)}} (l_i l_j I_2; I^{(2)} | \} I)^2 E_1(l_i l_j I_2) \tag{16.7.10}$$

where

$$E_1(l_i l_j I_2) = \langle \psi_0(l_i, l_j, I_2) | \frac{1}{r_{12}} | \psi_0(l_i, l_j, I_2) \rangle \tag{16.7.11}$$

We have now expressed the first order energy for an N-electron state as a weighted sum of first order energies for 2-electron states. These first order energies for the 2-electron states are known from (16.6.4) and it follows that a knowledge of the 2-electron CFP suffice to determine N-electron first order energies. As an example we have for the ground state of Li the result

$$E_1(1s^2\, 2s\, {}^2S) = E_1(1s^2\, {}^1S) + \tfrac{1}{2} E_1(1s2s\, {}^1S) + \tfrac{3}{2} E_1(1s2s\, {}^3S)$$

The ground state of Be is more complicated. Here the terms $1s^2\, 2s^2\, {}^1S_g$

and $1s^2\,2p^2\,{}^1S_g$ are of the same symmetry and have the same zeroth order energy. According to degenerate perturbation theory the zeroth order function for the ground state of Be is given by

$$\Psi_0(1\,{}^1S) = a\Psi_0(1s^2\,2s^2\,{}^1S) + b\Psi_0(1s^2\,2p^2\,{}^1S)$$

This is essentially configuration interaction. The coefficients a and b are determined from the secular equations

$$a(E_1(1s^2\,2s^2\,{}^1S) - E_1) + bH_{12} = 0$$

$$aH_{12} + b(E_1(1s^2\,2p^2\,{}^1S) - E_1) = 0$$

where

$$H_{12} = \langle\Psi_0(1s^2\,2s^2\,{}^1S)|\,\hat{H}_1\,|\Psi_0(1s^2\,2p^2\,{}^1S)\rangle$$

The relevant FP expansions are

$$\Psi_0(1s^2\,2s^2\,{}^1S) = \frac{1}{\sqrt{6}}\,\psi_0(1s^2\,{}^1S;\,2s^2\,{}^1S;\,{}^1S) + \frac{1}{\sqrt{6}}\,\psi_0(2s^2\,{}^1S;\,1s^2\,{}^1S;\,{}^1S)$$
$$-\frac{1}{\sqrt{6}}\,\psi_0(1s2s\,{}^1S;\,1s2s\,{}^1S;\,{}^1S) - \frac{1}{\sqrt{2}}\,\psi_0(1s2s\,{}^3S;\,1s2s\,{}^3S;\,{}^1S)$$

and

$$\Psi_0(1s^2\,2p^2\,{}^1S) = \frac{1}{\sqrt{6}}\,\psi_0(1s^2\,{}^1S;\,2p^2\,{}^1S;\,{}^1S) + \frac{1}{\sqrt{6}}\,\psi_0(2p^2\,{}^1S;\,1s^2\,{}^1S;\,{}^1S)$$
$$-\frac{1}{\sqrt{6}}\,\psi_0(1s2p\,{}^1P;\,1s2p\,{}^1P;\,{}^1S) - \frac{1}{\sqrt{2}}\,\psi_0(1s2p\,{}^3P;\,1s2p\,{}^3P;\,{}^1S)$$

According to (16.7.10) we immediately have

$$E_1(1s^2\,2s^2\,{}^1S) = E_1(1s^2\,{}^1S) + E_1(2s^2\,{}^1S) + E_1(1s2s\,{}^1S) + 3E_1(1s2s\,{}^3S)$$

and

$$E_1(1s^2\,2p^2\,{}^1S) = E_1(1s^2\,{}^1S) + E_1(2p^2\,{}^1S) + E_1(1s2p\,{}^1P) + 3E_1(1s2p\,{}^3P)$$

If we use (16.6.4) we find

$$E_1(1s^2\,{}^1S) = F_0(1s1s) = \tfrac{5}{8}$$
$$E_1(2s^2\,{}^1S) = F_0(2s2s) = \tfrac{77}{512}$$
$$E_1(1s2s\,{}^1S) = F_0(1s2s) + G_0(1s2s) = \tfrac{169}{729}$$
$$E_1(1s2s\,{}^3S) = F_0(1s2s) - G_0(1s2s) = \tfrac{137}{729}$$
$$E_1(1s2s\,{}^1P) = F_1(1s2p) + \tfrac{1}{3}G_1(1s2p) = \tfrac{1705}{6561}$$
$$E_1(1s2p\,{}^3P) = F_1(1s2p) - \tfrac{1}{3}G_1(1s2p) = \tfrac{1481}{6561}$$
$$E_1(2p^2\,{}^1S) = F_0(2p2p) + \tfrac{2}{5}F_2(2p2p) = \tfrac{111}{512}$$

In order to evaluate H_{12} we use (16.6.6). Thus

$$H_{12} = \langle \Psi_0(2s^2\,{}^1S)| \frac{1}{r_{12}} |\Psi_0(2p^2\,{}^1S)\rangle = -\frac{1}{\sqrt{3}}\, G_1(2s2p) = -\frac{15\sqrt{3}}{512}$$

The eigenvalues of the secular matrix can now be calculated. The eigenfunction corresponding to the lowest eigenvalue is found to be

$$\Psi_0(1\,{}^1S) = 0.97432\,\Psi_0(1s^2\,2s^2\,{}^1S) + 0.22517\,\Psi_0(1s^2\,2p^2\,{}^1S)$$

This shows that the ground state of Be is predominantly $1s^22s^2$ in character.

An interesting case from the group theoretical point of view concerns the two ${}^2D$ terms of the configuration $(3d)^3$. If we use the FP expansions (15.5.7) and (15.5.8) we find the first order energies

$$E_1(d^3\,{}^2_1D) = \tfrac{4}{5}E_1(d^2\,{}^1S) + \tfrac{1}{4}E_1(d^2\,{}^1D) + \tfrac{9}{20}E_1(d^2\,{}^1G)$$
$$+\tfrac{9}{20}E_1(d^2\,{}^3P) + \tfrac{21}{20}E_1(d^2\,{}^3F)$$
$$E_1(d^3\,{}^2_3D) = \tfrac{135}{140}E_1(d^2\,{}^1D) + \tfrac{75}{140}E_1(d^2\,{}^1G) + \tfrac{147}{140}E_1(d^2\,{}^3P)$$
$$+\tfrac{63}{140}E_1(d^2\,{}^3F)$$

Introducing the Racah parameters

$$A = F_0 - \tfrac{1}{9}F_4 \qquad B = \tfrac{1}{441}(9F_2 - 5F_4) \qquad C = \tfrac{5}{63}F_4$$

and using (16.6.5) we obtain

$$\begin{aligned} E_1(d^3\,{}^2_1D) &= 3A + 7B + 7C = H_{11} \\ E_1(d^3\,{}^2_3D) &= 3A + 3B + 3C = H_{22} \end{aligned} \qquad (16.7.12)$$

Since the functions $\Psi_0(d^3\,{}^2_1D)$ and $\Psi_0(d^3\,{}^2_3D)$ have the same zeroth order energy they are not necessarily eigenfunctions. In order to obtain the eigenfunctions we again consider degenerate perturbation theory and write

$$\Psi_0(d^3\,{}^2D) = a\Psi_0(d^3\,{}^2_1D) + b\Psi_0(d^3\,{}^2_3D)$$

The secular determinantal equation is

$$\begin{vmatrix} H_{11} - E & H_{12} \\ H_{12} & H_{22} - E \end{vmatrix} = 0$$

where H_{11} and H_{22} are given by (16.7.12) and where

$$H_{12} = \langle \Psi_0(d^3\,{}^2_1D)|\,\hat{H}_1\,|\Psi_0(d^3\,{}^2_3D)\rangle = -3\sqrt{21}\,B$$

It remains to evaluate the radial integrals and this, although tedious, is quite straightforward. When the calculation is carried out we find the eigenfunctions to be

$$\Psi_0(d^3\,1\,{}^2D) = 0.89\,\Psi_0(d^3\,{}^2_1D) - 0.46\,\Psi_0(d^3\,{}^2_3D)$$

$$\Psi_0(d^3\,2\,{}^2D) = 0.46\,\Psi_0(d^3\,{}^2_1D) + 0.89\,\Psi_0(d^3\,{}^2_3D)$$

The result of this calculation demonstrates that seniority is not a good quantum number. This however does not remove the utility of seniority as a means of state classification.

We end this chapter by noting that to obtain greater accuracy we should also calculate the second-order correction to the energy. This book is not the place to go into details but the method of fractional parentage can be used to obtain second-order energies for atomic states (see Chisholm and Dalgarno, 1966).

References

Chisholm, C. D. H. and Dalgarno, A. (1966) *Proc. Roy. Soc. (Lond.)*, **A290,** 264.

Edmonds, A. R. (1960) "Angular Momentum in Quantum Mechanics", Princeton University Press.

Fano, U. and Racah, G. (1959) "Irreducible Tensorial Sets", Academic Press.

Judd, B. R. (1963) "Operator Techniques in Atomic Spectroscopy", McGraw Hill.

Rotenberg, M., Bivens, R., Metropolis, N. and Wooten, J. K. (1959) "The 3*j* and 6*j* symbols", The Technology Press, M.I.T., Cambridge, Mass.

General Results in the Theory of Lie Groups

A1.1 Lie groups and Lie algebras

Let $\mathscr{G}$ be a group in which the elements of the group manifold can be labelled by a finite set of continuously varying independent real parameters. By an r-parameter Lie group of transformations $\mathscr{G}$ we mean a set of transformations

$$x_i' = f_i(x_1, x_2, \ldots, x_n; a_1, a_2, \ldots, a_r) \qquad i = 1, 2, \ldots, n \quad \text{(A1.1.1)}$$

for which the f_i are analytic functions of the parameters a_j and for which the transformations satisfy the group postulates. (The case of a 1-parameter group in a single variable is discussed in Section 9.1). By generalization of (9.1.9) we can consider infinitesimal transformations

$$x_i' + dx_i' = f_i(x_1', \ldots, x_n'; \delta a_1, \ldots, \delta a_r) \qquad i = 1, 2, \ldots, n \quad \text{(A1.1.2)}$$

where the parameters δa_i are infinitesimal. We denote the identity transformation $x_i' = x_i$ by

$$x_i' = x_i = f_i(x_1, \ldots, x_n, 0, 0, \ldots, 0)$$

It is convenient in the general case to introduce several conventions which simplify the writing of equations. Thus we use x to denote the collection $x_1, x_2, \ldots, x_n$ and a to denote the collection $a_1, a_2, \ldots, a_r$. We also employ the summation convention in which repeated indices in a product are to be summed over their range of values. Unless otherwise stated the summation convention is implicit in all equations that follow.

If we refer to Hamermesh (1962) we find that there is associated with an r-parameter Lie group of transformations a set of r independent infinitesimal operators (IO's). We shall call these natural IO's for the group (see Section 9.4). These natural IO's are given by

$$\hat{X}_k = \sum_{i=1}^{n} u_{ik}(x) \frac{\partial}{\partial x_i} \equiv u_{ik}(x) \frac{\partial}{\partial x_i} \qquad k = 1, 2, \ldots, r \qquad \text{(A1.1.3)}$$

where

$$u_{ik}(x) = \left|\frac{\partial f_i(x; a)}{\partial a_k}\right|_{a=0} \qquad \text{(A1.1.4)}$$

The natural IO's for a specific group are always easy to determine (see

Chapter 9). The r natural IO's can be regarded as forming a basis for a real r-dimensional linear space $\mathscr{L}$. We shall therefore refer to this basis as the natural basis for $\mathscr{L}$.

Now it can be shown (see Hamermesh, 1962) that in the general case the IO's satisfy the following important commutation relations (CR's)

$$[\hat{X}_\rho, \hat{X}_\sigma] \equiv \hat{X}_\rho \hat{X}_\sigma - \hat{X}_\sigma \hat{X}_\rho = c^\tau_{\rho\sigma} \hat{X}_\tau \quad \text{(A1.1.5)}$$

for all choices ρ, σ, τ in the set $1, 2, \ldots, r$. The coefficients $c^\tau_{\rho\sigma}$ in the linear combination (A1.1.5) are scalars called the structure constants for the group. Again for a specific group the structure constants are readily found (see Chapter 9). We also note that the IO's satisfy the Jacobi identity

$$[[\hat{X}_\rho, \hat{X}_\sigma], \hat{X}_\tau] + [[\hat{X}_\sigma, \hat{X}_\tau], \hat{X}_\rho] + [[\hat{X}_\tau, \hat{X}_\rho], \hat{X}_\sigma] = 0 \quad \text{(A1.1.6)}$$

The binary relation of the commutator $[\hat{X}_\rho, \hat{X}_\sigma]$ defines a product in the linear space $\mathscr{L}$. Further we see from (A1.1.5) that $\mathscr{L}$ is closed with respect to this binary relation. According to Section 2.14 the linear space $\mathscr{L}$ is now an r-dimensional algebra. This algebra is called the Lie algebra of the Lie group $\mathscr{G}$.

We now introduce the important concept of a simple Lie algebra (compare with Section 2.14 in what follows). Let $\mathscr{K}$ be an s-dimensional subspace ($s < r$) of $\mathscr{L}$ with basis $\hat{Y}_1, \hat{Y}_2, \ldots, \hat{Y}_s$. If for all $\hat{Y}_\rho$, $\hat{Y}_\sigma$ in $\mathscr{K}$ we have

$$[\hat{Y}_\rho, \hat{Y}_\sigma] = b^\tau_{\rho\sigma} \hat{Y}_\tau$$

then we say that $\mathscr{K}$ is an s-dimensional subalgebra of $\mathscr{L}$. If all the constants $b^\tau_{\rho\sigma}$ are zero then we say that $\mathscr{K}$ is an abelian subalgebra. Next if for all $\hat{Y}_\rho$ in $\mathscr{K}$ and all $\hat{X}_\sigma$ in $\mathscr{L}$ we have

$$[\hat{Y}_\rho, \hat{X}_\sigma] = a^\tau_{\rho\sigma} \hat{Y}_\tau$$

then we say that $\mathscr{K}$ is an invariant subalgebra or an ideal in $\mathscr{L}$. Apart from the void space and $\mathscr{L}$ itself all other ideals are called proper ideals. Now a Lie algebra is said to be simple if it contains no proper ideals and semi-simple if it contains no proper abelian ideals.

Next let $\mathscr{L}_1$ and $\mathscr{L}_2$ be two Lie algebras of dimensions n and m respectively. Let $\hat{X}_\rho$ ($\rho = 1, 2, \ldots, n$) and $\hat{Y}_\alpha$ ($\alpha = 1, 2, \ldots, m$) be basis operators in $\mathscr{L}_1$ and $\mathscr{L}_2$ respectively which satisfy the commutation relations (CR's)

$$\begin{aligned} [\hat{X}_\rho, \hat{X}_\sigma] &= c^\tau_{\rho\sigma} \hat{X}_\tau \\ [\hat{Y}_\alpha, \hat{Y}_\beta] &= d^\gamma_{\alpha\beta} \hat{Y}_\gamma \end{aligned} \quad \text{(A1.1.7)}$$

We define the direct sum $\mathscr{L}$ of the two algebras to be the linear space spanned by all the basis operators $\hat{X}_\rho$ and $\hat{Y}_\alpha$ which satisfy (A1.1.7) and for which

$$[\hat{X}_\rho, \hat{Y}_\alpha] = 0 \qquad \begin{aligned} \rho &= 1, 2, \ldots, n \\ \alpha &= 1, 2, \ldots, m \end{aligned}$$

We write

$$\mathscr{L} = \mathscr{L}_1 \oplus \mathscr{L}_2$$

This concept can clearly be generalized to give the direct sum of any number of algebras. It can be shown that the direct sum of simple Lie algebras is a semi-simple Lie algebra and conversely that any semi-simple Lie algebra can be written as a direct sum of simple Lie algebras (see Jacobson, 1962).

The importance of simple or semi-simple Lie algebras is that, unlike an arbitrary Lie algebra, all simple Lie algebras can be classified and treated together in a very elegant way. Furthermore nearly all the Lie algebras of importance in quantum chemistry are simple or semi-simple.

A1.2 The Lie algebra as a vector space

As we shall see later it is extremely useful to investigate the possibility of converting a Lie algebra $\mathscr{L}$ into a vector space. This requires the definition of a scalar product in $\mathscr{L}$. Such a definition can be obtained by the following argument.

Consider a simple Lie algebra $\mathscr{L}$ with natural basis $\hat{X}_1, \hat{X}_2, \ldots, \hat{X}_r$. Let $\hat{B}$ denote any element of $\mathscr{L}$ and let $\hat{A}$ be a fixed element of $\mathscr{L}$. We have

$$\hat{A} = a^\rho \hat{X}_\rho \qquad \hat{B} = b^\sigma \hat{X}_\sigma \tag{A1.2.1}$$

where a^ρ and b^σ are the components of $\hat{A}$ and $\hat{B}$ respectively in the natural basis. Since

$$[\hat{X}_\rho, \hat{X}_\sigma] = c^\tau_{\rho\sigma} \hat{X}_\tau$$

it follows that $[\hat{A}, \hat{B}]$ is an element of $\mathscr{L}$. We can therefore consider, for a fixed $\hat{A}$, a mapping of $\mathscr{L}$ onto itself given by

$$\hat{B} \to [\hat{A}, \hat{B}]$$

If we denote this mapping by $\hat{F}(\hat{A})$ we have

$$\hat{F}(\hat{A})\hat{B} = [\hat{A}, \hat{B}] \tag{A1.2.2}$$

In particular for the basis elements $\hat{X}_\sigma$ we have

$$\hat{F}(\hat{B})\hat{X}_\sigma = [\hat{B}, \hat{X}_\sigma] = b^\mu [\hat{X}_\mu, \hat{X}_\sigma] = b^\mu c^\tau_{\mu\sigma} \hat{X}_\tau$$

We now consider the product mapping

$$\hat{P} = \hat{F}(\hat{A})\hat{F}(\hat{B})$$

We have

$$\hat{F}(\hat{A})\hat{F}(\hat{B})\hat{X}_\sigma = b^\mu c^\tau_{\mu\sigma}[\hat{A}, \hat{X}_\tau]$$
$$= a^\lambda b^\mu c^\tau_{\mu\sigma}[\hat{X}_\lambda, \hat{X}_\tau] = a^\lambda b^\mu c^\tau_{\mu\sigma} c^\rho_{\lambda\tau} \hat{X}_\rho$$

and within the natural basis the mapping P has a matrix representation $\mathbf{P} = [P_{\rho\sigma}]$ given by

$$P_{\rho\sigma} = a^\lambda b^\mu c^\tau_{\mu\sigma} c^\rho_{\lambda\tau} \tag{A1.2.3}$$

The trace of $\mathbf{P}$ is

$$P_{\sigma\sigma} = a^\lambda b^\mu c^\tau_{\mu\sigma} c^\sigma_{\lambda\tau}$$

If we write

$$g_{\mu\lambda} = c^\tau_{\mu\sigma} c^\sigma_{\lambda\tau} \tag{A1.2.4}$$

then

$$P_{\sigma\sigma} = g_{\mu\lambda} a^\lambda b^\mu$$

and we have produced the required definition of a scalar product $\langle \hat{A} \mid \hat{B} \rangle$ in $\mathscr{L}$ given by

$$\langle \hat{A} \mid \hat{B} \rangle = g_{\mu\lambda} a^\lambda b^\mu \tag{A1.2.5}$$

It can be shown (see Hamermesh, 1962) that $g_{\mu\lambda}$ transforms like a rank two symmetric tensor and that $\langle \hat{A} \mid \hat{B} \rangle$ is indeed invariant under a basis transformation. We shall call the metric tensor $g_{\mu\lambda}$ the *Cartan* tensor for the Lie algebra $\mathscr{L}$.

The properties of the matrix $\mathbf{g}$ are very important. Thus it can be shown that the necessary and sufficient conditions for a Lie group to be semi-simple is that $\det \mathbf{g} \neq 0$ (see Racah, 1965). For most of the Lie groups of interest to us the range of variation of the parameters a_j is finite. When this is so the group manifold is said to be *closed* and the Lie group itself is said to be *compact.* It can be shown that the necessary and sufficient condition for a semi-simple Lie group to be compact is that $\mathbf{g}$ be negative definite (see Hermann, 1966).

A1.3 The classification of simple or semi-simple Lie algebras

The CR's in the natural basis are given by (A1.1.5). We are not however confined to using the natural basis. There may exist a new basis, obtained by taking linear combinations of the natural IO's, in which the CR's have a simpler structure than (A1.1.5). By examining the possibility of introducing a new basis Cartan (1894) found that there exists a standard basis in which the CR's for all simple or semi-simple Lie algebras can be written in a general classified form. These CR's are called the canonical CR's for the Lie algebra.

In order to find this standard basis we begin by returning to the mapping

(A1.2.2). Now associated with any linear operator $\hat{F}$ there exists a corresponding eigenvalue problem

$$\hat{F}\psi = \rho\psi$$

where ρ is an eigenvalue and ψ an eigenfunction of $\hat{F}$. Let $\hat{X} = x^\nu \hat{X}_\nu$ be any element of a simple or semi-simple Lie algebra of dimension r. Here the $\hat{X}_\upsilon$ are the natural IO's. Associated with the mapping $\hat{F}(\hat{A})$ (see A1.2.2) there is a corresponding (operator) eigenvalue problem

$$\hat{F}(\hat{A})\hat{X} = \rho\hat{X} \tag{A1.3.1}$$

which from (A1.2.2) becomes

$$[\hat{A}, \hat{X}] = \rho\hat{X} \tag{A1.3.2}$$

Now $\hat{A} = a^\mu \hat{X}_\mu$ and in terms of components (A1.3.2) gives

$$[a^\mu \hat{X}_\mu, x^\nu \hat{X}_\nu] = a^\mu x^\nu [\hat{X}_\mu, \hat{X}_\nu] = a^\mu x^\nu c^\tau_{\mu\nu} \hat{X}_\tau = \rho x^\tau \hat{X}_\tau \tag{A1.3.3}$$

Since the $\hat{X}_\tau$ are linearly independent we obtain from (A1.3.3) r secular equations

$$(a^\mu c^\tau_{\mu\nu} - \rho\, \delta^\tau_\nu)x^\nu = 0 \qquad \tau = 1, 2, \ldots, r \tag{A1.3.4}$$

where δ^τ_ν is the Kronecker delta. For non-trivial solutions of (A1.3.4) we require

$$\det |a^\mu c^\tau_{\mu\nu} - \rho\, \delta^\tau_\nu| = 0 \tag{A1.3.5}$$

Cartan (1894) made an exhaustive study of this eigenvalue problem and he deduced that for any simple or semi-simple Lie algebra there exists a particular element $\hat{A}$ for which (A1.3.5) has the maximum number of different roots. Furthermore the only degenerate root is $\rho = 0$. The degeneracy l of the eigenvalue $\rho = 0$ is called the *rank* of the Lie algebra. Cartan also showed that the l linearly independent eigenoperators $\hat{H}_1, \hat{H}_2, \ldots, \hat{H}_l$ corresponding to $\rho = 0$ are mutually commuting. The remaining $(r - l)$ eigenoperators corresponding to the non-degenerate eigenvalues are denoted by $\hat{E}_\alpha, \hat{E}_\beta, \ldots, \hat{E}_\nu$. The above results are contained in the following theorem.

Cartan's Theorem (see Cartan, 1894; and Jacobson, 1962)

Any simple or semi-simple r-dimensional Lie algebra $\mathscr{L}$ can be decomposed into a direct sum of subalgebras

$$\mathscr{L} = \mathscr{H} \oplus \mathscr{E}^{(\alpha)} \oplus \mathscr{E}^{(\beta)} \oplus \ldots \oplus \mathscr{E}^{(\nu)} \tag{A1.3.6}$$

The $(r - l)$ subalgebras $\mathscr{E}^{(\alpha)}, \mathscr{E}^{(\beta)}, \ldots, \mathscr{E}^{(\nu)}$ are each of dimension one and are spanned respectively by the IO's $\hat{E}_\alpha, \hat{E}_\beta, \ldots, \hat{E}_\nu$. The subalgebra $\mathscr{H}$ is called the Cartan subalgebra. In general $\mathscr{H}$ is of dimension $l > 1$ and is spanned by the mutually commuting IO's $\hat{H}_1, \hat{H}_2, \ldots, \hat{H}_l$.

In order to demonstrate how the Cartan eigenvalue problem works in practice we consider the trivial example of the Lie algebra B_1 of $R(3)$. The natural IO's are given by (9.4.6). We thus have

$$\hat{X} = x_1\hat{X}_1 + x_2\hat{X}_2 + x_3\hat{X}_3$$
$$\hat{A} = a_1\hat{X}_1 + a_2\hat{X}_2 + a_3\hat{X}_3$$

Using the CR's (9.4.8) we find the secular equations (A1.3.4) to be

$$\rho x_1 + a_3x_2 - a_2x_3 = 0$$
$$-a_3x_1 + \rho x_2 + a_1x_3 = 0$$
$$a_2x_1 - a_1x_2 + \rho x_3 = 0$$

Expansion of (A1.3.5) for this case gives

$$\rho[\rho^2 + (a_1^2 + a_2^2 + a_3^2)] = 0$$

and we have the eigenvalues

$$\rho = 0 \qquad \rho = \pm i\sqrt{a_1^2 + a_2^2 + a_3^2}$$

We can now choose $a_1 = a_2 = 0$ and $a_3 = \alpha i$. The roots are $\rho = 0$, α, $-\alpha$ and Cartan's result is established by choosing $\hat{A} = \alpha i\hat{X}_3$. For $\rho = 0$ we have $x_1 = x_2 = 0$ and the eigenoperator is

$$\hat{H}_1 = x_3\hat{X}_3 \tag{A1.3.7}$$

For $\rho = \pm\alpha$ we have $x_3 = 0$; $x_2 = \pm ix_1$ and the eigenoperators are

$$\hat{E}_{\pm\alpha} = x_1(\hat{X}_1 \pm i\hat{X}_2) \tag{A1.3.8}$$

Thus a standard basis for B_1 is given by the IO's $\hat{H}_1$, $\hat{E}_{\pm\alpha}$ and B_1 is a rank one Lie algebra.

We now return to the general case. Corresponding to the l roots $\rho = 0$ we find from the secular equations l solutions say

$$x^1 = b_i^1, x^2 = b_i^2, \ldots, x^r = b_i^r \qquad i = 1, 2, \ldots, l$$

which give rise to l eigenoperators

$$\hat{H}_i = b_i^\nu\hat{X}_\nu \qquad i = 1, 2, \ldots, l$$

Corresponding to each of the $(r-l)$ non-degenerate roots $\rho = \varepsilon \neq 0$ we find from the secular equations the solution say

$$x^1 = \varepsilon^1 \qquad x^2 = \varepsilon^2, \ldots, x^r = \varepsilon^r$$

which gives rise to the single eigenoperator

$$\hat{E}_\varepsilon = \varepsilon^\nu\hat{X}_\nu$$

There will be $(r-l)$ such eigenoperators $\hat{E}_\varepsilon$ ($\varepsilon = \alpha, \beta, \ldots, \nu$). We have thus replaced the natural basis $\hat{X}_1, \hat{X}_2, \ldots, \hat{X}_r$ by a standard basis $\hat{H}_1, \hat{H}_2, \ldots, \hat{H}_l, \hat{E}_\alpha, \hat{E}_\beta, \ldots, \hat{E}_\nu$ in which the standard IO's are some particular linear combinations of the natural IO's.

We now go on to construct canonical CR's which came from the standard IO's. If $\hat{A}$ is any element of the Lie algebra then from Cartan's theorem we have

$$[\hat{A}, \hat{H}_i] = 0 \qquad i = 1, 2, \ldots, l$$
$$[\hat{A}, \hat{E}_\alpha] = \alpha \hat{E}_\alpha \qquad \text{(A1.3.9)}$$
$$[\hat{H}_i, \hat{H}_k] = 0$$

where α is a non-degenerate eigenvalue. In (A1.3.9) and in what follows we shall not use the summation convention for greek indices $\alpha, \beta, \ldots, \nu$ but we shall retain it for all other indices. Since $[\hat{A}, \hat{A}] = 0$ and $[\hat{A}, \hat{H}_i] = 0$ it follows that $\hat{A}$ can be written as

$$\hat{A} = \lambda^i \hat{H}_i$$

for some scalars λ^i. Consider next the Jacobi identity

$$[\hat{A}, [\hat{H}_i, \hat{E}_\alpha]] + [\hat{H}_i, [\hat{E}_\alpha, \hat{A}]] + [\hat{E}_\alpha, [\hat{A}, \hat{H}_i]] = 0$$

we have

$$[\hat{A}, [\hat{H}_i, \hat{E}_\alpha]] = \alpha [\hat{H}_i, \hat{E}_\alpha]$$

which implies that (see A1.3.9) $[\hat{H}_i, \hat{E}_\alpha]$ is an eigenoperator corresponding to the non-degenerate eigenvalue α. Thus $[\hat{H}_i, \hat{E}_\alpha]$ must be proportional to $\hat{E}_\alpha$ i.e.

$$[\hat{H}_i, \hat{E}_\alpha] = \alpha_i \hat{E}_\alpha \qquad \text{(A1.3.10)}$$

for some scalars α_i. Since $\hat{A} = \lambda^i \hat{H}_i$ we have

$$[\lambda^i \hat{H}_i, \hat{E}_\alpha] = \lambda^i \alpha_i \hat{E}_\alpha = \alpha \hat{E}_\alpha$$

and it follows that

$$\alpha = \lambda^i \alpha_i$$

The set of l scalars α_i are called *simple roots*. They can be regarded as forming a basis for an l-dimensional linear space called simple root space and denoted by Π. The root α is then called a *root vector* and has contravariant components (see (2.9.1)) λ^i in the space Π. Clearly there exists $(r-l)$ non-zero root vectors $\alpha, \beta, \ldots, \nu$.

Consider next the Jacobi identity

$$[\hat{A}, [\hat{E}_\alpha, \hat{E}_\beta]] + [\hat{E}_\alpha, [\hat{E}_\beta, \hat{A}]] + [\hat{E}_\beta, [\hat{A}, \hat{E}_\alpha]] = 0$$

From (A1.3.9) this gives

$$[\hat{A}, [\hat{E}_\alpha, \hat{E}_\beta]] = (\alpha + \beta)[\hat{E}_\alpha, \hat{E}_\beta]$$

If $\alpha+\beta$ is not a root then $\hat{E}_\alpha$ and $\hat{E}_\beta$ must commute. If $\alpha+\beta$ is a non-zero root it follows that

$$[\hat{E}_\alpha, \hat{E}_\beta] = N_{\alpha\beta}\hat{E}_{\alpha+\beta} \tag{A1.3.11}$$

where $N_{\alpha\beta}$ is a structure constant. If $\alpha+\beta=0$ we have

$$[\hat{A}, [\hat{E}_\alpha, \hat{E}_{-\alpha}]] = 0$$

Since only the $\hat{H}_i$ are such that $[\hat{A}, \hat{H}_i]=0$ it follows that

$$[\hat{E}_\alpha, \hat{E}_{-\alpha}] = c^{i}_{\alpha-\alpha}\hat{H}_i \tag{A1.3.12}$$

where $c^{i}_{\alpha-\alpha}$ are structure constants.

In order to obtain canonical CR's we turn our attention to the form of the Cartan tensor (A1.2.4) when the standard basis is used. Consider first an element g_{ik}. Since

$$[\hat{H}_i, \hat{E}_\alpha] = \sum_\beta c^{\beta}_{i\alpha}\hat{E}_\beta = c^{\alpha}_{i\alpha}\hat{E}_\alpha = \alpha_i\hat{E}_\alpha$$

we have

$$g_{ik} = \sum_{\alpha,\beta} c^{\beta}_{i\alpha}c^{\alpha}_{k\beta} = \sum_\alpha c^{\alpha}_{i\alpha}c^{\alpha}_{k\alpha} = \sum_\alpha \alpha_i\alpha_k \tag{A1.3.13}$$

and the totality of g_{ik} can be regarded as a metric tensor for the Cartan subalgebra $\mathscr{H}$. Consider next the element $g_{\alpha\lambda}$ of row α of the Cartan tensor. From (A1.3.10)–(A1.3.12) we have

$$c^{\rho}_{\alpha i} = -\alpha_i\,\delta^{\rho}_{\alpha} \qquad c^{\rho}_{\alpha\beta} = N_{\alpha\beta}\,\delta^{\rho}_{\alpha+\beta} \qquad c^{i}_{\alpha-\alpha} = c^{i}_{\alpha-\alpha}$$

and thus

$$g_{\alpha\lambda} = c^{\tau}_{\alpha\sigma}c^{\sigma}_{\lambda\tau} = -\alpha_i c^{i}_{\lambda\alpha} + \sum_{\beta\neq-\alpha} N_{\alpha\beta}c^{\beta}_{\lambda\alpha+\beta} + c^{i}_{\alpha-\alpha}c^{-\alpha}_{\lambda i} \tag{A1.3.14}$$

Again referring to (A1.3.10)–(A1.3.12) we see that each term in (A1.3.14) will vanish unless $\lambda=-\alpha$. Now for a semi-simple group $\det \mathbf{g}\neq 0$ and it follows that if α is a root for a semi-simple Lie algebra then $-\alpha$ is also a root. Also the only element of row α in $g_{\alpha\lambda}$ different from zero is $g_{\alpha-\alpha}$. It is customary to normalize the $(r-l)$ operators $\hat{E}_\alpha$ such that

$$g_{\alpha-\alpha} = 1 \tag{A1.3.15}$$

We introduce next the inverse matrix $[g^{ik}]$ of $[g_{ik}]$. Thus

$$g^{jk}g_{ki} = \delta^{j}_{i} \tag{A1.3.16}$$

According to (2.9.13) if α_k are regarded as the covariant components of the root vector α then the contravariant components α^j are given by

$$\alpha^j = g^{jk}\alpha_k \tag{A1.3.17}$$

Also from (2.9.14) the scalar product $\langle\alpha\mid\beta\rangle$ of two root vectors α, β is given by

$$\langle\alpha \mid \beta\rangle = \alpha^i\beta_i = g^{jk}\alpha_j\beta_k = g_{jk}\alpha^j\beta^k \tag{A1.3.18}$$

We now define a tensor $c_{\rho\sigma\tau}$ by

$$c_{\rho\sigma\tau} = g_{\alpha\tau}c^{\alpha}_{\rho\sigma} \tag{A1.3.19}$$

It can be shown from (A1.2.4) and (A1.1.6) that $c_{\rho\sigma\tau}$ is invariant under cyclic permutations of the indices. In particular using (A1.3.15) and (A1.3.10) we have

$$\begin{aligned} c^{i}_{\alpha-\alpha} &= g^{ij}c_{\alpha-\alpha j} = g^{ij}c_{j\alpha-\alpha} = g^{ij}g_{-\alpha\beta}c^{\beta}_{j\alpha} \\ &= g^{ij}\,\delta_{\alpha\beta}\alpha_j\,\delta_{\alpha\beta} = g^{ij}\alpha_j = \alpha^i \end{aligned} \tag{A1.3.20}$$

This result enables us to write (A1.3.12) in the form

$$[\hat{E}_\alpha, \hat{E}_{-\alpha}] = \alpha^i\hat{H}_i \tag{A1.3.21}$$

We now have the canonical CR's given in compact general form by

$$[\hat{H}_i, \hat{H}_k] = 0 \tag{A1.3.22}$$

$$[\hat{H}_i, \hat{E}_\alpha] = \alpha_i\hat{E}_\alpha \tag{A1.3.23}$$

$$[\hat{E}_\alpha, \hat{E}_\beta] = N_{\alpha\beta}\hat{E}_{\alpha+\beta} \qquad \beta \neq \alpha \tag{A1.3.24}$$

$$[\hat{E}_\alpha, \hat{E}_{-\alpha}] = \alpha^i\hat{H}_i \tag{A1.3.25}$$

where $N_{\alpha\beta}$ are chosen such that (A1.3.15) is satisfied.

Having introduced the concept of root vectors we now consider some of their more important properties. Firstly if α and β are simple roots then (see Rowlatt, 1966; and Gourdin, 1967)

$$-2\frac{\langle\alpha \mid \beta\rangle}{\langle\alpha \mid \alpha\rangle} = N \qquad -2\frac{\langle\alpha \mid \beta\rangle}{\langle\beta \mid \beta\rangle} = M \tag{A1.3.26}$$

where N and M are non-negative integers. This result can be used to show that there is a restriction on both the angles between the simple root vectors and their relative lengths (see Rowlatt, 1966). Thus the angle between any two simple root vectors can only be 90°, 120°, 135° or 150°. Also the simple root vectors either all have the same length or all have one of two possible lengths.

Next let Σ denote the set of all root vectors of a simple Lie algebra. The set Π of simple roots is a subset of Σ and as we have seen the root vectors can be expressed in terms of the simple roots as

$$\alpha = \lambda^i\alpha_i \tag{A1.3.27}$$

If a is an element of Π then (A1.3.27) can be written as

$$\alpha = \sum_{a\in\Pi} \lambda^{\alpha}_{a}a \tag{A1.3.28}$$

where α is an element of Σ. We now define a quantity $\sigma(\alpha)$ by

$$\sigma(\alpha) = \sum_{a \in \Pi} \lambda_a^\alpha \tag{A1.3.29}$$

If $\sigma(\alpha) > 0$ we call α a positive root while if $\sigma(\alpha) < 0$ we call α a negative root. Since $-\alpha$ is a root whenever α is a root it follows that we need only determine all the positive roots to obtain the full set of roots.

Another useful property of the roots states that if α and β are any two roots then the vector

$$\gamma = \beta - 2\frac{\langle \alpha \mid \beta \rangle}{\langle \alpha \mid \alpha \rangle} \cdot \alpha \tag{A1.3.30}$$

is also a root (see Racah, 1965).

For rank two semi-simple Lie algebras the graphical representation of the root vectors is called a *root diagram* or Schouten diagram. A neat method for specifying the roots of a simple Lie algebra is by means of a Dynkin diagram. This Dynkin diagram is obtained as follows. Each simple root is represented by a circle, ○ if the root is a short one and ● if the root is a long one. When the angle between the roots is 90°, 120°, 135° or 150° the circles are joined by zero, one, two or three lines respectively. As a result of (A1.3.26) together with other considerations (see Jacobson, 1962) it can be shown that there exist only four infinite series of 'classical' simple Lie algebras and five exceptional simple Lie algebras. This classification of simple Lie algebras is given in terms of the Dynkin diagrams in Table 41.

TABLE 41. Classification of simple Lie algebras

Lie group	Simple Lie algebra	
$SU(n+1)$	A_n; $n = 1, 2, 3, \ldots, \infty$	
$R(2n+1)$	B_n; $n = 1, 2, 3, \ldots, \infty$	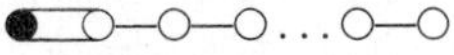
$Sp(n)$	C_n; $n = 2, 3, \ldots, \infty$	
$R(2n)$	D_n; $n = 3, 4, \ldots, \infty$	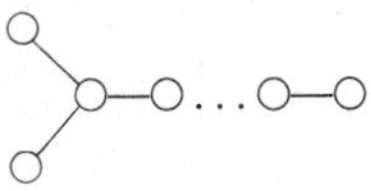
	G_2	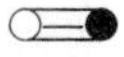
	F_4	
	E_6	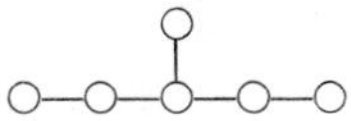
	E_7	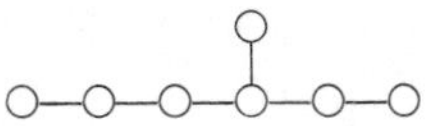
	E_8	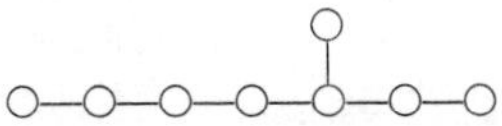

The Lie algebra C_1 does not exist while D_2 is omitted as it is semi-simple (see Chapter 9). Of the five exceptional Lie algebras only G_2 has found any use in quantum chemistry. This is in connection with f orbitals (see Judd, 1963).

A1.4 Some detailed examples for particular Lie algebras

Of specific importance in this book are the Lie groups $R(3)$, $R(4)$, $SU(2)$ and $SU(3)$. We consider in some detail standard IO's for the corresponding Lie algebras B_1, D_2, A_1 and A_2 respectively.

(a) The Lie algebra B_1 which is isomorphic to A_1.

From (A1.3.7) and (A1.3.8) we have

$$[\hat{H}_1, \hat{E}_\alpha] = -ix_3\hat{E}_\alpha = \alpha_1\hat{E}_\alpha$$

$$[\hat{E}_\alpha, \hat{E}_{-\alpha}] = -\frac{2ix_1^2}{x_3}\hat{H}_1 = \alpha^1\hat{H}_1$$

We can choose α_1 such that $g_{11} = 1$ which gives $\alpha_1 = 1/\sqrt{2}$. Likewise we choose α^1 such that $g_{\alpha-\alpha} = 1$ which gives $\alpha^1 = 1/\sqrt{2}$. It follows that

$$\hat{H}_1 = \frac{i}{\sqrt{2}}\hat{X}_3 = \frac{1}{\sqrt{2}}\hat{l}_z \qquad \hat{E}_{\pm\alpha} = \frac{i}{2}(\hat{X}_1 \pm i\hat{X}_2) = \tfrac{1}{2}\hat{l}_\pm \tag{A1.4.1}$$

These results should be compared with (9.4.13) and (9.4.14).

(b) The Lie algebra D_2 of $R(4)$.

Canonical CR's are given by (9.6.8) which shows that D_2 is a rank two semi-simple Lie algebra. We immediately see that there are two simple root vectors α and β whose covariant and contravariant components are identical.

(c) The Lie algebra A_2 of $SU(3)$.

By setting up the CR's between the operators (9.7.6) we see that they are in canonical form and further that $\gamma = \alpha + \beta$. Thus A_2 is a rank two simple Lie algebra the full system of roots being given by

$$\rho = 0, 0, \alpha, -\alpha, \beta, -\beta, (\alpha+\beta), -(\alpha+\beta) \tag{A1.4.2}$$

Now irrespective of the values of λ, μ, κ, ν we find

$$g_{\alpha-\alpha} = -24N_\alpha^2$$

If we normalize the $\hat{E}_\alpha$ such that $g_{\alpha-\alpha} = 1$ we find $N_\alpha = 1/i2\sqrt{6}$. Similarly $N_\beta = N_\gamma = N_\alpha$. By examining the CR's between $\hat{H}_1$, $\hat{H}_2$ and $\hat{E}_\alpha$'s we see that the simplest possible components for the simple root vectors α and β are obtained by making the choice

$$\lambda + 2\mu = 0 \qquad \kappa - \nu = 0$$

with $\lambda = \frac{2}{3}$ and $\kappa = \frac{1}{3}$. Of course other choices are possible but we shall not be concerned in this book with the definition of an absolute set of canonical CR's (see e.g. Rowlatt, 1966; Gourdin, 1967). With the above choice of λ, μ, κ, ν the covariant and contravariant components of the simple roots are

$$\begin{array}{llll} \alpha_1 = 1 & \alpha_2 = 0 & \beta_1 = 0 & \beta_2 = 1 \\ \alpha^1 = \frac{1}{3} & \alpha^2 - -\frac{1}{6} & \beta^1 = -\frac{1}{6} & \beta^2 = \frac{1}{3} \end{array} \qquad \text{(A1.4.3)}$$

It follows that all the root vectors have the same magnitude and the angle between α and β is

$$\theta = \arccos \frac{\langle \alpha \mid \beta \rangle}{|\alpha|\,|\beta|} = 120°$$

Thus the Schouten diagram (root diagram) for A_2 is:

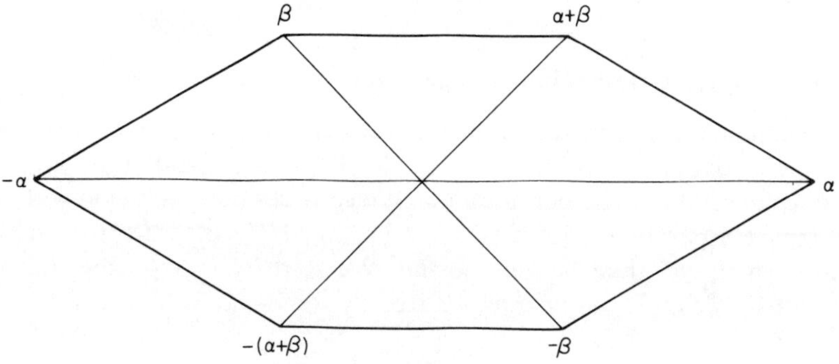

and the corresponding Dynkin diagram is O—O.

A1.5 Canonical parameters

We begin this section by considering the trivial Lie group $R(2)$. According to Section 9.1 $R(2)$ is a one-parameter Lie group. The parameter ω has the property that if $R(\omega_1)$ and $R(\omega_2)$ are two successive rotations then the resultant is given by

$$R(\omega_3) = R(\omega_2)R(\omega_1) = R(\omega_2 + \omega_1) \qquad \text{(A1.5.1)}$$

This shows that $\omega_3 = \omega_1 + \omega_2$. When a parameter has such an additive property we refer to it as a *canonical* parameter. For the canonical parameter ω of $R(2)$ we have from (9.1.7) the result

$$d\omega = \delta\omega \tag{A1.5.2}$$

The IO associated with ω is simply given by

$$\hat{X} = \frac{d}{d\omega} \tag{A1.5.3}$$

We now generalize the above findings and consider an r-parameter Lie group of transformations $\mathcal{G}$ given by (A1.1.1). We can associate with each of the r parameters a_i a one-parameter subgroup $\mathcal{G}_i$ of $\mathcal{G}$ in which only a_i varies. If we now refer to Hamermesh (1962) we see that for a one-parameter group it is always possible to replace a_i by a parameter t which is canonical. For such a canonical parameter results analogous to (A1.5.1)–(A1.5.3) can readily be shown to hold. The result analogous to (9.1.21) can also readily be established. Thus if $R(t)$ is an element of a one-parameter group then

$$\hat{R}(t) = \exp(t\hat{X}) \tag{A1.5.4}$$

where $\hat{X}$ is the IO corresponding to t.

A1.6 Conjugate classes in linear groups

In Section 9.4 we saw that class properties of the 3-parameter Lie group R(3) depend upon only one parameter. Thus in (9.4.30) the character depends upon the single parameter ω. We consider here, to begin with, the less trivial 8-parameter rank two Lie group $SU(3)$ in order to establish results which can then be generalized. We start by recognizing that any element $\mathbf{U}$ of $SU(3)$ can be diagonalized by some matrix $\mathbf{V}$. Thus

$$\mathbf{V}^{-1}\mathbf{U}\mathbf{V} = \mathbf{\Lambda}$$

It follows that the elements of $SU(3)$ are equivalent to the diagonal matrices

$$\mathbf{\Lambda} = \text{diag}(\lambda_1, \lambda_2, \lambda_3)$$

For unitarity we require $\lambda_p^*\lambda_p = 1$. If we write

$$\lambda_p = r_p e^{i\phi_p}$$

then $r_p = 1$ and

$$\mathbf{\Lambda} = \text{diag}(e^{i\phi_1}, e^{i\phi_2}, e^{i\phi_3})$$

For unimodularity we require

$$e^{i(\phi_1+\phi_2+\phi_3)} = 1$$

and it follows that

$$\phi_3 = -(\phi_1 + \phi_2)$$

For given values of the parameters ϕ_1 and ϕ_2 the elements of $SU(3)$ belong to the same conjugate class. Class properties such as the character are thus dependent upon only two parameters. Now

$$\begin{aligned}\mathbf{\Lambda}(\phi_1, \phi_2) &= \text{diag}\,(e^{i\phi_1}, e^{i\phi_2}, e^{-(\phi_1+\phi_2)}) \\ &= \text{diag}\,(e^{i\phi_1}, 1, e^{-i\phi_1})\,\text{diag}\,(1, e^{i\phi_2}, e^{-i\phi_2}) \\ &= \mathbf{\Lambda}_1(\phi_1)\mathbf{\Lambda}_2(\phi_2) \qquad \text{(A1.6.1)}\end{aligned}$$

The group element $\mathbf{\Lambda}(\phi_1, \phi_2)$ is thus given by

$$\mathbf{\Lambda}(\phi_1, \phi_2) = \mathbf{\Lambda}_1(\phi_1)\mathbf{\Lambda}_2(\phi_2) \qquad \text{(A1.6.2)}$$

The elements $\mathbf{\Lambda}_1(\phi_1)$ and $\mathbf{\Lambda}_2(\phi_2)$ form one-parameter subgroups of $SU(3)$ which are given by

$$\begin{aligned} x' &= e^{i\phi_1}x \\ z' &= e^{-i\phi_1}z \end{aligned} \quad \text{and} \quad \begin{aligned} y' &= e^{i\phi_2}y \\ z' &= e^{-i\phi_2}z \end{aligned}$$

respectively. The parameters ϕ_1, ϕ_2 are clearly canonical and the IO's associated with them are

$$\hat{X}_1 = \frac{\partial}{\partial \phi_1} \quad \text{and} \quad \hat{X}_2 = \frac{\partial}{\partial \phi_2} \qquad \text{(A1.6.3)}$$

respectively. We note that

$$[\hat{X}_1, \hat{X}_2] = 0 \qquad \text{(A1.6.4)}$$

The IO's can also be written as

$$\begin{aligned}\hat{X}_1 &= i\left(x\frac{\partial}{\partial x} - z\frac{\partial}{\partial z}\right) = -i(\hat{H}_1 + \hat{H}_2) \qquad \text{(A1.6.5)} \\ \hat{X}_2 &= i\left(y\frac{\partial}{\partial y} - z\frac{\partial}{\partial z}\right) = -i(2\hat{H}_2 - \hat{H}_1)\end{aligned}$$

where $\hat{H}_1$, $\hat{H}_2$ are given by (9.7.6). According to (A1.5.4) we have

$$\begin{aligned}\hat{\Lambda}_1(\phi_1) &= \exp\,(\phi_1\hat{X}_1) \\ \hat{\Lambda}_2(\phi_2) &= \exp\,(\phi_2\hat{X}_2)\end{aligned}$$

and it follows that

$$\hat{\Lambda}(\phi_1, \phi_2) = \exp\,(\phi_1\hat{X}_1 + \phi_2\hat{X}_2) \qquad \text{(A1.6.6)}$$

since $\hat{X}_1$ and $\hat{X}_2$ commute.

We now generalize the above findings. Let $\mathscr{G}$ be a semi-simple r-parameter Lie group of linear transformations. We denote the group elements by $R(a_1, a_2, \ldots, a_r)$ which correspond to the linear transformations

$$x_i' = R_{ij}(a_1, a_2, \ldots, a_r)x_j \qquad i = 1, 2, \ldots, n$$

According to Weyl (1956) if the rank of $\mathcal{G}$ is l then in any conjugate class there exists elements of the form $R(\phi_1, \phi_2, \ldots, \phi_l)$ which depend upon only l parameters ϕ_i. Furthermore $R(\phi_1, \phi_2, \ldots, \phi_l)$ can be expressed in the form (compare (A1.6.2))

$$R(\phi_1, \phi_2, \ldots, \phi_l) = R_1(\phi_1)R_2(\phi_2) \ldots R_l(\phi_l) \qquad \text{(A1.6.7)}$$

where $R_i(\phi_i)$ corresponds to a transformation of the one-parameter subgroup specified by canonical parameter ϕ_i. The IO's $\hat{X}_1, \hat{X}_2, \ldots, \hat{X}_l$ which are associated with the one-parameter subgroups are mutually commuting (compare (A1.6.4)) and can be written in terms of the standard basis IO's $\hat{H}_1, \hat{H}_2, \ldots, \hat{H}_l$ (compare (A1.6.5)). From (A1.5.4) we now have (compare (A1.6.6))

$$\hat{R}(\phi_1, \phi_2, \ldots, \phi_l) = \exp(\phi_1\hat{X}_1 + \phi_2\hat{X}_2 + \ldots + \phi_l\hat{X}_l) \qquad \text{(A1.6.8)}$$

since

$$[\hat{X}_i, \hat{X}_j] = 0$$

Equation (A1.6.8) establishes a vital link between the elements of the Lie group and the IO's of its Lie algebra, at least for groups of linear transformations. This exponential mapping takes a full neighbourhood of the zero of the Lie algebra onto a full neighbourhood of the unit of the Lie group. Only the structure of the Lie group in the neighbourhood of the unit (the so-called local Lie group) is necessarily determined by (A1.6.8). In particular we can use (A1.6.8) to obtain the IR's of a Lie group from the IR's of its Lie algebra whose derivation is usually very much easier. Thus if we have an IR $\mathbf{X}_i$ of the Lie algebra then (A1.6.8) gives an IR $\mathbf{R}(\phi_1, \phi_2, \ldots, \phi_l)$ of the (local) universal covering Lie group.

A1.7 Representations of Lie algebras and Lie groups

Consider a semi-simple r-dimensional Lie algebra $\mathcal{L}$ of rank l with a standard basis given by the IO's

$$\hat{H}_1, \hat{H}_2, \ldots, \hat{H}_l, \hat{E}_\alpha, \hat{E}_\beta, \ldots, \hat{E}_\nu$$

We say that these IO's have an n-dimensional matrix representation if there exists a set of $n \times n$ matrices

$$\mathbf{H}_1, \mathbf{H}_2, \ldots, \mathbf{H}_l, \mathbf{E}_\alpha, \mathbf{E}_\beta, \ldots, \mathbf{E}_\nu$$

which satisfy the canonical CR's (A1.3.22)–(A1.3.25).

Since the matrices $\mathbf{H}_i$ are mutually commuting it follows that there exists vectors which are simultaneous eigenvectors of $\mathbf{H}_1, \mathbf{H}_2, \ldots, \mathbf{H}_l$. We shall denote these eigenvectors by $|m_1, m_2, \ldots, m_l\rangle$. Thus we have

$$\mathbf{H}_i\,|m_1, m_2, \ldots, m_l\rangle = m_i\,|m_1, m_2, \ldots, m_l\rangle \qquad i = 1, 2, \ldots, l \quad \text{(A1.7.1)}$$

The sets of eigenvalues $\{m_1\}, \{m_2\}, \ldots, \{m_l\}$ play an extremely important role in the representation theory of semi-simple Lie algebras. It is convenient to regard a set of eigenvalues $m_1, m_2, \ldots, m_l$ as the covariant components of a vector $m = (m_1, m_2, \ldots, m_l)$ in the l-dimensional simple root space Π. It is customary to call the vector m a *weight* vector or simply a weight.

It is instructive at this stage to consider an illustration. For the Lie algebra A_2 we find from (9.7.6)

$$\hat{H}_1 = -\tfrac{2}{3}x\frac{\partial}{\partial x} + \tfrac{1}{3}y\frac{\partial}{\partial y} + \tfrac{1}{3}z\frac{\partial}{\partial z}$$

$$\hat{H}_2 = -\tfrac{1}{3}x\frac{\partial}{\partial x} - \tfrac{1}{3}y\frac{\partial}{\partial y} + \tfrac{2}{3}z\frac{\partial}{\partial z}$$

where we have used the choice of λ, μ, κ, ν given in Section A1.4. In the representation space (x, y, z) we readily find

$$\mathbf{H}_1 = \text{diag}\,(-\tfrac{2}{3}, \tfrac{1}{3}, \tfrac{1}{3}) \qquad \mathbf{H}_2 = \text{diag}\,(-\tfrac{1}{3}, -\tfrac{1}{3}, \tfrac{2}{3}) \tag{A1.7.2}$$

In a similar way the matrices $\mathbf{E}_{\pm\alpha}$, $\mathbf{E}_{\pm\beta}$, $\mathbf{E}_{\pm(\alpha+\beta)}$ can be constructed and using Schur's Lemma it can be checked that the representation is irreducible. From (A1.7.1) we have

$$m_1 = \tfrac{1}{3}, \tfrac{1}{3}, -\tfrac{2}{3} \qquad m_2 = \tfrac{2}{3}, -\tfrac{1}{3}, -\tfrac{1}{3}$$

and the weight vectors are thus given by

$$a = (\tfrac{1}{3}, \tfrac{2}{3}) \qquad b = (\tfrac{1}{3}, -\tfrac{1}{3}) \qquad c = (-\tfrac{2}{3}, -\tfrac{1}{3})$$

In terms of the simple roots α and β (see (A1.4.3)) we have

$$a = \tfrac{1}{3}(\alpha + 2\beta) \qquad b = \tfrac{1}{3}(\alpha - \beta) \qquad c = -\tfrac{1}{3}(2\alpha + \beta)$$

The weight vectors are drawn in Fig. 14 with respect to the root diagram. With the above example in mind we now return to the general theory. We begin with some important definitions. A weight is said to be *positive* if its

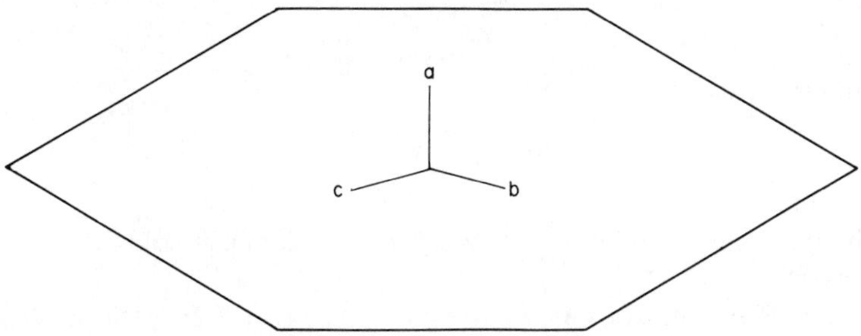

FIG. 14

first non-vanishing component is positive. One weight is said to be *higher* than another if the difference between them is positive. The highest (or greatest) weight is of particular importance in the representation theory of semi-simple Lie algebras.

We now state without proof some important results from the general representation theory of semi-simple Lie algebras (see Jacobson, 1962; Racah, 1965).

I. If α is a simple root then the possible highest weights $\mu = (\mu_1, \mu_2, \ldots, \mu_l)$ are given from the equations

$$2\frac{\langle \mu \mid \alpha \rangle}{\langle \alpha \mid \alpha \rangle} = p \qquad \text{(A1.7.3)}$$

where p is a non-negative integer.

II. Corresponding to each highest weight μ there exists an IR of the Lie algebra. The symbol

$$(\mu_1 \mu_2 \ldots \mu_l) \qquad \text{(A1.7.4)}$$

can be used as a label for the various IR's. Sometimes the symbol $(\mu^1 \mu^2 \ldots \mu^l)$ where μ^i are the contravariant components of μ is used to label the IR's.

III. An IR of a semi-simple Lie algebra is completely specified by its highest weight. In particular the dimension $n_{(\mu)}$ of the IR labelled by highest weight μ is given by

$$n_{(\mu)} = \prod_{\alpha \text{ in } \Sigma^+} \frac{\langle \mu + g \mid \alpha \rangle}{\langle g \mid \alpha \rangle} \qquad \text{(A1.7.5)}$$

where α is a positive root and

$$g = \tfrac{1}{2} \sum_{\alpha \text{ in } \Sigma^+} \alpha \qquad \text{(A1.7.6)}$$

IV. If m is a weight and α is a root then

$$2\frac{\langle m \mid \alpha \rangle}{\langle \alpha \mid \alpha \rangle}$$

is an integer and

$$m - 2\frac{\langle m \mid \alpha \rangle}{\langle \alpha \mid \alpha \rangle} \cdot \alpha \qquad \text{(A1.7.7)}$$

is a weight.

V. The maximum number of weights for an n-dimensional representation is precisely n.

VI. In an IR the matrices $\mathbf{H}_i$ are diagonal (see (A1.7.2)) with the components of the weights along the diagonals.

The results I–VI can be used to obtain the IR's and the characters for the groups $SU(2)$ and $R(3)$ in a very elegant manner. The results obtained are of course identical to those found by the alternative direct method in Sections 9.5 and 9.4. The interested reader can check this very quickly. We shall consider the application of results I–VI for the less trivial but important groups $R^*(4)$ and $SU(3)$ for which the alternative direct method is cumbersome and not self-evident.

A1.8 The irreducible representations of $R^*(4)$

The Lie algebra D_2 of $R(4)$ is a rank 2 semi-simple algebra with simple roots α, β given by (A1.4.1). The covering group of D_2 is $R^*(4)$.

Let $j' = (j'_1, j'_2)$ be a highest weight. According to (A1.7.3) we have

$$2\frac{\langle j' \mid \alpha\rangle}{\langle \alpha \mid \alpha\rangle} = p \qquad 2\frac{\langle j' \mid \beta\rangle}{\langle \beta \mid \beta\rangle} = q \tag{A1.8.1}$$

where p and q are non-negative integers. If we write

$$j'_1 = \frac{j_1}{\sqrt{2}} \qquad j'_2 = \frac{j_2}{\sqrt{2}}$$

then from (A1.8.1) and (A1.4.1) we immediately obtain

$$\begin{aligned} j_1 &= 0, \tfrac{1}{2}, 1, \tfrac{3}{2}, \ldots \\ j_2 &= 0, \tfrac{1}{2}, 1, \tfrac{3}{2}, \ldots \end{aligned} \tag{A1.8.2}$$

According to (A1.7.4) the IR's of D_2 can be labelled by a symbol $(j_1 j_2)$ with j_1 and j_2 given by (A1.8.2). From (A1.7.5) and (A1.7.6) the dimension of the IR $(j_1 j_2)$ is given by

$$n_{(j_1 j_2)} = \frac{\langle j' + g \mid \alpha\rangle\langle j' + g \mid \beta\rangle}{\langle g \mid \alpha\rangle\langle g \mid \beta\rangle}$$

where

$$g = \tfrac{1}{2}(\alpha + \beta) = \left(\frac{1}{2\sqrt{2}}, \frac{1}{2\sqrt{2}}\right)$$

Thus we find

$$n_{(j_1 j_2)} = (2j_1 + 1)(2j_2 + 1) \tag{A1.8.3}$$

Following (A1.6.7) and using (9.6.12) we write

$$\mathbf{U}(\psi, \theta) = \mathbf{U}_1(\psi)\mathbf{U}_2(\theta)$$

where the matrices of $\mathbf{U}_1$ and $\mathbf{U}_2$ are given by

$$\begin{aligned} \mathbf{U}_1(\psi) &= \text{diag}\,[e^{i\psi/2}, e^{-i\psi/2}, 1, 1] \\ \mathbf{U}_2(\theta) &= \text{diag}\,[1, 1, e^{i\theta/2}, e^{-i\theta/2}] \end{aligned}$$

The IO's corresponding to the canonical parameters ψ and θ are

$$\frac{i}{2}\left(x\frac{\partial}{\partial x}-y\frac{\partial}{\partial y}\right)=\sqrt{2}i\hat{H}_1$$

and

$$\frac{i}{2}\left(z\frac{\partial}{\partial z}-t\frac{\partial}{\partial t}\right)=\sqrt{2}i\hat{H}_2$$

respectively. From (A1.6.8) we have

$$\hat{U}(\psi,\theta)=\exp\left(\sqrt{2}i\psi\hat{H}_1+\sqrt{2}i\theta\hat{H}_2\right)$$

and from this result it is a simple matter to show that the character of $U(\psi, \theta)$ in the IR $(j_1 j_2)$ is given by

$$\chi^{(j_1 j_2)}(\psi,\theta)=\frac{\sin\left[(j_1+\frac{1}{2})\psi\right]\sin\left[(j_2+\frac{1}{2})\theta\right]}{\sin\left(\frac{1}{2}\psi\right)\sin\left(\frac{1}{2}\theta\right)} \tag{A1.8.4}$$

A1.9 The irreducible representations of $SU(3)$

The Lie algebra of $SU(3)$ is A_2 and the covering group of A_2 is $SU(3)$.

Let $M=(M_1, M_2)$ be a highest weight. If we use (A1.7.3) and (A1.4.3) we find

$$M_1=\frac{2n+m}{3}\qquad M_2=\frac{n+2m}{3}$$

where n and m are non-negative integers. It follows that we can label the IR's of A_2 by a symbol (nm) with

$$\begin{aligned} n&=0,1,2,3,\ldots\\ m&=0,1,2,3,\ldots \end{aligned} \tag{A1.9.1}$$

From (A1.4.3) and (A1.3.13) we find

$$g_{ij}=\begin{bmatrix}4 & 2\\ 2 & 4\end{bmatrix}\qquad g^{ij}=\begin{bmatrix}\frac{1}{3} & -\frac{1}{6}\\ -\frac{1}{6} & \frac{1}{3}\end{bmatrix} \tag{A1.9.2}$$

and the contravariant components of the highest weights are thus $(n/6, m/6)$. According to (A1.7.6) we have

$$g=\tfrac{1}{2}(\alpha+\beta+\gamma)=\tfrac{1}{2}(\alpha+\beta+\alpha+\beta)=(1,1)$$

in terms of covariant components and

$$g=\left(\tfrac{1}{6},\tfrac{1}{6}\right)$$

in terms of contravariant components. The dimension of the IR (nm) is now given by (A1.7.5) as

$$n_{(nm)} = \frac{\langle (n+1, m+1) \mid (1, 0)\rangle\langle (n+1, m+1) \mid (0, 1)\rangle\langle (n+1, m+1) \mid (1, 1)\rangle}{\langle (1, 1) \mid (1, 0)\rangle\langle (1, 1) \mid (0, 1)\rangle\langle (1, 1) \mid (1, 1)\rangle}$$

$$= \tfrac{1}{2}(n+1)(m+1)(n+m+2) \qquad \text{(A1.9.3)}$$

It is quite easy to find the character of the IR (nm) in $SU(3)$ by methods analogous to those used for $R^*(4)$ in the last section. As we do not use the characters for $SU(3)$ in this book we shall not obtain them here.

A1.10 The Casimir operator

Let $\mathscr{L}$ be an r-dimensional rank l semi-simple Lie algebra whose natural IO's are $\hat{X}_\rho$ and let $g^{\rho\sigma}$ be the Cartan tensor given by (A1.3.16). We define an operator $\hat{C}$ by

$$\hat{C} = g^{\rho\sigma}\hat{X}_\rho\hat{X}_\sigma \qquad \text{(A1.10.1)}$$

This operator can readily be shown to commute with all the IO's $\hat{X}_\tau$ and it is called the *Casimir* operator for $\mathscr{L}$. In terms of a standard basis the Casimir operator takes the form

$$\hat{C} = g^{ik}\hat{H}_i\hat{H}_k + \sum_\alpha \hat{E}_\alpha\hat{E}_{-\alpha} \qquad \text{(A1.10.2)}$$

due to the normalization condition (A1.3.15).

Consider now an IR of $\mathscr{L}$ specified by highest weight $\mu = (\mu_1, \mu_2, \ldots, \mu_l)$. In terms of operators we have (see (A1.7.1))

$$\hat{H}_i\psi = \mu_i\psi \qquad i = 1, 2, \ldots, l \qquad \text{(A1.10.3)}$$

where ψ is a simultaneous eigenfunction of $\hat{H}_1, \hat{H}_2, \ldots, \hat{H}_l$. Now

$$\hat{H}_i\hat{E}_\alpha = \hat{E}_\alpha\hat{H}_i + [\hat{H}_i, \hat{E}_\alpha]$$

and it follows from (A1.3.23) that

$$\hat{H}_i\hat{E}_\alpha = \hat{E}_\alpha\hat{H}_i + \alpha_i\hat{E}_\alpha$$

We now have

$$\hat{H}_i\hat{E}_\alpha\psi = (\mu_i + \alpha_i)\hat{E}_\alpha\psi \qquad \text{(A1.10.4)}$$

and we see that if $\hat{E}_\alpha\psi \neq 0$ then $\hat{E}_\alpha\psi$ is an eigenfunction of $\hat{H}_i$ corresponding to a weight $(\mu + \alpha)$. Since μ is a highest weight $(\mu + \alpha)$ cannot be a weight if

$\alpha > 0$ in which case $\hat{E}_\alpha \psi = 0$. For the Casimir operator we have

$$\hat{C}\psi = g^{ik}\hat{H}_i\hat{H}_k\psi + \sum_\alpha \hat{E}_\alpha\hat{E}_{-\alpha}\psi$$

$$= g^{ik}\mu_i\mu_k\psi + \sum_{\alpha \text{ in } \Sigma^+} [\hat{E}_\alpha, \hat{E}_{-\alpha}]\psi$$

$$= \langle\mu \mid \mu\rangle\psi + \sum_{\alpha \text{ in } \Sigma^+} \alpha^i\mu_i\psi$$

$$= \left[\langle\mu \mid \mu\rangle + \sum_{\alpha \text{ in } \Sigma^+} \langle\alpha \mid \mu\rangle\right]\psi$$

where the sum is over the set of all positive roots. If we define a vector

$$k = \mu + g \tag{A1.10.5}$$

where g is given by (A1.7.6) then

$$\hat{C}\psi = \lambda\psi \tag{A1.10.6}$$

with

$$\lambda = k^2 - g^2 \tag{A1.10.7}$$

In a given IR of $\mathscr{L}$ it is clear that λ (the eigenvalue of $\hat{C}$) has a definite value. This single scalar cannot in general be used to characterize the IR since, as we have seen, the IR can only be completely specified by its highest weight and this is a set of l scalars. However the concept of an invariant operator, such as $\hat{C}$, can be generalized and the following result is obtained (see Racah, 1965).

For any semi-simple Lie algebra of rank l there exists a set of invariant operators $\hat{F}_i(\hat{X}_1, \hat{X}_2, \ldots, \hat{X}_r)$; $i = 1, 2, \ldots, l$ which commute with all the IO's $\hat{X}_\rho$. The set of l eigenvalues $\lambda_1, \lambda_2, \ldots, \lambda_l$ serve to characterize the various IR's of the Lie algebra.

As a first example we take the Lie algebra A_1 (or B_1). This Lie algebra is of rank one and it follows that the IR's can be specified by the eigenvalues of a single invariant operator. This operator can be taken to be the Casimir operator. From (A1.10.2) we have

$$\hat{C} = \hat{H}_1^2 + \hat{E}_\alpha\hat{E}_{-\alpha} + \hat{E}_{-\alpha}\hat{E}_\alpha \tag{A1.10.8}$$

The eigenvalues of $\hat{C}$ are given by (A1.10.7). In this case

$$g = \frac{1}{2}\frac{1}{\sqrt{2}} \qquad \mu = \frac{j}{\sqrt{2}}$$

and $k = \frac{1}{2}(j + \frac{1}{2})$. It follows that

$$\lambda = \tfrac{1}{2}j(j+1) \qquad j = 0, \tfrac{1}{2}, 1, \tfrac{3}{2}, 2, \ldots \tag{A1.10.9}$$

and we write (A1.10.6) as

$$\hat{C}\,|j, m\rangle = \tfrac{1}{2}j(j+1)\,|j, m\rangle \qquad \text{(A1.10.11)}$$

For the Lie algebra D_2 of the rotation group $R(4)$ there exists two invariant operators $\hat{F}_1$ and $\hat{F}_2$ whose eigenvalues serve to characterize the IR's. From (9.6.6) and (A1.10.8) it can be seen that we may take $\hat{F}_1$ and $\hat{F}_2$ to be

$$\begin{aligned}\hat{F}_1 &= \hat{H}_1^2 + \hat{E}_\alpha\hat{E}_{-\alpha} + \hat{E}_{-\alpha}\hat{E}_\alpha \\ \hat{F}_2 &= \hat{H}_2^2 + \hat{E}_\beta\hat{E}_{-\beta} + \hat{E}_{-\beta}\hat{E}_\beta\end{aligned} \qquad \text{(A1.10.12)}$$

By comparison with (A1.10.11) we can immediately write

$$\begin{aligned}\hat{F}_1\,|j_1, j_2, m_1, m_2\rangle &= \tfrac{1}{2}j_1(j_1+1)\,|j_1, j_2, m_1, m_2\rangle \\ \hat{F}_2\,|j_1, j_2, m_1, m_2\rangle &= \tfrac{1}{2}j_2(j_2+1)\,|j_1, j_2, m_1, m_2\rangle\end{aligned} \qquad \text{(A1.10.13)}$$

The Casimir operator for D_2 is clearly

$$\hat{C} = \hat{F}_1 + \hat{F}_2 \qquad \text{(A1.10.14)}$$

The eigenvalues of $\hat{C}$ are determined in Section 9.6.

Finally we consider the Lie algebra A_2 of the group $SU(3)$. According to (A1.9.2) and (A1.10.2) the Casimir operator for A_2 is

$$\begin{aligned}\hat{C} = {}&\tfrac{1}{3}(\hat{H}_1^2 - \hat{H}_1\hat{H}_2 + \hat{H}_2^2) + \hat{E}_\alpha\hat{E}_{-\alpha} + \hat{E}_{-\alpha}\hat{E}_\alpha \\ &+ \hat{E}_\beta\hat{E}_{-\beta} + \hat{E}_{-\beta}\hat{E}_\beta + \hat{E}_\gamma\hat{E}_{-\gamma} + \hat{E}_{-\gamma}\hat{E}_\gamma\end{aligned} \qquad \text{(A1.10.15)}$$

The eigenvalues of this Casimir operator are given by (A1.10.7). Using the results of Section A1.9 we have

$$g = (g^1, g^2) = (\tfrac{1}{6}, \tfrac{1}{6}) \qquad g = (g_1, g_1) = (1, 1)$$

$$\mu = (\mu^1, \mu^2) = \left(\frac{n}{6}, \frac{m}{6}\right) \qquad \mu = (\mu_1, \mu_2) = \left(\frac{2n+m}{3}, \frac{n+2m}{3}\right)$$

$$k = (k^1, k^2) = \left(\frac{n+1}{6}, \frac{m+1}{6}\right) \qquad k = (k_1, k_2) = \left(\frac{2n+m+3}{3}, \frac{n+2m+3}{3}\right)$$

and

$$g^2 = g^i g_i = \tfrac{1}{3}$$

$$k^2 = k^i k_i = \tfrac{1}{9}(n^2+m^2+nm) + \tfrac{1}{3}(n+m) + \tfrac{1}{3}$$

Thus the eigenvalues λ of $\hat{C}$ are given by

$$\lambda = \tfrac{1}{9}(n^2+m^2+nm) + \tfrac{1}{3}(n+m) \qquad \text{(A1.10.16)}$$

References

Cartan, E. (1952) These Paris (1894) reprinted in E. Cartan "Oeuvres Completes", Gauthier-Villars, Paris.

Gourdin, M. (1967) "Unitary Symmetries", North Holland.
Hamermesh, M. (1962) "Group Theory and its Application to Physical Problems", Addison-Wesley.
Hermann, R. (1966) "Lie groups for Physicists", Benjamin.
Jacobson, N. (1962) "Lie algebras", Interscience.
Judd, B. R. (1963) "Operator Techniques in Atomic Spectroscopy", McGraw-Hill.
Racah, G. (1965) Ergebnisse der Exacten Naturwissenschaften 37, (in English).
Rowlatt, P. A. (1966) "Group Theory and Elementary Particles", Longmans.
Weyl, H. (1956) Selecta (Birkhauser Verlag, Basel und Stuttgart), p. 262.

Some Basic Tables Relating to Classification of States

TABLE 35. Terms arising from p^N

N	S	$[\lambda]$	$[\tilde{\lambda}]$	L	Terms
1	$\frac{1}{2}$	[1]	[10]	P	^{2}P
2	0	[11]	[20]	S, D	$^{1}S, ^{1}D$
	1	[20]	[11]	P	^{3}P
3	$\frac{1}{2}$	[21]	[21]	P, D	$^{2}P, ^{2}D$
	$\frac{3}{2}$	[3]	[111]	S	^{4}S

TABLE 36. Terms arising from d^N

N	S	$[\lambda]$	$[\tilde{\lambda}]$	L	Terms
1	$\frac{1}{2}$	[10]	[10]	D	^{2}D
2	0	[11]	[20]	S, D, G	$^{1}S, ^{1}D, ^{1}G$
	1	[20]	[11]	P, F	$^{3}P, ^{3}F$
3	$\frac{1}{2}$	[21]	[21]	$P, 2D, F, G, H$	$^{2}P, ^{2}D, ^{2}D, ^{2}F, ^{2}G, ^{2}H$
	$\frac{3}{2}$	[3]	[111]	P, F	$^{4}P, ^{4}F$
4	0	[22]	[22]	$2S, 2D, F, 2G, I$	$^{1}S, ^{1}S, ^{1}D, ^{1}D, ^{1}F, ^{1}G, ^{1}G, ^{1}I$
	1	[31]	$[21^2]$	$2P, D, 2F, G, H$	$^{3}P, ^{3}P, ^{3}D, ^{3}F, ^{3}F, ^{3}G, ^{3}H$
	2	[4]	$[1^4]$	D	^{5}D
5	$\frac{1}{2}$	[32]	$[21^2]$	$S, P, 3D, 2F, 2G, H, I$	$^{2}S, ^{2}P, ^{2}D, ^{2}D, ^{2}D, ^{2}F, ^{2}F, ^{2}G, ^{2}G, ^{2}H, ^{2}I$
	$\frac{3}{2}$	[41]	$[21^3]$	P, D, F, G	$^{4}P, ^{4}D, ^{4}F, ^{4}G$
	$\frac{5}{2}$	[5]	$[1^5]$	S	^{6}S

TABLE 37. Seniority in the terms from d^N

N	S	$[\tilde{\lambda}]$	(w_2w_2)	L	v
1	$\frac{1}{2}$	[10]	(10)	D	1
2	0	[20]	(00)	S	0
			(20)	DG	2
	1	[11]	(11)	PF	2
3	$\frac{1}{2}$	[21]	(10)	D	1
			(21)	$PDFGH$	3
	$\frac{3}{2}$	$[1^3]$	(11)	PF	3
4	0	$[2^2]$	(00)	S	0
			(20)	DG	2
			(22)	$SDFGI$	4
	1	$[21^2]$	(11)	PF	2
			(21)	$PDFGH$	4
	2	$[1^4]$	(10)	D	4
5	$\frac{1}{2}$	$[21^2]$	(10)	D	1
			(21)	$PDFGH$	3
			(22)	$SDFGI$	5
	$\frac{3}{2}$	$[21^3]$	(11)	PF	3
			(20)	DG	5
	$\frac{5}{2}$	$[1^5]$	(00)	S	5

TABLE 38. Classification of States of s^mp^{n-m}

n	$SU(4)$	State	Configurations
1	[1]	$(\frac{1}{2}\frac{1}{2})^2S$	s
		$(\frac{1}{2}\frac{1}{2})^2P$	p
2	[2]	$(00)^1S$	s^2, p^2
		$(11)^1S$	s^2, p^2
		$(11)^1P$	sp
		$(11)^1D$	p^2
	[11]	$(10)^3P$	sp, p^2
		$(01)^3P$	sp, p^2
3	$[1^3]$	$(\frac{1}{2}\frac{1}{2})^4S$	p^3
		$(\frac{1}{2}\frac{1}{2})^4P$	sp^2
	[21]	$(\frac{3}{2}\frac{1}{2})^2D$	p^3, sp^2

TABLE 39. Multiplets from $(j)^N$ configurations

N	[λ]	J
		$j = \frac{1}{2}$
N	[λ]	J
1	[1]	$\frac{1}{2}$
		$j = 1$
N	[λ]	J
1	[1]	1
		$j = \frac{3}{2}$
N	[λ]	J
1	[1]	$\frac{3}{2}$
2	[11]	0, 2
		$j = 2$
N	[λ]	J
1	[1]	2
2	[11]	1, 3
		$j = \frac{5}{2}$
N	[λ]	J
1	[1]	$\frac{5}{2}$
2	[11]	0, 2, 4
3	[111]	$\frac{3}{2}, \frac{5}{2}, \frac{9}{2}$
		$j = 3$
N	[λ]	J
1	[1]	3
2	[11]	1, 3, 5
3	[111]	0, 2, 3, 4, 6
		$j = \frac{7}{2}$
N	[λ]	J
1	[1]	$\frac{7}{2}$
2	[11]	0, 2, 4, 6
3	[111]	$\frac{3}{2}, \frac{5}{2}, \frac{7}{2}, \frac{9}{2}, \frac{11}{2}, \frac{15}{2}$
4	[1111]	0, 2, 2, 4, 4, 5, 6, 8

Subject Index

A

Accidental degeneracy, 44, 166
Adjoint representation, 113
Algebra, 28
  Lie, 131, 144, 243
  semi-simple, 29
  simple, 29
Ambivalent class, 42
Angular momentum, 98, 229
  coupling 174, 200
  recoupling, 181
  spin, 98
Antisymmetric product representation, 67
Antisymmetrizer, 94
Associated Legendre polynomial, 46
Axial group, 132

B

Basis vectors, 11
  orthonormal, 20
Bilinear form, 125
Branching rules, 72, 156, 169, 188, 202, 207, 209
  for symmetric group, 84

C

Cartan subalgebra, 246, 249
  tensor, 245, 249
  theorem, 246
Casimir operator, 145, 157, 160, 261
Character, 37
  table, 50 169
Clebsch–Gordon coefficients, 174, 200
  theorem, 148, 174
Cogredience, 17
Complex conjugate representation, 52
Conjugate class, 4
Contraction, 27, 113
Contragredience, 17
Contragredient representation, 111
Contravariant components, 22
  set, 17
  tensor, 25, 111
  vector, 17
Coset, 6, 95
Coupling coefficients, 90, 174, 175, 177
  scheme, 181
Covariant components, 22
  set, 18
tensor, 25, 108
vector, 18
Cubic harmonics, 49
Cycle, 76
  structure, 77

D

Decomposable representation, 38
  set of mappings, 15
Dictionary order of tableaux, 81
Direct sum of linear subspaces, 12
  representations, 38
Direct product of linear subspaces, 12
  representations, 66, 89, 107, 129, 173
Dual basis, 16
  representation, 79, 88
  space, 16
Dynkin diagram, 251

E

Electron configuration, 103, 104, 199, 208
Equivalent representations, 15, 35, 37
Euclidean 3-space, 29

F

Fractional parentage, 213
  coefficients of, 216
  expansion, 216

G

Group, Abelian, 3
  cubic, 7
  direct product of, 5, 8, 94
  double, 151, 167
  finite, 2

full linear, 106
generators, 4
isomorphic, 6, 9
Lie, 131, 242,
manifold, 28, 106, 139, 242
molecular symmetry, 3, 137
octahedral, 54
orthogonal, 119
rotation, 119, 142, 153, 247, 252, 259
semi-simple, 5
simple, 5
simply reducible, 177
special orthogonal, 119
special unitary, 116, 148, 157, 252, 254, 257, 260
symmetric, 5, 75
symplectic, 126
unitary, 116

H

Hamiltonian, 43, 51, 72, 101, 161, 164, 198, 205, 208
Harmonic oscillator, 161
Hilbert space, 27, 43
Homomorphison, 6
Hydrogen atom, 164

I

Ideal, 29
Infinitesimal operator, 135, 144, 242
transformation, 131
Invariant integration, 139
subspace, 15
Irreducible representation, 39, 50
tensor, 108
tensor operator, 230

J

jj coupling, 198, 205

L

Lagrange theorem, 6
Levi–Civita tensor, 116
Lie algebra, 131, 144, 243
rank of, 246
semi-simple, 243
simple, 243
Lie group, 131, 242
Linear independence, 10
operator, 14
space, 10
subspace, 12

M

Mapping, 13
homomorphic, 35
Matrix, 11
block diagonal, 13
contragredient, 17, 111
direct product of, 13, 66
direct sum of, 13
representation of a group, 35
orthogonal, 18
unitary, 18
Metric, 19
matrix, 20
tensor, 27, 249
Multiplet, 204, 206

O

Operator, 14
equivalent representations of, 15
irreducible tensor, 230
linear, 14
matrix representation of, 14
projection, 61
unitary, 21
Orbital, atomic, 58, 141, 148, 199, 234
molecular, 62, 208
symmetry, 62
Orthogonal transformation, 18, 119
Outer product representation, 92, 99, 107

P

Pair trace, 113
Partition, 77
associated, 122
permissible, 121
shape of, 78
Pauli principle, 102, 201, 206
spin postulate, 98
Permutation, 75
parity of, 76
Perturbation theory, 103, 199, 237, 240
Principal parent, 221
Projection operator, 61, 80, 88

R

Racah coefficient, 182
Realization of a group, 35

Recoupling coefficient, 183
Reducible representation, 39
Representation space, 35
Rigid rotator, 164
Root diagram, 251
  simple, 248
  vector, 248
Rotation, 119, 132
Russell-Saunders coupling, 198

S

Scalar product, 18, 124
Schrödinger equation, 42, 102, 199, 237
  spin-free, 102
Schur's Lemma, 15
Selection rules, 71
Seniority, 202, 207, 240
Skew product, 125
Spherical harmonics, 46, 47
Spin, 98
Spinor, 98
Spin-free formalism, 102
Spin-orbit interaction, 198, 204
Standard basis functions, 58
  representation, 35, 36, 85
  Young tableau, 79
Young operator, 88
Subalgebra, 29
Subgroup, 4
  invariant, 5
Symmetric group, 75
  standard irreducible representations of, 85
Symmetric product representation, 67
Symmetry element, 1
  orbital, 62, 141
  species, 58, 140
  transformation, 1
Symplectic transformation, 126

T

Tensor, contravariant, 25, 111
  covariant, 25, 108
  metric, 27
  mixed, 26, 112
  operator, 230
  space, 25
Term, atomic, 200
  molecular, 208
Transformation coefficient, 182, 185
Transformed function, 31
Transposition, 76

U

Unitary operator, 21
  space, 20
  transformation, 18, 116, 148

V

Vector, 10
  complex conjugate, 24
  components, 11
  image, 13
  space, 18

W

Weight, highest, 258
  vector, 257
Wigner 3j symbol, 176
  3Γ symbol, 179
  6j symbol, 182
  9j symbol, 186
  6Γ symbol, 186
  9Γ symbol, 186
Wigner-Eckart theorem, 231

Y

Yamanouchi symbol, 85
Young operator, 80
  standard operator, 88
  tableau, 79
  theorem, 79